EMBRYOLOGY – UPDATES AND HIGHLIGHTS ON CLASSIC TOPICS

Edited by **Luís Antonio Violin Pereira**

Embryology – Updates and Highlights on Classic Topics
Edited by Luís Antonio Violin Pereira

Published by InTech

Janeza Trdine 9, 51000 Rijeka, Croatia

Publishing Process Manager Romina Skomersic

Technical Editor Teodora Smiljanic

Cover Designer InTech Design Team

Additional hard copies can be obtained from orders@intechopen.com

Embryology – Updates and Highlights on Classic Topics, Edited by Luís Antonio Violin Pereira
p. cm.
ISBN 978-953-51-0465-0

For EU product safety concerns:
IN TECH d.o.o., Prolaz Marije Krucifikse Kozulić 3, 51000 Rijeka, Croatia,
info@intechopen.com or visit our website at intechopen.com.

Meet the editor

Dr Luis Antonio Violin Pereira received his M.D. in 1991 from the Federal University of Santa Maria, Brazil. This was followed in 1994 by a M.Sc. in Cell Biology and in 1998 by a Ph.D. in Pathology, from the State University of Campinas (UNICAMP) and the University of São Paulo (USP), Brazil, respectively. Dr. Pereira initiated his teaching career at UNICAMP in 1994 when he started lecturing on human embryology to medical students. In 1998, Dr. Pereira was promoted to Assistant Professor and, since 2007, he has been an Associate Professor in the Department of Histology and Embryology of the Institute of Biology at UNICAMP. Dr. Pereira has developed methods for teaching human embryology based on active learning (Annals of Anatomy 192 (2010) 388-395). His research interests include the role of the epididymis in the mechanisms of fertilization, experimental models of birth defects, and teaching methods in human embryology.

Contents

Preface

Embryology is a branch of science concerned with the morphological aspects of organismal development. The genomic and molecular revolution of the second half of the 20th century, together with the classic descriptive aspects of this science have allowed greater integration in our understanding of many developmental events. Current studies in embryology and developmental biology are not restricted to gamete formation, fertilization (*in vivo* or *in vitro*), zygote formation, early growth or the development of living organisms, but also involve investigation of the genetic control of these processes and of development itself (so called morphogenesis). Modern embryology seeks to provide practical knowledge that can be applied to assisted reproduction, stem cell therapy, birth defects, fetal surgery and other fields.

This book focuses on human embryology and aims to provide an up-to-date source of information on a variety of selected topics. The book consists of nine chapters organized into three sections, namely: 1) gametes and infertility, 2) implantation, placentation and early development, and 3) perspectives in embryology.

1. Gametes and Infertility

Development begins with fertilization. However, the success of fertilization depends on the ability of the gametes involved. The molecular alterations that occur in oocytes during female reproductive aging represent a controversial area of clinical interest since these changes can markedly affect human female fecundity by 40 years of age or less. This topic is discussed in chapter 1 by Drs. Imai, Qin, Yamakawa, Miyado, Umezawa and Takahashi, who review their studies and recent knowledge on female reproductive aging, as well as the possibility of preventing age-associated infertility.

Infertility affects approximately 15% of all couples trying to conceive. The important subject of male fertility and the biology of male gametes are dealt with in chapters two and three. Reduced semen quality contributes to ~50% of the cases of male infertility, while there is no clear explanation for infertility in the remaining 50% of cases. In chapter 2, Dr. Bungum discusses the techniques for assessing the intactness of sperm DNA since recent molecular studies have shown that factors such as breaks in sperm DNA can contribute to male infertility. In chapter 3, Drs. Arrotéia, Garcia, Barbieri, Justino and Pereira discuss sperm epididymal maturation, i.e., the process by which spermatozoa acquire their post-testicular ability to fertilize. The embryology, structure, function and role of the epididymis in fertilization and in male infertility are also emphasized.

2. Implantation, Placentation and Early Development

Fertilization is followed by pre-implantational events that lead to formation of the blastocyst which is subsequently implanted in the endometrium. The onset of endometrial receptivity to the blastocyst, trophoblast invasion and placental development involves the expression of specific genes in particular cell types. As discussed by Dr. Tapia in chapter 4, this expression is temporally regulated, with some genes being turned on or showing enhanced expression while others are down-regulated or completely switched off. Likewise, when implantation has occurred, another set of genes facilitates the continuing trophoblast invasion and placentation.

In chapter 5, Drs. Gridelet, Gaspard, Polese, Ruggeri, Ravet, Munaut, Geenen, Foidart, Lédée and Perrier d'Hauterive point out that implantation of the blastocyst and placentation remain the black boxes of fertility since only 20-25% of embryos transferred to the uterus after *in vitro* fertilization result in a birth. As pointed out by these authors, successful implantation requires that a good quality oocyte meet a normal sperm, leading to the development of a functionally normal blastocyst able to interact with the maternal endometrium.

Placental macrophages play a central role in the establishment and maintenance of pregnancy through their ability to produce a variety of endogenous mediators involved in pregnancy, as well as in parturition, lactation, local immune reactions and maternal-fetal tolerance. In chapter 6, Drs. Pinhal-Enfield, Vasan and Leibovich show that the study of placental macrophages can provide insights into normal embryonic development and the possible causes of embryo loss.

DNA methylation during early embryonic development has received considerable attention. In chapter 7, Drs. Pan, Smith and Zakar describe the influence of DNA methylation in cell-lineage determination, genomic imprinting and the genesis of germ cells, as well as its role in a group of diseases related to the Developmental Origins of Health and Disease (DOHaD).

3. Perspectives in Embryology

Stem cells are undifferentiated cells that can proliferate and give rise to various types of differentiating cell lines. Stem cells were discovered through a combination of studies involving early embryonic development, genetics, cell surface immunology and tissue culture. The ability of these cells to differentiate into various cell types has created numerous perspectives for their use in many fields. However, the presence of these cells raises important questions. For example, if stem cells occur in humans then why does the human body not regenerate completely? And what about stem cell therapy? Chapter 8 by Drs. Amat, Becerra, Medina, Quesada and Marí-Beffa provides an introduction to the concept, nature and diversity of stem cells. The authors summarize current research in this field and describe the potential applications of stem cells in a variety of areas.

The final chapter (9) by Dr. Beloussov provides a useful summary of what we still do not know and do not understand about organism development. This chapter clearly shows that there are more black boxes or "dark areas" in our understanding than one might initially suppose based on the numerous advances in embryology in recent years. One avenue for future progress suggested by Dr. Beloussov is the application of a self-organizational theory for developmental events, an approach that is generally not addressed in conventional text-books of embryology.

The contents of this book should be of interest to biology and medical students, clinical embryologists, laboratory researchers, obstetricians and urologists, developmental biologists, molecular geneticists and anyone who wishes to know more about contemporary topics of human development. I hope that this book will help to highlight embryology by approaching classic topics of development from a modern perspective.

Luis Antonio Violin Pereira, MD, PhD
Department of Histology and Embryology, State University, Campinas
Brazil

Part 1

Gametes and Infertility

1

Molecular Alterations During Female Reproductive Aging: Can Aged Oocytes Remind Youth?

Misa Imai[1], Junwen Qin[2], Naomi Yamakawa[3], Kenji Miyado[4], Akihiro Umezawa[4] and Yuji Takahashi[4]

[1]Department of Biochemistry, Tufts University School of Medicine
[2]Institute of Reproductive Immunology, College of Life Science and Technology, Jinan University
[3]Research Team for Geriatric Disease, Tokyo Metropolitan Institute of Gerontology
[4]Department of Reproductive Biology, National Center for Child Health and Development
[1]USA
[2]China
[3,4]Japan

1. Introduction

Aging is a multifunctional disorder that leads to cell death, tumors and the other diseases. Accumulation of improper molecular information during aging results in loss of functions in cells and tissues. The ovary is one of the first organs to age; women lose their fertility in their middle age (around 35 years old) and their fecundity expires soon thereafter (at more than 40 years old) (Baird *et al.* 2005; Alviggi *et al.* 2009). Although the exact mechanism underlying female reproductive aging remains unclear, common features among species, including loss of the ovarian follicle pool, disability of chromosome segregation leading to aneuploidy, and increasing mitochondrial dysfunctions, have been reported (Djahanbakhch *et al.* 2007). These changes are largely associated with the unique mechanism of oogenesis. Oocytes that mitotically proliferate during fetal development are stored in the ovaries without further proliferation and are repeatedly ovulated after they enter meiosis. Accordingly, oocytes that are stored for a longer duration gradually lose their functions because the ovarian microenvironment changes with aging.

Ovulation is known to produce reactive oxygen species (ROS) in the ovaries. Although ROS are toxic and sometimes lethal for any cell types, they are even necessary for proper ovulation because direct administration of ROS scavengers, N-acetylcysteine and dibutylhydroxytoluene, into mouse bursa blocked ovulation and hydrogen peroxide-assisted ovulation by functioning like luteinizing hormone (LH) (Shkolnik *et al.* 2011). Nevertheless, repeated exposure of stored oocytes to ROS at each ovulation results in loss of the integrity of these stored oocytes (Chao *et al.* 2005; Miyamoto *et al.* 2010). Oxidative stress is well known to damage macromolecules and cellular components, e.g., mitochondrial desensitization, mitochondrial DNA mutation, irregular DNA methylation, and improper chromosome segregation. In addition, these changes affect the hormonal regulation; losing ovarian endocrine cells by both ovulation and oxidative damages alters the hormonal feedback system in the pituitary-gonad axis.

From a clinical viewpoint, age-associated infertility is not a small part in all the infertile patients. However, lack of knowledge regarding the aging mechanism hampers clinical approaches for treatment of aged women. Here we review recent findings on female reproductive aging and propose possible treatment options for age-associated infertility.

2. The prenatal and postnatal pathways of oogenesis

The debate on the duration of oogenesis in the whole life of females had been sealed for decades. However, recent reports on postnatal oogenesis and germline stem cells have resumed this debate. Herein we describe the prenatal and postnatal pathways of oogenesis from the viewpoint of reproductive aging.

2.1 The prenatal pathway of oogenesis

In the early stage of embryogenesis, primordial germ cells (PGCs) - from which oocytes originate - migrate from the dorsal yolk sac into the genital ridge where gonads would be formed (De Felici & Siracusa 1985). In mice, the germ cells originated from the proximal epiblast of the egg cylinder at embryonic day 5.5 to 6 in response to Bmp4 and Bmp8b signaling (Ying *et al.* 2001). In humans, this process occurs during the first month of gestation (Djahanbakhch *et al.* 2007). Then, the cells undergo mitosis; however the number of PGCs is highly limited at this time point. The PGCs proliferate rapidly, and approximately 7 × 10^6 oogonia are eventually formed at 6–8 weeks of gestation in humans. During this process, transforming growth factor-β (TGF-β) family members, including activins, BMPs, and TGF-β1, support the proliferation of PGCs (Godin & Wylie 1991; Richards *et al.* 1999; Farini *et al.* 2005; Childs *et al.* 2010). Activins and their receptors are highly expressed in human oogonia at later stages of gestation and activin A supports the proliferation of oogonia *in vitro* (Martins da Silva *et al.* 2004). The oogonia then enter meiosis at 11–12 weeks of gestation in humans (Gondos *et al.* 1986).

After oogonia are enclosed by the granulosa cells and primordial follicles are formed, a number of oocytes are destined to die without contributing to reproduction during meiotic prophase I (Hussein 2005; Ghafari *et al.* 2007). More than one-third of all pachytene oocytes are proapoptotic, and a high frequency of atresia is observed between midterm and birth in the human ovaries (Speed 1988; De Pol *et al.* 1997). The large-scale loss of the ovarian follicle pool has been estimated to be more than 80% in humans (Martins da Silva *et al.* 2004).

Several paracrine factors that affect oocyte survival have been reported (Fig. 1). For example, growth factors including KIT ligand, leukocyte inhibitory factor (LIF), BMP-4, SDF-1, and basic FGF have been shown to be able to sustain the survival and proliferation of PGCs in the absence of somatic cell support (Farini *et al.* 2005). In addition, SCF, insulin-like growth factor I (IGF-I), and LIF have been found to assist the survival of germ cells in mice (Morita *et al.* 1999; Gu *et al.* 2009). In contrast, tumor necrosis factor-α (TNF-α) promotes apoptosis at the neonatal stage in rats (Marcinkiewicz *et al.* 2002; Morrison & Marcinkiewicz 2002). In addition, intracellular factors determine the fate of oocytes. Members of the B cell lymphoma/leukemia (BCL) protein family, including BCL-2 and BAX, have been suggested to be involved in this process (Felici *et al.* 1999); BCL-2 is expressed in oocytes undergoing meiosis, and its expression is stable during meiotic prophase I, whereas upregulation of BAX is observed in oocytes undergoing apoptosis. Genetic inactivation (knockout) of BAX in mice resulted in higher number of germ cells in peri-natal ovaries compared with wild-type or heterozygous mice (Alton & Taketo 2007). Moreover, NANOS3 and DND1 protect PGCs from apoptosis (Tsuda *et al.* 2003; Youngren *et al.* 2005). Although the biological basis of the oocyte selection has not been completely understood, prenatal loss of oocytes may occur because of exclusion of accumulated mutations in mitochondria, clearance of lethal errors arising during the mitosis or meiotic prophase, or increased survival of some oocytes within a particular sibling "nest" (Ghafari *et al.* 2007). An imbalance between cell death and survival signaling would result in an abnormal number of follicles that would be stored in the ovaries at this stage; higher frequency of oocyte death that is a result of atresia eventually leads to irreversible premature ovarian failure (Krysko *et al.* 2008). Therefore, the number of oocytes that are stored prenatally must be extremely important for the subsequent reproductive period.

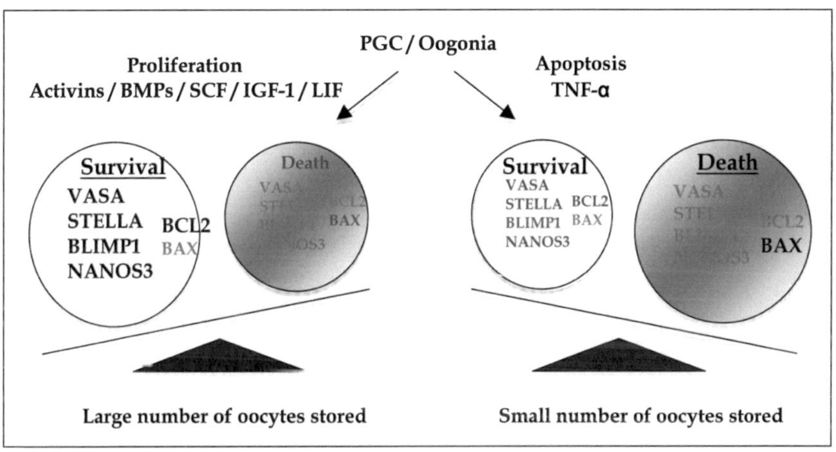

Fig. 1. Determinants of the maximum number of oocytes in the entire life of an organism. Several growth factors, including activins, BMPs, SCF, IGF-1, and LIF, promote proliferation of PGCs and oocyte growth, whereas TNF-α induces oocyte death. The number of follicles at this stage is determined by the balance between survival and death signaling. In addition, the intracellular balance between BCL-2 and BAX determines oocyte survival and death. NANOS3 is another anti-apoptotic molecule found in PGCs.

2.2 The postnatal pathway of oogenesis

Can oocytes be newly produced in adult ovaries? This ancient question arises with the observation of adult mice whose ovaries contain mitotically active germ cells (Johnson *et al.* 2004). This finding led to the hypothesis that oocytes can be generated from sources other than those prenatally stored in the ovaries. A possible origin of postnatal oocytes has been reported to be a specific set of bone marrow cells or peripheral blood cells expressing germline markers (Johnson *et al.* 2005). Johnson *et al.* reported that both bone marrow and peripheral blood transplantations restored oocytes in mice that lost all oocytes by chemotherapy. However, another study claimed that fresh mature oocytes could not be obtained when wild-type and GFP-transgenic mice were parabiotically jointed to establish blood crossover (Eggan *et al.* 2006). Later, this report was supported when transplanted bone marrow cells could be transformed only into immature oocytes (Lee *et al.* 2007; Tilly *et al.* 2009). Moreover, other reports emphasized the possibility that putative germline stem cells exist inside and outside the ovaries in some species. In pigs, fetal skin cells have been reported to contribute to the generation of oocytes (Dyce *et al.* 2006). The ovarian surface epithelium (OSE) cells are another candidate for the origin of oocyte in adult human and rat ovaries (Bukovsky *et al.* 2008; Parte *et al.* 2011), although the candidate cells in OSE may originate from bone marrow cells. In addition, a pancreatic stem cell line seems to differentiate into oocyte-like cells in rats (Danner *et al.* 2007). Unfortunately, none of these germline stem cells have contributed to the production of the next generation. Furthermore, the molecular mechanisms underlying postnatal oogenesis of putative germline stem cells continue to be a black box. Even so, these cells might be useful for clinical applications if a specific condition in which mature fertile oocytes are postnatally generated is elucidated. Although these findings are fascinating, the following questions are arising.

1. Can the germline stem cells participate in oogenesis over the entire life of females? 2. Do other germline stem cells exist? 3. Is there a specific condition in which the germline stem cells participate in oogenesis? 4. How many oocytes are generated in the ovaries through postnatal oogenesis? 5. Are the factors that assist postnatal oogenesis the same as those that assist in prenatal oogenesis? 6. What is the exact role of postnatal oogenesis? Unfortunately, we still have to wait many years to obtain sufficient data to answer these questions.

3. Modification of oocyte quantities and qualities during aging

The common physiology of the ovaries during aging among species includes loss of the ovarian follicle pool, chromosomal abnormalities and cytoplasmic abnormalities (Fig. 2). All these changes may be inevitable and are associated with declining oocyte quality. Here recent findings regarding the alterations in oocyte quantities and qualities are discussed.

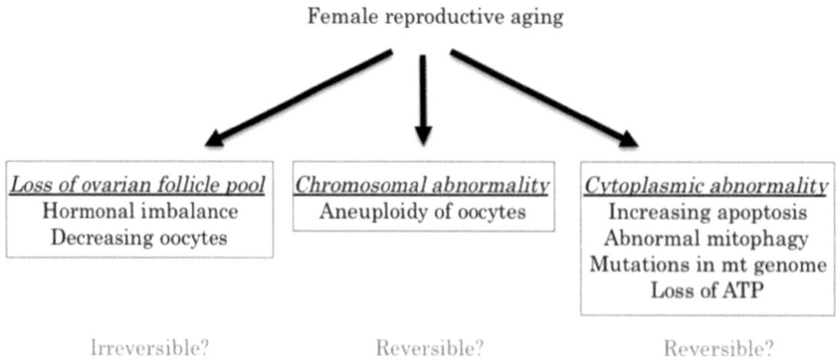

Fig. 2. Common features of female reproductive aging in mammals. Loss of the ovarian follicle pool is caused partly by characteristic oogenesis and partly by ovulation, thereby leading to hormonal imbalance. Chromosomal aberrancy and cytoplasmic abnormalities are possibly induced by longer exposures of stored oocytes to oxidative stress. Recently, these abnormalities have been shown to be reversible by calorie restriction (Selesniemi *et al.* 2011).

3.1 Age-associated decrease of the ovarian follicle pool

After puberty, the number of oocytes steadily decreases due to repeated ovulation. A lot of oocytes are consumed in 1 estrous cycle, but only a few oocytes are eventually ovulated. In humans, only 400 oocytes are dedicated for ovulation throughout the lifespan. During aging, the ovarian follicle pool declines continuously, and this is partly because of atresia. Resting follicles in humans enter atresia through a necrotic process during the initial recruitment phase of folliculogenesis because the ooplasm in those follicles contains increasing numbers of multivesicular bodies and lipid droplets, dilation of the smooth endoplasmic reticulum and the Golgi apparatus, and irregular mitochondria with changed matrix densities (de Bruin *et al.* 2002; de Bruin *et al.* 2004). The atresia of ovarian follicles during aging may be induced by dysfunctions of proteosomes and the endoplasmic reticulum (Matsumine *et al.* 2008).

Although most primordial follicles that are initiated to grow are destined to cell death, the recruitment of follicles from the resting ovarian follicle pool is the sole method to salvage the follicles. FSH is a strong trophic factor that supports both the cyclic recruitment of antral follicles and the growth of follicular somatic cells (Chun *et al.* 1996).

3.2 Age-associated aberrancies of oocyte chromosomes

The most deleterious damage in oocytes is often observed in chromosomes. The relationship between maternal age and the increased incidence of oocyte aneuploidies has been studied in several epidemiological studies (Hassold & Jacobs 1984; McFadden & Friedman 1997; Pellestor *et al.* 2003). In women in their early 20's, the risk of trisomy in a clinically recognized pregnancy is only approximately 2%, whereas it increases up to 35% in women in their 40's (Hassold & Chiu 1985). Supportively, more than half of oocytes from patients of advanced age exhibit aneuploidy (Kuliev & Verlinsky 2004; Pellestor *et al.* 2005). This abnormality is believed to occur because of chromosomal nondisjunction during either meiosis I or II (Nicolaidis & Petersen 1998; Hassold *et al.* 2007). The incidence of aneuploidy is not random; abnormalities of chromosomes 16 and 22 originate more frequently in meiosis II than in meiosis I, and those of chromosomes 13, 18, and 21 occur more frequently in meiosis I than in meiosis II (Kuliev *et al.* 2005). In addition to this nondisjunction theory, the premature separation of chromatids during meiosis is suggested to be responsible for aneuploidy; the age-associated degradation of cohesins or the other molecules sustaining chromatids during metaphase I may contribute to the age-related increase in aneuploidy (Watanabe & Nurse 1999).

3.3 Age-associated decline of mitochondrial activities

The mitochondria alter their organization, shape, and size, depending on various signals (Bereiter-Hahn & Voth 1994). Mitochondrial turnover is the most important process to maintain a healthy state of the mitochondria. During this process, they undergo biogenesis and degradation (Seo *et al.* 2010). Mitochondrial biogenesis is enhanced in muscle cells under certain physiological conditions such as myogenesis, exercise, cold exposure, hypoxia, and calorie restriction (CR) (Freyssenet *et al.* 1996; Nisoli *et al.* 2003; Kraft *et al.* 2006; Civitarese *et al.* 2007; Zhu *et al.* 2010). Damaged or incompetent mitochondria are removed by macroautophagy (Wang & Klionsky 2011).

Although whether mitochondrial dynamics are important for the maintenance of oocyte integrity during aging remains unclear, the copy number of the mitochondria is one of the factors that affect the developmental capacity of oocytes after implantation in mice (Wai *et al.* 2010). In addition, mouse oocytes with artificial mitochondrial damages lost their ability to be matured *in vitro* (Takeuchi *et al.* 2005). Thus, the oocyte quality is largely dependent on mitochondrial health.

Progressive and accumulative damages to mitochondrial DNA (mtDNA) have been postulated to be responsible for the aging process. In aged human fibroblasts, point mutations are likely to occur at specific positions in the replication-controlling region (Michikawa *et al.* 1999). Although these specific mutations have never been reported in aged human oocytes, several mutations in mtDNA were responsible for the decreased ability of oocytes to develop (Barritt *et al.* 2000). In the report, ooplasmic transfer from young oocytes to aged oocytes improved the quality of aged oocytes, indicating that the decreased mitochondrial activities in aged oocytes were complemented.

4. Molecular events during aging

Oocyte quality declines during aging in a complicated process involving several events (Fig. 3). Oxidative stress affects both the size of the ovarian follicle pool and oocyte quality. The reduced follicle pool accelerates hormonal dysregulation. This, in turn, promotes the decrease in the size of the ovarian follicle pool and oocyte quality. In this section, the recent findings regarding the molecular events that occur during reproductive aging are discussed.

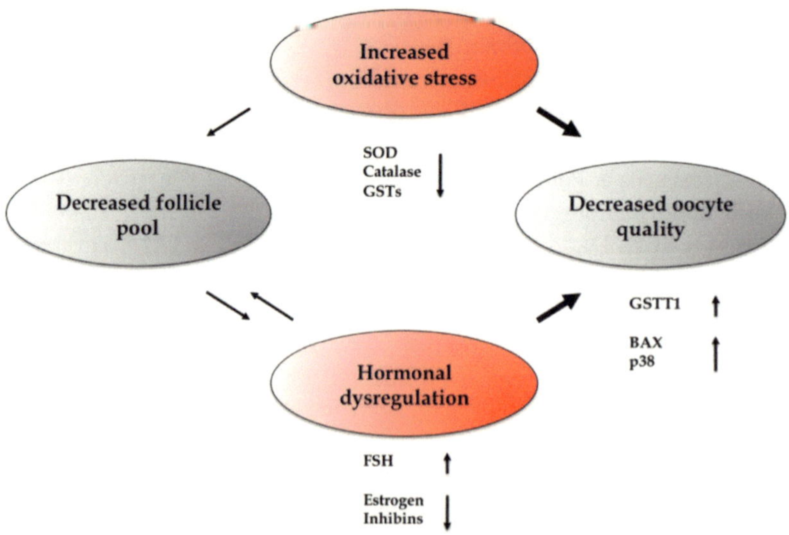

Fig. 3. A negative loop leading to the age-associated decline of oocyte quality. Because of repeated ovulation and the loss of antioxidants including SOD, catalase, glutathione S-transferases (GSTs) etc., excess oxidative stress accelerates the decrease of both oocyte quality and the size of the follicle pool. The decreased follicle pool results in the insufficient secretion of ovarian estrogens and inhibins and the rise of FSH. These changes accelerate the decrease of the follicle pool and directly affect oocyte quality. Both oxidative stress and aberrant hormones induce the molecular alterations (GSTT1, p-p38 etc.) in oocytes and granulosa cells.

4.1 Serum hormone levels

Full competence of oocytes to develop to term is acquired depending on the proper timing of hormonal activation. The decrease in follicle numbers because of aging ensures aberrant hormonal regulation as a result of incomplete feedback mechanisms. This hormonal dysregulation will, in turn, accelerate the loss of follicles at the advanced age (McTavish *et al.* 2007).

The most well-known hormones that affect aging are FSH and LH. Under the normal conditions, relatively high levels of FSH promote the synthesis of estradiol, inhibin A, and inhibin B in granulosa cells (Tonetta & diZerega 1989). On the other hand, LH regulates the production of androgens in theca cells of small antral follicles and promotes the conversion of androgens to estradiol by aromatization (Erickson *et al.* 1985). These factors, in turn, decrease the serum levels of FSH and LH. With luteal regression, the downregulation of estradiol and inhibin A in luteinized granulosa cells allows the rise of FSH again at the onset of the subsequent menstrual cycle (Broekmans *et al.* 2009). Therefore, the negative feedback system between the pituitary and the ovary enables follicles to grow properly.

The dysfunction of the hypothalamic GnRH pulse generators results in an abnormal release of FSH and LH from the pituitary around menopause (Wise *et al.* 1996). However, the factor that is critical to induce improper hormonal regulation is the loss of the follicle pool (Fig. 3). Decreasing numbers of follicles in the ovaries result in decreasing concentrations of circulating estrogens and inhibins during aging (Broekmans *et al.* 2009). Accordingly, the serum concentration of FSH is elevated because of aging (Klein *et al.* 1996). The hormonal changes, especially the decrease in the levels of inhibins, are highly associated with oocyte quality (Chang *et al.* 2002).

Anti-Mullerian inhibitory hormone (AMH) is expressed in granulosa cells of nonatretic preantral and small antral follicles (Baarends *et al.* 1995). AMH has been postulated to regulate the entry of primordial follicles into the growing pool (Durlinger *et al.* 2002). As the number of antral follicles decreases with age, the serum amount of AMH diminishes (van Rooij *et al.* 2004).

Because these changes are largely associated with the unique features of oogenesis, age-associated hormonal changes are inevitable. Abnormal levels of hormones can be a risk factor for certain diseases other than infertility. For example, elevated FSH levels stimulate TNF-α synthesis directly from bone marrow granulocytes and macrophages and promote osteoporosis in mice (Iqbal *et al.* 2006; Sun *et al.* 2006). In addition, the single nucleotide polymorphism (SNP) rs6166 of the FSHR gene significantly influences bone mineral density in postmenopausal women (Rendina *et al.* 2010).

4.2 Oxidative stress and cellular scavengers

Oxidative stress is generally accepted as the major cause of aging. The major source of ROS is believed to be the mitochondria, because ROS are generated as byproducts of electron transport during respiration. Although about 1 - 2 % of oxygen in the heart is converted into ROS under physiological conditions, ROS generation increases under pathological conditions (O'Rourke *et al.* 2005; Valko *et al.* 2007). ROS are removed rapidly through multiple pathways to protect cells and tissues in normal young individuals.

A similar system must be present in the ovaries. Free radical activities in human follicular fluid have been shown to be increased during aging (Wiener-Megnazi *et al.* 2004). However, the levels of free radical scavengers, including SOD1, SOD2, and catalase, were significantly decreased in the granulosa cells from older women compared with those in the granulosa cells from younger women (Tatone *et al.* 2006). In addition, oxidative damages measured by the expression of 8-hydroxydeoxyguanosine were observed in oocytes after ovulation (Chao *et al.* 2005; Miyamoto *et al.* 2010). Although the ovarian levels of oxidative stress during aging remain unclear, excess ROS induced by ovulation may affect the quality of oocytes that are stored in the ovaries (Fig. 3). In fact, ovulation induced ROS generation in the ovaries, resulting in oxidative damage of genomic DNA and mitochondrial DNA mutations (Agarwal *et al.* 2005; Chao *et al.* 2005).

Glutathione (GSH) – a direct ROS scavenger – protects cells from deleterious attacks of ROS. Accordingly, it is highly correlated with oocyte quality in terms of viability (Zuelke *et al.* 2003; Luberda 2005). However, whether the level of GSH in oocytes from aged females is decreased compared with that from young females is unclear.

GSTs are well-known detoxification factors that excrete genotoxins by conjugation of GSH directly to the genotoxins (Sheehan *et al.* 2001). In addition to this characteristic, some GSTs have been shown to play important roles in ROS scavenging by affecting JNK stress signaling (Adler *et al.* 1999; Cheng *et al.* 2001). As expected from the known functions, GST activities in aged oocytes were lower than those in young oocytes (Tarin *et al.* 2004). We previously reported that GSTT1 was upregulated in aged granulosa cells, although the other isoforms of GSTs were downregulated (Ito *et al.* 2008). GSTT1 is known to have bilateral features in that it removes toxins and oxidative stress from cells and tissues and it produces harmful formaldehyde using halogenated substrates (Sherratt *et al.* 1998; Landi 2000). Although it remains uncertain whether GSTT1 is positively or negatively involved in reproductive aging, GSTT1-depleted granulosa cells exhibit mitochondrial hyperpolarization, suggesting that GSTT1 plays a role in controlling mitochondrial activities (Ito *et al.*, 2011).

4.3 Genes-related to apoptosis during reproductive aging

BCL family members are closely related to apoptosis in ovarian cells as well as in other cell types (Tilly *et al.* 1997). Overexpression of BCL-2 in mouse ovaries leads to decreased follicular apoptosis (Hsu *et al.* 1996). A prominent decrease of BCL-2 was also observed in eggs aged *in vitro* (Gordo *et al.* 2002). In addition, the upregulation of BIM in cumulus cells seems to accelerate oocyte aging (Wu *et al.* 2011). More impressively, ovarian functions in mice with genetically engineered BAX were prolonged (Perez *et al.* 1999). Supporting this report, damaged oocytes in mice exhibited higher expression level of BAX (Kujjo *et al.* 2010). Hence, BAX may be a therapeutic target for oocyte rejuvenation.

4.4 Cellular signaling involved in oocyte survival and death

The crosstalk of the signal kinases is important for oocyte survival. For the survival of primordial follicles, 3-phosphoinositide-dependent protein kinase-1 (PDK1) in oocytes preserves the lifespan by maintaining the ovarian follicle pool (Reddy *et al.* 2009). PDK1 and PTEN have been reported to be critical regulators in the phosphatidylinositol 3-kinase (PI3K) signaling pathway (Iwanami *et al.* 2009). Therefore, the loss of PTEN results in the activation and depletion of the primordial follicle pool in early adulthood (Reddy *et al.* 2008).

Higher levels of M-phase promoting factors and mitogen-activated protein kinases (MAPKs) have been observed in ovulated oocytes from aged mice (Tarin *et al.* 2004). Although how these signaling molecules are involved in oocyte activities remains uncertain, JNK, but not p38 MAPK, was found to participate in oocyte fragmentation and parthenogenetic activation in aged oocytes (Petrova *et al.* 2009).

In our previous report, p38 MAPK in human granulosa cells showed a unique pattern of activation and localization during aging (Ito *et al.* 2010). Because p38 has been shown to translocate between the nucleus and cytoplasm upon stimulation (Gong *et al.* 2010), the changes in the subcellular localization of active p38 during aging reflect the microenvironmental status around oocytes and granulosa cells. The nuclear localization of p38 in young granulosa cells may be due to the proper transactivation of genes in response to hormones. On the other hand, the cytoplasmic localization of p38 in aged granulosa cells may be resulted from exposure to oxidative stress. Although it is unclear whether or not such kind of age-associated changes occurs in oocytes, regulation of activation and localization of p38 may contribute to oocyte juvenescence.

5. Is it possible to rescue age-related infertility?

5.1 Anti-aging effects of calorie restriction

Although aged oocytes adapt to the stressful environment and endure multiple disorders that occur inside and outside oocytes, they eventually lose their ability to develop to term. Is it possible to rejuvenate aged oocytes? Although oocyte rejuvenation should be considerably attractive for treatment of age-related infertility, it has not succeeded so far. However, it may be possible to prolong ovarian functions (or to delay ovarian aging). Physical juvenescence can be achieved with several methods, including calorie restriction (CR), moderate fitness and nutritional supply (Prokopov & Reinmuth 2010). These treatments are believed to maximize mitochondrial performance and lower the incidence of mitochondrial dysfunction (Lopez-Lluch *et al.* 2006; McKiernan *et al.* 2007).

CR has been suggested to elongate the lifespan of many organisms (Wolf 2006). CR influences energy metabolism, oxidative damage, inflammation, and insulin sensitivity. CR is reported to activate SIRT1, a key factor that regulates longevity (Allard *et al.* 2009). An important role of SIRT1 may be refreshment of damaged mitochondria by inducing macroautophagy (Kume *et al.* 2010). Therefore, CR enhances cellular homeostasis. Apart from the beneficial effects of CR on longevity, it was believed to be the cost for reproductive capacities (Holliday 1989). Harsh CR (30% or more) is unable to maintain stored oocytes enough for their subsequent development, whereas milder CR (20%) resulted in the loss of the negative effects on fecundity (Rocha *et al.* 2007). Recently, CR has been shown to be beneficial for maintaining the integrity of oocyte chromosomes in mice (Selesniemi *et al.* 2011). These effects were mimicked by the genetic loss of the metabolic regulator, peroxisome proliferator-activated receptor gamma coactivator-1 alpha (PGC-1α). Our preliminary data also reveal that CR (approximately 15% reduction of body mass after one year of treatment) did not reduce fertility in mice. Rather, the CR-treated mice bore more offspring. These data indicate that CR supports the oocyte quality in aged females.

Although the hypothesis that physical juvenescence correlates with the maintenance of oocyte integrity must be explored further, other treatments that lead to anti-aging may support oocyte integrity during aging.

5.2 Nutritional supports of fertility

Since CR has been demonstrated to be beneficial for the juvenescence of cells and tissues including germ cells, as described above, the daily nutritional intake must be crucial for cellular juvenescence. One of the most successful nutrients that affect anti-aging may be polyphenols. For example, turmeric-derived tetrahydrocurcumin and green tea polyphenols promote longevity of mice (Kitani *et al.* 2007). Of those polyphenols, resveratrol has been a potent therapeutic target for age-related diseases. Resveratrol is found in eucalyptus, peanuts, and grapevines (Soleas *et al.* 1997), and its functional properties are versatile probably because of its divergent targets. It has beneficial effects in terms of prevention of various cancers, cardiovascular diseases, neurodegenerative diseases, and diabetes in animal models (Vang *et al.* 2011), because of its antioxidant and anti-inflammatory properties (Schmitt *et al.* 2010). Moreover, resveratrol has been demonstrated to prolong lifespan in short-lived vertebrates (Valenzano *et al.* 2006), because it greatly enhances the activity of SIRT1 (Howitz *et al.* 2003; Knutson & Leeuwenburgh 2008). In addition to its anti-aging properties, resveratrol has been reported to function as estrogen through direct association with estrogen-responsive element (ERE) (Klinge *et al.* 2003). Although the beneficial effects of resveratrol in humans remain to be determined, it is expected to be a mimetic of CR and have the potential to preserve oocyte quality in aged females. Supportively, resveratrol assisted the increase of the ovarian follicle pool in both neonatal and aged rat ovaries (Kong *et al.* 2011). Genistein, one of isoflavones, also seems to increase the ovarian follicle pool by inhibiting atresia in aged rats (Chen *et al.* 2010).

Royal jelly (RJ) was reported to contribute to the prolongation of longevity in mice (Inoue *et al.* 2003). RJ reduced the damages of DNA by acting as an antioxidant. Similar to resveratrol, RJ contains an estrogen-like component that associates with the ERE (Mishima *et al.* 2005). Traditionally, it has been used to treat menopausal symptoms, although the detailed mechanism by which RJ treats menopausal disorders is yet to be determined. RJ seems to rebalance the hormonal concentration in the blood; it decreased the FSH concentrations and increased the estrogen concentrations in aged mice (Fig. 4). These changes, in turn, increased the number of ovulated oocytes. This may improve the oocyte quality in aged body, although further investigation is required.

Recently, a probiotic strain, LKM512, present in yogurt was shown to prolong the longevity of mice (Matsumoto *et al.* 2011). LKM512 has been suggested to act on the polyamines circulating bodies and result in the unexpected prolongation of longevity. Although whether these beneficial effects on longevity can sustain the maintenance of oocyte quality remains unclear, some of these nutritional elements may alleviate female reproductive aging.

The exact mechanism by which some anti-aging treatments improve female reproductive capacity remains unknown. However, hormonal regulation in aged females becomes similar to that in young females with anti-aging treatments, and this is probably due to the prevention of follicle loss or the enhancement of hormonal secretion from aged ovaries (Fig. 5).

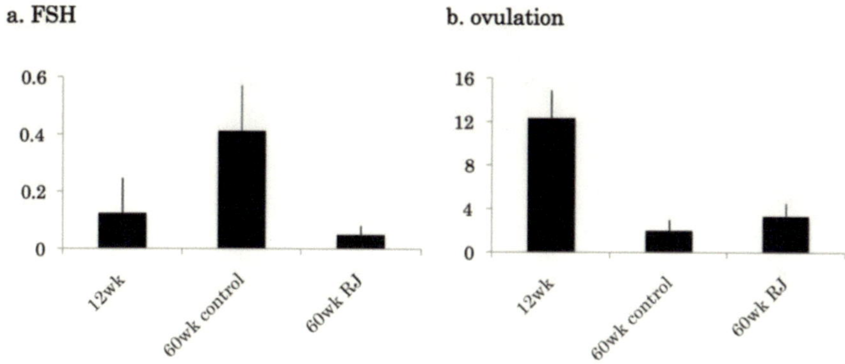

Fig. 4. Effects of RJ on female reproductive capacities. The administration of RJ in drinking water to aged female mice (60 weeks old) for 2 months markedly decreased the serum concentrations of FSH (a). RJ slightly increased the number of ovulated oocytes (b).

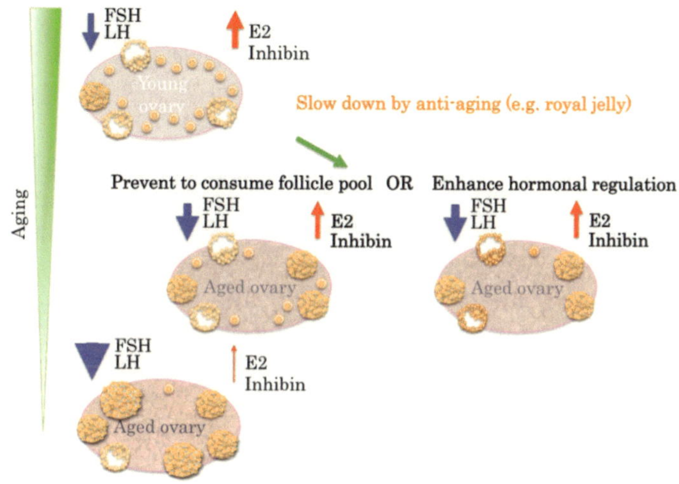

Fig. 5. Age-dependent hormonal regulation in the ovaries. In young females, the negative feedback system in the pituitary-gonadal axis is active, and the amount of FSH and LH is regulated by estrogens (E2) and inhibins secreted from the ovaries. However, inadequate amounts of E2 and inhibins synthesized from aged ovaries cannot decrease the serum levels of FSH and LH. These changes can be treated by anti-aging therapies, such as the supplemental administration of RJ that may prevent the waste of follicles during ovulation or enhance the synthesis of ovarian hormones.

5.3 Other compounds that affect oocyte aging

Because oxidative stress promotes reproductive aging, antioxidants can be effective in regaining reproductive juvenescence. In fact, oral administration of vitamins C and E could prevent age-related ovarian disorders in mice (Tarin *et al.* 1998a; Tarin *et al.* 2002). In addition, some antioxidants are used to treat infertility (Visioli & Hagen 2011).

L-cystine, a component of GSH, and β-mercaptoethanol decreases oocyte quality to develop to the blastocyst stage, whereas dithiothreitol (DTT) enhances the fertilization rate and the developmental capacity of oocytes (Tarin *et al.* 1998b). Moreover, DTT supports embryonic integrity regarding cell number in inner cell mass cells (Rausell *et al.* 2007). Therefore, reagents that assist redox may be effective in enhancing oocyte quality. However, all these compounds were tested in oocytes that were aged *in vitro*, and thus, their reported effects may not be observed in oocytes from aged females.

On the other hand, nitric oxide (NO) seems to be a strong candidate for the treatment of the declining oocyte quality in aged females, because the exposure of aged oocytes to NO decreased the loss of cortical granules and the frequency of spindle abnormalities (Goud *et al.* 2005).

6. Future perspectives to achieve the juvenescence of female fertility

The impact of reproduction on the maternal longevity has been postulated by numerous epidemiological and historical studies (Westendorp & Kirkwood 1998; Le Bourg 2007; Mitteldorf 2010), and the tradeoff between fertility and longevity may occur through genetic or metabolic factors. However, some studies reported a positive correlation between maternal age at reproduction and female longevity (Muller *et al.* 2002; Helle *et al.* 2005). Although the outcomes of those surveys varied, maternal age at the time of the first childbirth seems to be positively correlated with female longevity.

Therefore, the most important factor affecting their fecundity must be physical juvenescence. Because the quality of oocytes from women who successfully give birth at advanced ages is somehow integral, physical juvenescence can increase oocyte quality (chromosomes and cytoplasm), although it may be difficult to increase the number of follicles stored in aged ovaries. Regarding longevity, fitness, or CR is successful in several species, including humans (Lahdenpera *et al.* 2004). In addition, some nutritional supplements could assist the longevity of animals, as described above. Although it remains uncertain whether or not those supplements can improve the integrity of oocytes even after aging, these kinds of treatment may help age-associated infertility in the future.

7. Acknowledgement

We thank Dr. Megumu Ito and Ms. Miho Muraki for their technical assistances. This work was supported partly by a grant from Honeybee Research Foundation of Yamada Bee Farm and partly by a Grant-in Aid for Scientific Research (C) from the Ministry of Education, Culture, Sports, Science and Technology of Japan (18591818).

8. References

Adler V, Yin Z, Fuchs SY, Benezra M, Rosario L, Tew KD, Pincus MR, Sardana M, Henderson CJ, Wolf CR, Davis RJ, Ronai Z (1999). Regulation of JNK signaling by GSTp. *EMBO J*. 18, 1321-1334.

Agarwal A, Gupta S, Sharma RK (2005). Role of oxidative stress in female reproduction. *Reprod Biol Endocrinol*. 3, 28.

Allard JS, Perez E, Zou S, de Cabo R (2009). Dietary activators of Sirt1. *Mol Cell Endocrinol*. 299, 58-63.

Alton M, Taketo T (2007). Switch from BAX-dependent to BAX-independent germ cell loss during the development of fetal mouse ovaries. *J Cell Sci.* 120, 417-424.

Alviggi C, Humaidan P, Howles CM, Tredway D, Hillier SG (2009). Biological versus chronological ovarian age: implications for assisted reproductive technology. *Reprod Biol Endocrinol.* 7, 101.

Baarends WM, Uilenbroek JT, Kramer P, Hoogerbrugge JW, van Leeuwen EC, Themmen AP, Grootegoed JA (1995). Anti-mullerian hormone and anti-mullerian hormone type II receptor messenger ribonucleic acid expression in rat ovaries during postnatal development, the estrous cycle, and gonadotropin-induced follicle growth. *Endocrinology.* 136, 4951-4962.

Baird DT, Collins J, Egozcue J, Evers LH, Gianaroli L, Leridon H, Sunde A, Templeton A, Van Steirteghem A, Cohen J, Crosignani PG, Devroey P, Diedrich K, Fauser BC, Fraser L, Glasier A, Liebaers I, Mautone G, Penney G, Tarlatzis B (2005). Fertility and ageing. *Hum Reprod Update.* 11, 261-276.

Barritt JA, Brenner CA, Willadsen S, Cohen J (2000). Spontaneous and artificial changes in human ooplasmic mitochondria. *Hum Reprod.* 15 Suppl 2, 207-217.

Bereiter-Hahn J, Voth M (1994). Dynamics of mitochondria in living cells: shape changes, dislocations, fusion, and fission of mitochondria. *Microsc Res Tech.* 27, 198-219.

Broekmans FJ, Soules MR, Fauser BC (2009). Ovarian aging: mechanisms and clinical consequences. *Endocr Rev.* 30, 465-493.

Bukovsky A, Gupta SK, Virant-Klun I, Upadhyaya NB, Copas P, Van Meter SE, Svetlikova M, Ayala ME, Dominguez R (2008). Study origin of germ cells and formation of new primary follicles in adult human and rat ovaries. *Methods Mol Biol.* 450, 233-265.

Chang CL, Wang TH, Horng SG, Wu HM, Wang HS, Soong YK (2002). The concentration of inhibin B in follicular fluid: relation to oocyte maturation and embryo development. *Hum Reprod.* 17, 1724-1728.

Chao HT, Lee SY, Lee HM, Liao TL, Wei YH, Kao SH (2005). Repeated ovarian stimulations induce oxidative damage and mitochondrial DNA mutations in mouse ovaries. *Ann N Y Acad Sci.* 1042, 148-156.

Chen ZG, Luo LL, Xu JJ, Zhuang XL, Kong XX, Fu YC (2010). Effects of plant polyphenols on ovarian follicular reserve in aging rats. *Biochem Cell Biol.* 88, 737-745.

Cheng JZ, Singhal SS, Sharma A, Saini M, Yang Y, Awasthi S, Zimniak P, Awasthi YC (2001). Transfection of mGSTA4 in HL-60 cells protects against 4-hydroxynonenal-induced apoptosis by inhibiting JNK-mediated signaling. *Arch Biochem Biophys.* 392, 197-207.

Childs AJ, Kinnell HL, Collins CS, Hogg K, Bayne RA, Green SJ, McNeilly AS, Anderson RA (2010). BMP signaling in the human fetal ovary is developmentally regulated and promotes primordial germ cell apoptosis. *Stem Cells.* 28, 1368-1378.

Chun SY, Eisenhauer KM, Minami S, Billig H, Perlas E, Hsueh AJ (1996). Hormonal regulation of apoptosis in early antral follicles: follicle-stimulating hormone as a major survival factor. *Endocrinology.* 137, 1447-1456.

Civitarese AE, Carling S, Heilbronn LK, Hulver MH, Ukropcova B, Deutsch WA, Smith SR, Ravussin E (2007). Calorie restriction increases muscle mitochondrial biogenesis in healthy humans. *PLoS Med.* 4, e76.

Danner S, Kajahn J, Geismann C, Klink E, Kruse C (2007). Derivation of oocyte-like cells from a clonal pancreatic stem cell line. *Mol Hum Reprod*. 13, 11-20.

de Bruin JP, Dorland M, Spek ER, Posthuma G, van Haaften M, Looman CW, te Velde ER (2002). Ultrastructure of the resting ovarian follicle pool in healthy young women. *Biol Reprod*. 66, 1151-1160.

de Bruin JP, Dorland M, Spek ER, Posthuma G, van Haaften M, Looman CW, te Velde ER (2004). Age-related changes in the ultrastructure of the resting follicle pool in human ovaries. *Biol Reprod*. 70, 419-424.

De Felici M, Siracusa G (1985). Adhesiveness of mouse primordial germ cells to follicular and Sertoli cell monolayers. *J Embryol Exp Morphol*. 87, 87-97.

De Pol A, Vaccina F, Forabosco A, Cavazzuti E, Marzona L (1997). Apoptosis of germ cells during human prenatal oogenesis. *Hum Reprod*. 12, 2235-2241.

Djahanbakhch O, Ezzati M, Zosmer A (2007). Reproductive ageing in women. *J Pathol*. 211, 219-231.

Durlinger AL, Visser JA, Themmen AP (2002). Regulation of ovarian function: the role of anti-Mullerian hormone. *Reproduction*. 124, 601-609.

Dyce PW, Wen L, Li J (2006). In vitro germline potential of stem cells derived from fetal porcine skin. *Nat Cell Biol*. 8, 384-390.

Eggan K, Jurga S, Gosden R, Min IM, Wagers AJ (2006). Ovulated oocytes in adult mice derive from non-circulating germ cells. *Nature*. 441, 1109-1114.

Erickson GF, Magoffin DA, Dyer CA, Hofeditz C (1985). The ovarian androgen producing cells: a review of structure/function relationships. *Endocr Rev*. 6, 371-399.

Farini D, Scaldaferri ML, Iona S, La Sala G, De Felici M (2005). Growth factors sustain primordial germ cell survival, proliferation and entering into meiosis in the absence of somatic cells. *Dev Biol*. 285, 49-56.

Felici MD, Carlo AD, Pesce M, Iona S, Farrace MG, Piacentini M (1999). Bcl-2 and Bax regulation of apoptosis in germ cells during prenatal oogenesis in the mouse embryo. *Cell Death Differ*. 6, 908-915.

Freyssenet D, Berthon P, Denis C (1996). Mitochondrial biogenesis in skeletal muscle in response to endurance exercises. *Arch Physiol Biochem*. 104, 129-141.

Ghafari F, Gutierrez CG, Hartshorne GM (2007). Apoptosis in mouse fetal and neonatal oocytes during meiotic prophase one. *BMC Dev Biol*. 7, 87.

Godin I, Wylie CC (1991). TGF beta 1 inhibits proliferation and has a chemotropic effect on mouse primordial germ cells in culture. *Development*. 113, 1451-1457.

Gondos B, Westergaard L, Byskov AG (1986). Initiation of oogenesis in the human fetal ovary: ultrastructural and squash preparation study. *Am J Obstet Gynecol*. 155, 189-195.

Gong X, Ming X, Deng P, Jiang Y (2010). Mechanisms regulating the nuclear translocation of p38 MAP kinase. *J Cell Biochem*. 110, 1420-1429.

Gordo AC, Rodrigues P, Kurokawa M, Jellerette T, Exley GE, Warner C, Fissore R (2002). Intracellular calcium oscillations signal apoptosis rather than activation in in vitro aged mouse eggs. *Biol Reprod*. 66, 1828-1837.

Goud AP, Goud PT, Diamond MP, Abu-Soud HM (2005). Nitric oxide delays oocyte aging. *Biochemistry*. 44, 11361-11368.

Gu Y, Runyan C, Shoemaker A, Surani A, Wylie C (2009). Steel factor controls primordial germ cell survival and motility from the time of their specification in the allantois, and provides a continuous niche throughout their migration. *Development*. 136, 1295-1303.

Hassold T, Chiu D (1985). Maternal age-specific rates of numerical chromosome abnormalities with special reference to trisomy. *Hum Genet*. 70, 11-17.

Hassold T, Hall H, Hunt P (2007). The origin of human aneuploidy: where we have been, where we are going. *Hum Mol Genet*. 16 Spec No. 2, R203-208.

Hassold TJ, Jacobs PA (1984). Trisomy in man. *Annu Rev Genet*. 18, 69-97.

Helle S, Lummaa V, Jokela J (2005). Are reproductive and somatic senescence coupled in humans? Late, but not early, reproduction correlated with longevity in historical Sami women. *Proc Biol Sci*. 272, 29-37.

Holliday R (1989). Food, reproduction and longevity: is the extended lifespan of calorie-restricted animals an evolutionary adaptation? *Bioessays*. 10, 125-127.

Howitz KT, Bitterman KJ, Cohen HY, Lamming DW, Lavu S, Wood JG, Zipkin RE, Chung P, Kisielewski A, Zhang LL, Scherer B, Sinclair DA (2003). Small molecule activators of sirtuins extend Saccharomyces cerevisiae lifespan. *Nature*. 425, 191-196.

Hsu SY, Lai RJ, Finegold M, Hsueh AJ (1996). Targeted overexpression of Bcl-2 in ovaries of transgenic mice leads to decreased follicle apoptosis, enhanced folliculogenesis, and increased germ cell tumorigenesis. *Endocrinology*. 137, 4837-4843.

Hussein MR (2005). Apoptosis in the ovary: molecular mechanisms. *Hum Reprod Update*. 11, 162-177.

Inoue S, Koya-Miyata S, Ushio S, Iwaki K, Ikeda M, Kurimoto M (2003). Royal Jelly prolongs the life span of C3H/HeJ mice: correlation with reduced DNA damage. *Exp Gerontol*. 38, 965-969.

Iqbal J, Sun L, Kumar TR, Blair HC, Zaidi M (2006). Follicle-stimulating hormone stimulates TNF production from immune cells to enhance osteoblast and osteoclast formation. *Proc Natl Acad Sci U S A*. 103, 14925-14930.

Ito M, Imai M, Muraki M, Miyado K, Qin J, Kyuwa S, Yoshikawa Y, Hosoi Y, Saito H, Takahashi Y (2011). GSTT1 is upregulated by oxidative stress through p38-MK2 signaling pathway in human granulosa cells: possible association with mitochondrial activity. *Aging (Albany NY)*. 3, 1213-1223.

Ito M, Miyado K, Nakagawa K, Muraki M, Imai M, Yamakawa N, Qin J, Hosoi Y, Saito H, Takahashi Y (2010). Age-associated changes in the subcellular localization of phosphorylated p38 MAPK in human granulosa cells. *Mol Hum Reprod*. 16, 928-937.

Ito M, Muraki M, Takahashi Y, Imai M, Tsukui T, Yamakawa N, Nakagawa K, Ohgi S, Horikawa T, Iwasaki W, Iida A, Nishi Y, Yanase T, Nawata H, Miyado K, Kono T, Hosoi Y, Saito H (2008). Glutathione S-transferase theta 1 expressed in granulosa cells as a biomarker for oocyte quality in age-related infertility. *Fertil Steril*. 90, 1026-1035.

Iwanami A, Cloughesy TF, Mischel PS (2009). Striking the balance between PTEN and PDK1: it all depends on the cell context. *Genes Dev*. 23, 1699-1704.

Johnson J, Bagley J, Skaznik-Wikiel M, Lee HJ, Adams GB, Niikura Y, Tschudy KS, Tilly JC, Cortes ML, Forkert R, Spitzer T, Iacomini J, Scadden DT, Tilly JL (2005). Oocyte generation in adult mammalian ovaries by putative germ cells in bone marrow and peripheral blood. *Cell*. 122, 303-315.

Johnson J, Canning J, Kaneko T, Pru JK, Tilly JL (2004). Germline stem cells and follicular renewal in the postnatal mammalian ovary. *Nature*. 428, 145-150.

Kitani K, Osawa T, Yokozawa T (2007). The effects of tetrahydrocurcumin and green tea polyphenol on the survival of male C57BL/6 mice. *Biogerontology*. 8, 567-573.

Klein NA, Battaglia DE, Fujimoto VY, Davis GS, Bremner WJ, Soules MR (1996). Reproductive aging: accelerated ovarian follicular development associated with a monotropic follicle-stimulating hormone rise in normal older women. *J Clin Endocrinol Metab*. 81, 1038-1045.

Klinge CM, Risinger KE, Watts MB, Beck V, Eder R, Jungbauer A (2003). Estrogenic activity in white and red wine extracts. *J Agric Food Chem*. 51, 1850-1857.

Knutson MD, Leeuwenburgh C (2008). Resveratrol and novel potent activators of SIRT1: effects on aging and age-related diseases. *Nutr Rev*. 66, 591-596.

Kong XX, Fu YC, Xu JJ, Zhuang XL, Chen ZG, Luo LL (2011). Resveratrol, an effective regulator of ovarian development and oocyte apoptosis. *J Endocrinol Invest*.

Kraft CS, LeMoine CM, Lyons CN, Michaud D, Mueller CR, Moyes CD (2006). Control of mitochondrial biogenesis during myogenesis. *Am J Physiol Cell Physiol*. 290, C1119-1127.

Krysko DV, Diez-Fraile A, Criel G, Svistunov AA, Vandenabeele P, D'Herde K (2008). Life and death of female gametes during oogenesis and folliculogenesis. *Apoptosis*. 13, 1065-1087.

Kujjo LL, Laine T, Pereira RJ, Kagawa W, Kurumizaka H, Yokoyama S, Perez GI (2010). Enhancing survival of mouse oocytes following chemotherapy or aging by targeting Bax and Rad51. *PLoS One*. 5, e9204.

Kuliev A, Cieslak J, Verlinsky Y (2005). Frequency and distribution of chromosome abnormalities in human oocytes. *Cytogenet Genome Res*. 111, 193-198.

Kuliev A, Verlinsky Y (2004). Meiotic and mitotic nondisjunction: lessons from preimplantation genetic diagnosis. *Hum Reprod Update*. 10, 401-407.

Kume S, Uzu T, Horiike K, Chin-Kanasaki M, Isshiki K, Araki S, Sugimoto T, Haneda M, Kashiwagi A, Koya D (2010). Calorie restriction enhances cell adaptation to hypoxia through Sirt1-dependent mitochondrial autophagy in mouse aged kidney. *J Clin Invest*. 120, 1043-1055.

Lahdenpera M, Lummaa V, Helle S, Tremblay M, Russell AF (2004). Fitness benefits of prolonged post-reproductive lifespan in women. *Nature*. 428, 178-181.

Landi S (2000). Mammalian class theta GST and differential susceptibility to carcinogens: a review. *Mutat Res*. 463, 247-283.

Le Bourg E (2007). Does reproduction decrease longevity in human beings? *Ageing Res Rev*. 6, 141-149.

Lee HJ, Selesniemi K, Niikura Y, Niikura T, Klein R, Dombkowski DM, Tilly JL (2007). Bone marrow transplantation generates immature oocytes and rescues long-term fertility in a preclinical mouse model of chemotherapy-induced premature ovarian failure. *J Clin Oncol*. 25, 3198-3204.

Lopez-Lluch G, Hunt N, Jones B, Zhu M, Jamieson H, Hilmer S, Cascajo MV, Allard J, Ingram DK, Navas P, de Cabo R (2006). Calorie restriction induces mitochondrial biogenesis and bioenergetic efficiency. *Proc Natl Acad Sci U S A*. 103, 1768-1773.

Luberda Z (2005). The role of glutathione in mammalian gametes. *Reprod Biol*. 5, 5-17.

Marcinkiewicz JL, Balchak SK, Morrison LJ (2002). The involvement of tumor necrosis factor-alpha (TNF) as an intraovarian regulator of oocyte apoptosis in the neonatal rat. *Front Biosci*. 7, d1997-2005.

Martins da Silva SJ, Bayne RA, Cambray N, Hartley PS, McNeilly AS, Anderson RA (2004). Expression of activin subunits and receptors in the developing human ovary: activin A promotes germ cell survival and proliferation before primordial follicle formation. *Dev Biol.* 266, 334-345.

Matsumine M, Shibata N, Ishitani K, Kobayashi M, Ohta H (2008). Pentosidine accumulation in human oocytes and their correlation to age-related apoptosis. *Acta Histochem Cytochem.* 41, 97-104.

Matsumoto M, Kurihara S, Kibe R, Ashida H, Benno Y (2011). Longevity in mice is promoted by probiotic-induced suppression of colonic senescence dependent on upregulation of gut bacterial polyamine production. *PLoS One.* 6, e23652.

McFadden DE, Friedman JM (1997). Chromosome abnormalities in human beings. *Mutat Res.* 396, 129-140.

McKiernan SH, Tuen VC, Baldwin K, Wanagat J, Djamali A, Aiken JM (2007). Adult-onset calorie restriction delays the accumulation of mitochondrial enzyme abnormalities in aging rat kidney tubular epithelial cells. *Am J Physiol Renal Physiol.* 292, F1751-1760.

McTavish KJ, Jimenez M, Walters KA, Spaliviero J, Groome NP, Themmen AP, Visser JA, Handelsman DJ, Allan CM (2007). Rising follicle-stimulating hormone levels with age accelerate female reproductive failure. *Endocrinology.* 148, 4432-4439.

Michikawa Y, Mazzucchelli F, Bresolin N, Scarlato G, Attardi G (1999). Aging-dependent large accumulation of point mutations in the human mtDNA control region for replication. *Science.* 286, 774-779.

Mishima S, Suzuki KM, Isohama Y, Kuratsu N, Araki Y, Inoue M, Miyata T (2005). Royal jelly has estrogenic effects in vitro and in vivo. *J Ethnopharmacol.* 101, 215-220.

Mitteldorf J (2010). Female fertility and longevity. *Age (Dordr).* 32, 79-84.

Miyamoto K, Sato EF, Kasahara E, Jikumaru M, Hiramoto K, Tabata H, Katsuragi M, Odo S, Utsumi K, Inoue M (2010). Effect of oxidative stress during repeated ovulation on the structure and functions of the ovary, oocytes, and their mitochondria. *Free Radic Biol Med.* 49, 674-681.

Morita Y, Manganaro TF, Tao XJ, Martimbeau S, Donahoe PK, Tilly JL (1999). Requirement for phosphatidylinositol-3'-kinase in cytokine-mediated germ cell survival during fetal oogenesis in the mouse. *Endocrinology.* 140, 941-949.

Morrison LJ, Marcinkiewicz JL (2002). Tumor necrosis factor alpha enhances oocyte/follicle apoptosis in the neonatal rat ovary. *Biol Reprod.* 66, 450-457.

Muller HG, Chiou JM, Carey JR, Wang JL (2002). Fertility and life span: late children enhance female longevity. *J Gerontol A Biol Sci Med Sci.* 57, B202-206.

Nicolaidis P, Petersen MB (1998). Origin and mechanisms of non-disjunction in human autosomal trisomies. *Hum Reprod.* 13, 313-319.

Nisoli E, Clementi E, Paolucci C, Cozzi V, Tonello C, Sciorati C, Bracale R, Valerio A, Francolini M, Moncada S, Carruba MO (2003). Mitochondrial biogenesis in mammals: the role of endogenous nitric oxide. *Science.* 299, 896-899.

O'Rourke B, Cortassa S, Aon MA (2005). Mitochondrial ion channels: gatekeepers of life and death. *Physiology (Bethesda).* 20, 303-315.

Parte S, Bhartiya D, Telang J, Daithankar V, Salvi V, Zaveri K, Hinduja I (2011). Detection, characterization, and spontaneous differentiation in vitro of very small embryonic-like putative stem cells in adult Mammalian ovary. *Stem Cells Dev.* 20, 1451-1464.

Pellestor F, Anahory T, Hamamah S (2005). Effect of maternal age on the frequency of cytogenetic abnormalities in human oocytes. *Cytogenet Genome Res*. 111, 206-212.

Pellestor F, Andreo B, Arnal F, Humeau C, Demaille J (2003). Maternal aging and chromosomal abnormalities: new data drawn from in vitro unfertilized human oocytes. *Hum Genet*. 112, 195-203.

Perez GI, Robles R, Knudson CM, Flaws JA, Korsmeyer SJ, Tilly JL (1999). Prolongation of ovarian lifespan into advanced chronological age by Bax-deficiency. *Nat Genet*. 21, 200-203.

Petrova I, Sedmikova M, Petr J, Vodkova Z, Pytloun P, Chmelikova E, Rehak D, Ctrnacta A, Rajmon R, Jilek F (2009). The roles of c-Jun N-terminal kinase (JNK) and p38 mitogen-activated protein kinase (p38 MAPK) in aged pig oocytes. *J Reprod Dev*. 55, 75-82.

Prokopov AF, Reinmuth J (2010). Affordable rejuvenation: a prototype facility in action. *Rejuvenation Res*. 13, 350-352.

Rausell F, Pertusa JF, Gomez-Piquer V, Hermenegildo C, Garcia-Perez MA, Cano A, Tarin JJ (2007). Beneficial effects of dithiothreitol on relative levels of glutathione S-transferase activity and thiols in oocytes, and cell number, DNA fragmentation and allocation at the blastocyst stage in the mouse. *Mol Reprod Dev*. 74, 860-869.

Reddy P, Adhikari D, Zheng W, Liang S, Hamalainen T, Tohonen V, Ogawa W, Noda T, Volarevic S, Huhtaniemi I, Liu K (2009). PDK1 signaling in oocytes controls reproductive aging and lifespan by manipulating the survival of primordial follicles. *Hum Mol Genet*. 18, 2813-2824.

Reddy P, Liu L, Adhikari D, Jagarlamudi K, Rajareddy S, Shen Y, Du C, Tang W, Hamalainen T, Peng SL, Lan ZJ, Cooney AJ, Huhtaniemi I, Liu K (2008). Oocyte-specific deletion of Pten causes premature activation of the primordial follicle pool. *Science*. 319, 611-613.

Rendina D, Gianfrancesco F, De Filippo G, Merlotti D, Esposito T, Mingione A, Nuti R, Strazzullo P, Mossetti G, Gennari L (2010). FSHR gene polymorphisms influence bone mineral density and bone turnover in postmenopausal women. *Eur J Endocrinol*. 163, 165-172.

Richards AJ, Enders GC, Resnick JL (1999). Activin and TGFbeta limit murine primordial germ cell proliferation. *Dev Biol*. 207, 470-475.

Rocha JS, Bonkowski MS, de Franca LR, Bartke A (2007). Effects of mild calorie restriction on reproduction, plasma parameters and hepatic gene expression in mice with altered GH/IGF-I axis. *Mech Ageing Dev*. 128, 317-331.

Schmitt CA, Heiss EH, Dirsch VM (2010). Effect of resveratrol on endothelial cell function: Molecular mechanisms. *Biofactors*. 36, 342-349.

Selesniemi K, Lee HJ, Muhlhauser A, Tilly JL (2011). Prevention of maternal aging-associated oocyte aneuploidy and meiotic spindle defects in mice by dietary and genetic strategies. *Proc Natl Acad Sci U S A*. 108, 12319-12324.

Seo AY, Joseph AM, Dutta D, Hwang JC, Aris JP, Leeuwenburgh C (2010). New insights into the role of mitochondria in aging: mitochondrial dynamics and more. *J Cell Sci*. 123, 2533-2542.

Sheehan D, Meade G, Foley VM, Dowd CA (2001). Structure, function and evolution of glutathione transferases: implications for classification of non-mammalian members of an ancient enzyme superfamily. *Biochem J*. 360, 1-16.

Sherratt PJ, Manson MM, Thomson AM, Hissink EA, Neal GE, van Bladeren PJ, Green T, Hayes JD (1998). Increased bioactivation of dihaloalkanes in rat liver due to induction of class theta glutathione S-transferase T1-1. *Biochem J.* 335 (Pt 3), 619-630.

Shkolnik K, Tadmor A, Ben-Dor S, Nevo N, Galiani D, Dekel N (2011). Reactive oxygen species are indispensable in ovulation. *Proc Natl Acad Sci U S A.* 108, 1462-1467.

Soleas GJ, Diamandis EP, Goldberg DM (1997). Resveratrol: a molecule whose time has come? And gone? *Clin Biochem.* 30, 91-113.

Speed RM (1988). The possible role of meiotic pairing anomalies in the atresia of human fetal oocytes. *Hum Genet.* 78, 260-266.

Sun L, Peng Y, Sharrow AC, Iqbal J, Zhang Z, Papachristou DJ, Zaidi S, Zhu LL, Yaroslavskiy BB, Zhou H, Zallone A, Sairam MR, Kumar TR, Bo W, Braun J, Cardoso-Landa L, Schaffler MB, Moonga BS, Blair HC, Zaidi M (2006). FSH directly regulates bone mass. *Cell.* 125, 247-260.

Takeuchi T, Neri QV, Katagiri Y, Rosenwaks Z, Palermo GD (2005). Effect of treating induced mitochondrial damage on embryonic development and epigenesis. *Biol Reprod.* 72, 584-592.

Tarin J, Ten J, Vendrell FJ, de Oliveira MN, Cano A (1998a). Effects of maternal ageing and dietary antioxidant supplementation on ovulation, fertilisation and embryo development in vitro in the mouse. *Reprod Nutr Dev.* 38, 499-508.

Tarin JJ, Gomez-Piquer V, Pertusa JF, Hermenegildo C, Cano A (2004). Association of female aging with decreased parthenogenetic activation, raised MPF, and MAPKs activities and reduced levels of glutathione S-transferases activity and thiols in mouse oocytes. *Mol Reprod Dev.* 69, 402-410.

Tarin JJ, Perez-Albala S, Cano A (2002). Oral antioxidants counteract the negative effects of female aging on oocyte quantity and quality in the mouse. *Mol Reprod Dev.* 61, 385-397.

Tarin JJ, Ten J, Vendrell FJ, Cano A (1998b). Dithiothreitol prevents age-associated decrease in oocyte/conceptus viability in vitro. *Hum Reprod.* 13, 381-386.

Tatone C, Carbone MC, Falone S, Aimola P, Giardinelli A, Caserta D, Marci R, Pandolfi A, Ragnelli AM, Amicarelli F (2006). Age-dependent changes in the expression of superoxide dismutases and catalase are associated with ultrastructural modifications in human granulosa cells. *Mol Hum Reprod.* 12, 655-660.

Tilly JL, Niikura Y, Rueda BR (2009). The current status of evidence for and against postnatal oogenesis in mammals: a case of ovarian optimism versus pessimism? *Biol Reprod.* 80, 2-12.

Tilly JL, Tilly KI, Perez GI (1997). The genes of cell death and cellular susceptibility to apoptosis in the ovary: a hypothesis. *Cell Death Differ.* 4, 180-187.

Tonetta SA, diZerega GS (1989). Intragonadal regulation of follicular maturation. *Endocr Rev.* 10, 205-229.

Tsuda M, Sasaoka Y, Kiso M, Abe K, Haraguchi S, Kobayashi S, Saga Y (2003). Conserved role of nanos proteins in germ cell development. *Science.* 301, 1239-1241.

Valenzano DR, Terzibasi E, Genade T, Cattaneo A, Domenici L, Cellerino A (2006). Resveratrol prolongs lifespan and retards the onset of age-related markers in a short-lived vertebrate. *Curr Biol.* 16, 296-300.

Valko M, Leibfritz D, Moncol J, Cronin MT, Mazur M, Telser J (2007). Free radicals and antioxidants in normal physiological functions and human disease. *Int J Biochem Cell Biol.* 39, 44-84.

van Rooij IA, Tonkelaar I, Broekmans FJ, Looman CW, Scheffer GJ, de Jong FH, Themmen AP, te Velde ER (2004). Anti-mullerian hormone is a promising predictor for the occurrence of the menopausal transition. *Menopause.* 11, 601-606.

Vang O, Ahmad N, Baile CA, Baur JA, Brown K, Csiszar A, Das DK, Delmas D, Gottfried C, Lin HY, Ma QY, Mukhopadhyay P, Nalini N, Pezzuto JM, Richard T, Shukla Y, Surh YJ, Szekeres T, Szkudelski T, Walle T, Wu JM (2011). What is new for an old molecule? Systematic review and recommendations on the use of resveratrol. *PLoS One.* 6, e19881.

Visioli F, Hagen TM (2011). Antioxidants to enhance fertility: Role of eNOS and potential benefits. *Pharmacol Res.*

Wai T, Ao A, Zhang X, Cyr D, Dufort D, Shoubridge EA (2010). The role of mitochondrial DNA copy number in mammalian fertility. *Biol Reprod.* 83, 52-62.

Wang K, Klionsky DJ (2011). Mitochondria removal by autophagy. *Autophagy.* 7, 297-300.

Watanabe Y, Nurse P (1999). Cohesin Rec8 is required for reductional chromosome segregation at meiosis. *Nature.* 400, 461-464.

Westendorp RG, Kirkwood TB (1998). Human longevity at the cost of reproductive success. *Nature.* 396, 743-746.

Wiener-Megnazi Z, Vardi L, Lissak A, Shnizer S, Reznick AZ, Ishai D, Lahav-Baratz S, Shiloh H, Koifman M, Dirnfeld M (2004). Oxidative stress indices in follicular fluid as measured by the thermochemiluminescence assay correlate with outcome parameters in in vitro fertilization. *Fertil Steril.* 82 Suppl 3, 1171-1176.

Wise PM, Krajnak KM, Kashon ML (1996). Menopause: the aging of multiple pacemakers. *Science.* 273, 67-70.

Wolf G (2006). Calorie restriction increases life span: a molecular mechanism. *Nutr Rev.* 64, 89-92.

Wu Y, Wang XL, Liu JH, Bao ZJ, Tang DW, Zeng SM (2011). BIM(EL)-mediated apoptosis in cumulus cells contributes to degenerative changes in aged porcine oocytes via a paracrine action. *Theriogenology.*

Ying Y, Qi X, Zhao GQ (2001). Induction of primordial germ cells from murine epiblasts by synergistic action of BMP4 and BMP8B signaling pathways. *Proc Natl Acad Sci U S A.* 98, 7858-7862.

Youngren KK, Coveney D, Peng X, Bhattacharya C, Schmidt LS, Nickerson ML, Lamb BT, Deng JM, Behringer RR, Capel B, Rubin EM, Nadeau JH, Matin A (2005). The Ter mutation in the dead end gene causes germ cell loss and testicular germ cell tumours. *Nature.* 435, 360-364.

Zhu L, Wang Q, Zhang L, Fang Z, Zhao F, Lv Z, Gu Z, Zhang J, Wang J, Zen K, Xiang Y, Wang D, Zhang CY (2010). Hypoxia induces PGC-1alpha expression and mitochondrial biogenesis in the myocardium of TOF patients. *Cell Res.* 20, 676-687.

Zuelke KA, Jeffay SC, Zucker RM, Perreault SD (2003). Glutathione (GSH) concentrations vary with the cell cycle in maturing hamster oocytes, zygotes, and pre-implantation stage embryos. *Mol Reprod Dev.* 64, 106-112.

2

Role of Sperm DNA Integrity in Fertility

Mona Bungum
Skånes University Hospital
Sweden

1. Introduction

The traditional semen analysis is a cornerstone in diagnosis and treatment of male fertility (World Health Organisation (WHO), 1999, 2010). The sperm parameters concentration, motility and morphology are, however, claimed to be poor predictors of a male´s fertility status, this in natural conception (Bonde et al., 1998; Guzick et al., 2001) as well as in assisted reproductive techniques (ART) (Wolf et al., 1996).

As a consequence of the limited predictive role of the traditional semen analysis there have for long been searched for better parameters, For the last decade an increased focus on sperm DNA is seen.

Infertile men are shown to have significantly more sperm DNA damage compared to fertile men (Evenson et al., 1999; Gandini et al., 2000; Irvine et al., 2000; Larson et al., 2000; Spanò et al., 2000; Carrell and Liu 2001; Hammadeh et al., 2001; Zini et al., 2001a; Zini et al., 2002) and time to spontaneous pregnancy is proved to be longer in couples where the male partner have an increased amount of sperm with DNA damage (Evenson et al., 1999; Spanò et al., 2000). Methods assessing sperm DNA integrity have shown a better predictivity of both in vivo and in vitro fertility than the WHO sperm parameters (Bungum et al., 2007). Moreover, studies have shown that sperm DNA integrity assessment can be applied in ART in order to find the most effective treatment in a given couple (Bungum et al., 2004, 2007; Boe-Hansen et al., 2006).

Semen quality is known to be influenced by a variety of lifestyle, environmental, and occupational factors. Although still much is unknown, the origins of sperm DNA damage are believed to be multi-factorial where defects during spermatogenesis, abortive apoptosis and oxidative stress may be possible causes of a defective sperm DNA (reviewed in (Erenpreiss et al., 2006a)).

During the last decades a variety of techniques to assess sperm DNA integrity have been developed. In the context of fertility the COMET, TUNEL, and Sperm Chromatin Structure assays (SCSA) as well as the sperm chromatin dispersion (SCD) test are the most frequently used (reviewed in (Erenpreiss et al., 2006a)). So far, SCSA is the test that is found to have the most stable threshold values in regard to fertility and therefore of best clinical value (Bungum et al., 2007) .

Accordingly, this chapter will first of all refer to available SCSA data in regard to fertility.

2. Male infertility

Infertility is a health problem affecting approximately 15% of all couples trying to conceive. It is now evident that in at least 50% of all cases, reduced semen quality is a factor contributing to the problem of the couple. In 20% of the couples, the main cause is solely male related, and in another 27%, both partners contribute to the problem (WHO 2000).

Male infertility can be the result of congenital and acquired urogenital abnormalities, infections of the genital tract, varicocele, endocrine disturbances, genetic or immunological factors (WHO, 2000). However, in at least 50% of the infertile men, no explanation to their reduced semen quality can be found (Seli and Sakkas, 2005; Matzuk and Lamb, 2008; O'Flynn O'Brien et al., 2010).

Recent studies have shown that also sperm factors at a molecular level can cause infertility. One example of this is DNA breaks (Evenson et al., 2002; Sharma et al., 2004; Lewis and Aitken, 2005; Lewis 2007; Aitken, 2006; Erenpreiss et al., 2006a; Evenson and Wixon, 2006; Muratori et al., 2006; Collins et al., 2008; Lewis and Agbaje, 2008; Lewis et al., 2008; Bungum et al., 2007; Zini and Sigman, 2009; Aitken and De Iuliis, 2010; Sakkas and Alvarez, 2010).

2.1 Diagnosis and treatment of male infertility

Traditionally, diagnosis of male infertility is based on the conventional sperm analysis where World Health Organisation (WHO) has set criteria for normality in regard to semen volume, sperm concentration, motility and morphology (WHO, 2010). The traditional semen analysis has, however, been criticized (Bonde et al., 1998; Giwercman et al., 1999; Auger et al., 2001; Guzick et al., 2001; Nallella et al., 2006; Swan 2006), in particular because of lack of power in regard to predict fertility. Human semen is a highly fluctuable fluid and all WHO parameters vary significantly between individuals, seasons, countries and regions and even between consecutive samples from one individual (Chia et al., 1998; WHO 1999; Auger et al., 2000; Jorgensen et al., 2001; Chen et al., 2003; Jorgensen et al., 2006). The traditional analysis is performed by light microscopy of 1-200 spermatozoa and this means a high grade of intra- and interlaboratory variation (Neuwinger et al., 1990; Cooper et al., 1992) and a considerable overlap in all three parameters; sperm concentration, motility and morphology between fertile and infertile is shown (Bonde et al., 1998; Guzick et al., 2001). One of the reasons behind the low status as fertility predictor may be that the WHO analysis only takes few sperm characteristics into consideration. Generally overlooked has been the fact that sperm carry DNA and that the DNA can be of a different quality. Such parameters describing sperm nuclear potential are not routinely assessed. However, there are ongoing debates whether sperm DNA integrity assessment should be introduced as a routine test in all or selected groups of infertile men (Evenson et al., 2000; Giwercman et al., 2010; ASRM, 2008, Makhlouf and Niederberger, 2006; Erenpreiss et al., 2006; Zini and Sigman, 2009).

Until the 1990s, the majority of cases of severe male factor subfertility were virtually untreatable, however, the introduction of ICSI revolutionized the treatment of male infertility (Palermo et al., 1992). However, ICSI is a subject of an ongoing debate regarding its indications and safety (Govaerts et al., 1996; Griffin et al., 2003; Kurinczuk 2003; Verpoest and Tournaye 2006; Varghese et al., 2007). One of the causes to this is that ICSI must be seen as a symptomatic treatment, not taking the underlying causes of infertility into account.

3. Sperm DNA structure

With a volume 40 times less than that of a somatic cell nucleus the genetic material of a spermatozoon is more compact packaged than in the nucleus of a somatic cell (Ward et al., 1991). During spermiogenesis histones are replaced by the more basic and small protamines (Fuentes-Mascorro et al. 2000). Each unit of mammalian sperm chromatin is a toroid containing 50–60 kb of DNA and individual toroids represent DNA loop-domains highly condensed by protamines and fixed at the nuclear matrix. The toroids are bound by disulfide crosslinks, formed by oxidation of sulfhydryl groups of cysteine present in the protamines and each chromosome represents a garland of toroids (Fuentes-Mascorro et al., 2000; Ward et al., 1993). While in most other species the protamines comprise as much as 95%, human protamines comprise 85% of the spermatozoal nucleoproteins. (Fuentes-Mascorro et al., 2000). This may explain why human sperm chromatin is less compacted and more frequently contains DNA breaks (Bench et al., 1993) compared to other species.

4. Sperm DNA damage

Human sperm DNA is often not so well packaged as meant to be (Sakkas et al., 1999a) and is therefore susceptible to DNA damage (Irvine et al., 2000). Whilst the mature sperm is a repair-deficient cell (Sega et al., 1978), oocytes and embryos are, to a certain degree, able to repair DNA damage (Matsuda and Tobari 1988; Ahmadi and Ng 1999b).

Sperm possess a variety of abnormalities at the nuclear level and that these anomalies can have an impact on fertility (Evenson et al., 1980; Hewitson 1999; Huszar 1999). The most common types of DNA damage include chemical modification of a base, inter- and intra-strand crosslinks, and single or double DNA strand breaks (Marchetti and Wyrobek 2005).

The origin of human sperm DNA damage is involving both testicular and post-testicular mechanisms. Testicular mechanisms include a) alterations in chromatin modelling during the process of spermiogenesis, and b) abortive apoptosis, whereas post-testicular factors are mostly related to the action of c) reactive oxygen species (ROS), and d) activation of caspases and endonucleases(Reviewed in (Aitken and De Iuliis, 2010 and Sakkas and Alvarez, 2010)).

Oxidative stress during sperm transport through the male reproductive tract is likely the most frequent cause of sperm DNA damage (Aitken and De Iuliis, 2010; Sakkas and Alvarez, 2010). Under normal conditions ROS are necessary for the functioning of sperm (reviewed by Aitken 2006), however, oxidative stress resulting from an over production or reduced antioxidant protection are thought to cause DNA damage (Aitken et al. 1998).

The risk of having ROS induced DNA damaged sperm increases by advanced age, abstinence time, influence of cancer treatment, varicocele and obesity (reviewed in Erenpreiss et al., 2006a; Aitken and De Iulius 2007). Moreover, several studies have reported a negative effect of cigarette smoking on sperm DNA, but data are not conclusive (Robbins et al., 1997; Sun et al., 1997; Rubes et al., 1998; Potts et al., 1999; Saleh et al., 2002b; Sepaniak et al., 2006). Other sources of ROS include organophosphorous pesticides (Sanchez-Pena et al., 2004) and other types of air pollution (Rubes et al., 1998; Selevan et al., 2000; Evenson and Wixon 2005). These agents possess estrogenic properties that are capable of inducing ROS production (Sanchez-Pena et al., 2004; Baker and Aitken 2005; Bennetts et al., 2008).

5. Sperm DNA integrity testing

Several tests developed to assess sperm DNA damage are available. The most frequently used is the Sperm Chromatin Structure assay (SCSA), the single-cell gel electrophoresis assay (COMET assay) in its alkaline, neutral, 2-tailed versions (Singh et al., 1997; Lewis and Agbaje, 2008; Enciso et al., 2009), the terminal deoxynucleotidyl transferase dUTP nick-end labeling (TUNEL) assay (Gorczyka et al., 1993) and the sperm chromatin dispersion (SCD) test (Fernandez et al., 2003). Although these tests correlate to each other (Aravindan et al. 1997; Zini et al. 2001; Perera et al. 2002; Erenpreiss et al. 2004, Donelly et al., 2000) and all measure single and double strand breaks, the methodologies are based on different principles and different aspects of sperm DNA damage (Makhlouf and Niederberger, 2006).

5.1 Sperm Chromatin Structure Assay (SCSA)

The SCSA® is a flow-cytrometic test based on the fact that damaged sperm chromatin denatures when exposed to a low pH-buffer, whereas normal chromatin remains stable (Evenson et al., 1980). The SCSA measures the denaturation of sperm DNA stained with acridine orange, which differentially stains double- and single stranded DNA. Five to ten thousand cells are analysed. Thereafter, the flow cytometric data is further analyzed using dedicated software (SCSASoft; SCSA Diagnostics, Brookings, SD, USA). Data appears in histo- and cytograms and results given as DAN fragmentation index (DFI) and High DNA stainability (HDS). It is still unclear which mechanisms and types of DNA damage that are lying behind DFI and HDS, however, it is believed that DFI are related to the percentage of sperm with both single strand breaks (SSB) and double strand breaks (DSB) or problems in the histone to protamine exchange. HDS is thought to represent immature sperm. The clear advantage of SCSA is the objectivity of the test as well as the high reproducibility (Giwercman et al., 2003) when ran after the standardised protocol (Evenson et al., 2002). Moreover, the clear cut-off levels in relation to fertility is maybe the most obvious benefit compared to other sperm DNA integrity tests (Bungum et al., 2007). A disadvantage is that an expensive flow cytometer is required to run the analysis. Moreover, the test irreversibly damage spermatozoa; after analysis they cannot be used for fertilisation purposes.

5.2 COMET assay

The COMET assay (Singh et al., 1997) is a single cell gel electrophoresis of immobilised sperm, which involves their encapsulation in agarose, lysis and electrophoresis. When the electric field is applied the negatively charged DNA will be drawn towards the positively charged anode. While undamaged DNA are too large and will remain in nucleus, the smaller broken DNA fragments move in a given period of time. The amount of DNA that leaves the nucleus is a measure of the DNA damage in the cell. The sperm are stained with a DNA-binding dye and the intensity of the fluorescence is measured by image analysis. The overall structure resembles a COMET with a circular head corresponding to the undamaged DNA that remains in the nucleus and a tail of damaged DNA. COMET assay can be ran under neutral or alkaline conditions. Under neutral conditions (pH 8–9), mainly DSB are detected (Collins et al. 2004). Under alkaline conditions, DSB and SSB (at pH 12.3) and additionally alkali labile sites (at pH≥13) can be visualised resulting in increased DNA migration in the electrophoretic field (Fairbairn et al., 1994). In COMET assay, normally only around 100 cells are analysed.

5.3 TUNEL assay

The terminal deoxynucleotidyl transferase (TdT)-mediated 2´-deoxyuridine 5´-triphosphate-nick end labelling (TUNEL) assay can be applied for both light microscopy and flow cytometry. The assay uses TdT to label the 3'-OH ends of double-stranded DNA breaks, but also works on the single strand 3'-OH (Gorczyka et al., 1993). The assay detects the DNA breaks directly, without any initial step of denaturation as in SCSA or by introducing acid or alkaline pH as in COMET assay. Whilst TUNEL assay based on light microscopy normally assess 2-500 cells the flow cytometry TUNEL assess 5-10 000 cells.

5.4 Sperm Chromatin Dispersion (SCD) test

The Sperm chromatin dispersion (SCD) test is a relatively simple test that can be applied either by fluorescence microscopy or by bright field microscopy. The test is based on the principle that sperm with fragmented DNA fails to produce the characteristic halo of dispersed DNA loops that are observed on sperm with non-fragmented DNA, when mixed with aqueous agarose following acid denaturation and removal of chromatin nuclear proteins (Fernandez et al., 2003). In the SCD test normally 500 spermatozoa are analysed.

6. Intra-individual variation of DFI

The WHO semen analysis is a golden standard in diagnosis of male infertility, this despite an intra-individual variation reported to be as high as up to 54% (Keel 2006). The first SCSA studies demonstrated that sperm chromatin parameters varied less within a man (Evenson et al., 1991; Spanò et al., 1998). In a study of 45 men who delivered monthly semen samples Evenson reported an average within-donor CV of the SCSA-parameter DFI around 23% (Evenson et al., 1991). These results were confirmed by another SCSA study of 277 men whose semen was measured two times during a period of six months (Spanò et al., 1998). Also Smit and co-workers in 100 men from an outpatient andrology clinic demonstrated a lower biological variation of sperm DNA fragmentation than the classical WHO sperm parameters (Smit et al., 2007). However, conflicting results were obtained in 282 patients undergoing ART with repeated (between 2 and 5) SCSA measurements. In this study, CV of DFI was as high as 29% (Erenpreiss et al., 2006b). In a more recent study these results were reproduced by Olechuk et al., (2011) who also found the mean CV for DFI to be around 30%. In this study 85% of the men that repeated their SCSA analysis remained in the same DFI category (DFI <30% or DFI >30%) from sample one to sample two.

7. Sperm chromatin damage and male infertility

It is evident that infertile men possess more sperm with DNA damage than fertile men (Evenson et al., 1999; Gandini et al., 2000; Host et al., 2000b; Irvine et al., 2000; Larson et al., 2000; Spanò et al., 2000; Carrell and Liu 2001; Hammadeh et al., 2001; Zini et al., 2001a; Sakkas et al., 2002; Saleh et al., 2002b; Zini et al., 2002; Erenpreisa et al., 2003; Muratori et al., 2003; Saleh et al., 2003a). However, very few infertile men are offered a sperm DNA integrity analysis during their fertility work-up, and are therefore not aware of the problem. This despite the fact that in 10-25% of men diagnosed with unexplained infertility sperm DNA damage can, at least partly explain their childlessness (Bungum et al., 2007; Erenpreiss et al.,2008; Smit et al., 2010; Giwercman et al., 2010).

8. Predictive role of DFI in spontaneous pregnancy

Only few studies have studied the role of sperm DNA integrity in relation to fertility in an unselected population. The US Georgetown study included 165 couples (Evenson et al., 1991) and the Danish first pregnancy planners study (Spanò et al., 2000) included 215 couples who tried to obtain pregnancy. Both studies analysed sperm DNA damage by the use of SCSA and demonstrated that the chance of spontaneous pregnancy, measured by the time-to-pregnancy (TTP), decreased when DFI, exceeded 20–30% and became infinite when DFI was more than 30%.

These results have been confirmed in a more recent SCSA case-control study of 127 men from infertile couples where female factors were excluded and 137 men with proven fertility. The risk of being infertile was increased when DFI rised above 20% in men with normal standard semen parameters (OR 5.1), whereas if one of the WHO parameters were abnormal, the OR for infertility was increased already at DFI above 10% (OR 16) (Giwercman et al., 2010). This above mentioned study demonstrated that SCSA can be used in prediction of the chance of spontaneous pregnancy, independently of the standard sperm parameters but also that combining WHO parameters with DFI can be beneficial. Giwercman and co-workers claimed that since a DFI >20% was found in 40% of men with otherwise normal standard sperm parameters, in almost half of the cases of unexplained infertility, sperm DNA defects are a contributing factor to the problem.

9. Predictive role of DFI in intrauterine insemination (IUI)

Several reports have studied DFI in prediction of fertility following intrauterine insemination (IUI). The first report used the TUNEL assay on prepared semen for sperm DNA integrity analysis (Duran et al., 2002). In 154 couples they found lack of pregnancy when DFI was above 12%. Other smaller SCSA studies have confirmed this (Saleh et al., 2003; Boe-Hanssen el al., 2006). In 2007, our group published a study based on 387 IUI cycles where DFI was assessed by SCSA (Bungum et al., 2007). DFI was shown to be a predictor of fertility independent of other sperm parameters. In men having a DFI level below 30% the proportion of children born per cycle was 19.0%. This was in contrast to those having a DFI value above 30% who only had a take-home-baby rate of 1.5 %. The chance of IUI pregnancy started to decrease already when the DFI value exceeded the 20%-level, but became close to zero when exceeding 30%.

10. Predictive role of DFI in IVF and ICSI

More contrasting data exist regarding role of sperm DNA damage in relation to fertilisation, embryo development and pregnancy outcome in IVF and ICSI.

10.1 IVF and ICSI pregnancy

The first SCSA studies based on a relatively limited number of couples indicated that DFI above 27% could be used as a cut-off value for infertility (Larsson et al., 2000; Larsson-Cook et al., 2003). However, in 2004 three independent SCSA reports demonstrated that one through the use of IVF and ICSI were able to compensate for poor sperm chromatin quality (Bungum et al., 2004; Virro et al., 2004; Gandini et al., 2004). Then in 2007 data based on 388

IVF and 223 ICSI cycles were published (Bungum et al., 2007). We observed no statistically significant differences between the outcomes of ICSI and ICSI when a DFI level of 30% was used.. However, in the DFI >30% group, the results of ICSI were significantly better than those of IVF. When comparing ICSI to IVF the odds ratios (ORs) for biochemical pregnancy, clinical pregnancy and delivery were 3.0 (95% CI: 1.4–6.2), 2.3 (5% CI: 1.1–4.6) and 2.2 (95% CI: 1.0–4.5), respectively. Also smaller reports using TUNEL or COMET assays shows that sperm DNA damage is more predictive in IVF than in ICSI (Hammadeh et al., 1998; Host et al., 2000; Simon et al., 2011). In contrast, one single SCSA study has reported that DFI threshold did not predict IVF outcome (Payne et al., 2005). However in this study the authors did not discriminate between IVF and ICSI.

10.2 Fertilisation and embryo development

In a mouse model Ahmadi and Ng demonstrated that spermatozoa with DNA damage were able to fertilise oocytes (Ahmadi and Ng 1999a). They also reported sperm DNA damage to be related to poor embryo development (Ahmadi and Ng 1999b), although the oocyte, to a certain degree, was able to repair the sperm DNA damage. The human data from ART populations regarding fertilisation and embryo development in relation to DNA damage is, however, conflicting.

In our large study of 611 IVF and ICSI couples we compared fertilisation rates between those having a DFI>30% and those having a DFI≤30%, however, no statistically significant differences were seen, neither for IVF nor for ICSI patients (Bungum et al., 2007). Our findings are in accordance with most other reports using the SCSA analysis (Larson et al., 2000; Larson-Cook et al., 2003; Gandini et al., 2004; Virro et al., 2004; Li et al., 2006) as well as other sperm DNA integrity testing methods (Sakkas et al., 1996; Hammadeh et al., 2001; Tomlinson et al., 2001; Morris et al., 2002; Tomsu et al., 2002; Henkel et al., 2004; Lewis et al., 2004; Huang et al., 2005; Nasr-Esfahani et al., 2005). In contrast, a negative correlation between sperm DNA fragmentation and IVF and ICSI fertilisation rates were reported by others (Sun et al., 1997; Lopes et al., 1998a; Saleh et al., 2003a; Payne et al., 2005). A Danish study (Host et al., 2000a) assessed sperm DNA breaks with the TUNEL- assay in infertile couples and found negative correlations between the proportion of spermatozoa with DNA damage and fertilisation in all groups except for those treated with ICSI.

In our ART-study from 2007 (Bungum et al., 2007), embryo development was compared between those having a SCSA-DFI>30% and those having a DFI≤30%, however, no statistically significant differences between the groups were seen. Several other authors have reported identical results (Larson et al., 2000; Larson-Cook et al., 2003; Gandini et al., 2004; Boe-Hansen et al., 2006). This, however, is contrasted by the findings of others (Sun et al., 1997; Morris et al., 2002; Saleh et al., 2003a), who showed that DFI levels were negatively correlated with embryo quality after IVF and ICSI. Seli et al. (2004) and Virro et al. (2004) reported that men with a high DFI had higher risk of low blastocyst formation rate compared to those with a low DFI.

10.3 DFI and risk of pregnancy loss

Traditionally, pregnancy loss has been explained by either genetic, structural, infective, endocrine or unexplained causes (reviewed in (Rai and Regan 2006)). However, recently also sperm DNA damage has been associated to recurrent pregnancy loss (Evenson et al., 1999; Carrell et al., 2003). In a population of first pregnancy planners Evenson et al. (1999) reported a miscarriage rate higher in fertile couples where the partner had a high SCSA-DFI compared to those with a low DFI. Also Carrell et al. (2003) by using the TUNEL assay on sperm from 24 couples with unexplained recurrent pregnancy loss showed that pregnancy loss was associated with DNA damage. As control group they used donors of known fertility and unscreened men from the general population.

Several studies have reported an increased risk of pregnancy loss after IVF and ICSI (Check et al., 2005; Zini et al. 2005; Lin et al., 2008), however data are conflicting (Bungum et al., 2007). In a recent meta-analysis Zini and co-workers (Zini et al., 2008) collected data from 11 studies involving 1549 IVF/ICSI treatments and 640 pregnancies. They demonstrated a combined odds ratio (OR) of 2.48 (95% CI; 1.52, 4.04, $p<0.0001$). The conclusions from this analysis should, however, be interpreted with care since different types of sperm DNA integrity tests and different sperm sources were included. We have demonstrated that washed semen is not predictive for the outcome of ART (Bungum et al., 2008). Further large-scale studies should be performed to get firm conclusions.

11. Prevention or treatment of sperm DNA damage

A more precise diagnosing would enable clinicians to better counsel the infertile couple and may also result in improvement and further development of cause-related therapy, which is very little used in today's clinical practice (Skakkebaek et al., 1994). The effects of therapy based on antioxidants on sperm DNA quality has been evaluated, however, the studies have been small and conflicting (Greco et al., 2005a; Greco et al., 2005b; Silver et al., 2005; Song et al., 2006; Menezo et al., 2007; Kefer et al., 2009) and so far no standardized, clinically well-proven treatment of sperm oxidative stress and DNA damage are established.

Data on how to best advise couples in what they self can do to prevent against sperm DNA damage are lacking. Several life-style factors are suggested to negatively influence sperm DNA, however, data are conflicting. Examples are smoking and obesity (Reviewed in (DuPlessis et al., 2010 and Agarwal et al., 2008)). Interestingly, the first published intervention study in severe obese men demonstrated improvement in WHO parameters but not in sperm DNA integrity after weight reduction (Hakonsen et al., 2011).

12. Conclusion and future perspectives

Sperm DNA damage is a frequent problem in infertile men. Although not yet routine, sperm DNA integrity is a good tool in investigation and treatment of infertility. Sperm DNA integrity is a marker of male fertility, alone or in combination with the WHO semen parameters, this in natural conception as well as in ART.

The role of DFI as a predictor of fertilisation, embryo development and pregnancy in IVF and ICSI is still discussed. Although not conclusive, also a relation between sperm DNA damage and risk of pregnancy loss have been suggested.

Among the different sperm DNA integrity assays available SCSA is currently the only method that has provided clear and stable clinical cut-off levels and therefore can be recommended for a clinical sperm DNA damage evaluation. The normality ranges and thresholds for male fertility potential of the other assays still need to be clarified.

It seems clear that ART, especially ICSI, are able to overcome the natural barriers of sperm DNA damage levels not compatible with fertilisation under natural circumstances. The consequences of this for the progeny are still not clear, however, it is also reason to concern regarding possible consequences of achieving a pregnancy using spermatozoa possessing DNA damage. So far no proven treatment of sperm DNA damage is available. Adequately powered, placebo-controlled trials of antioxidants in the prevention of sperm oxidative stress should be performed.

13. References

Agarwal, A., Desai,. NR., Ruffoli, R. & Carpi A. (2008). Lifestyle and testicular dysfunction: a brief update. Biomed Pharmacother 62(8):550-553.

Ahmadi, A. & Ng, S. C. (1999a). Fertilizing ability of DNA-damaged spermatozoa. J Exp Zool 284, 696-704.

Ahmadi, A. & Ng, S. C. (1999b). Developmental capacity of damaged spermatozoa. Hum Reprod 14, 2279-2285.

Aitken, J. & De Iulius, G. N. (2007). Origins and concequences of DNA damage in male germ cells. Reprod Biomed Online 14, 727-733

Aitken, R. J. & De Iuliis, G.N. (2010). On the possible origins of DNA damage in human spermatozoa. Mol Hum Reprod 16:3-13.

Aitken, R. J. (2006). Sperm function tests and fertility. Int J Androl 29, 69-75; discussion 105-108.

Aitken, R. J. , De Iuliis, G.N. & McLachlan R.I. (2009). Biological and clinical significance of DNA damage in the male germ line. Int J Androl 32:46-56.

Aravindan, G. R., Bjordahl, J., Jost, L. K. & Evenson, D. P. (1997). Susceptibility of human sperm to in situ DNA denaturation is strongly correlated with DNA strand breaks identified by single-cell electrophoresis. Exp Cell Res 236, 231-237

Auger, J., Eustache, F., Andersen, A. G., Irvine, D. S., Jorgensen, N., Skakkebaek, N. E., Suominen, J., Toppari, J., Vierula, M. & Jouannet, P. (2001). Sperm morphological defects related to environment, lifestyle and medical history of 1001 male partners of pregnant women from four European cities. Hum Reprod 16, 2710-2717.

Baker, M. A. & Aitken, R. J. (2005). Reactive oxygen species in spermatozoa: methods for monitoring and significance for the origins of genetic disease and infertility. Reprod Biol Endocrinol 3, 67.

Bench, G.S., Friz, A.M., Corzett, M.H., Morse, D.H. & Balhorn R. (1996). DNA and total protamine masses in individual sperm from fertile mammalian subjects. Cytometry 23(4):263-271.

Bennetts, L. E., De Iuliis, G. N., Nixon, B., Kime, M., Zelski, K., McVicar, C. M., Lewis, S. E. & Aitken, R. J. (2008). Impact of estrogenic compounds on DNA integrity in human spermatozoa: Evidence for cross-linking and redox cycling activities. Mutat Res 641, 1-11.

Boe-Hansen, G. B., Fedder, J., Ersboll, A. K. & Christensen, P. (2006) The sperm chromatin structure assay as a diagnostic tool in the human fertility clinic. Hum Reprod 21, 1576-1582.

Bonde, J. P., Ernst, E., Jensen, T. K., Hjollund, N. H., Kolstad, H., Henriksen, T. B., Scheike, T., Giwercman, A., Olsen, J. & Skakkebaek, N. E. (1998). Relation between semen quality and fertility: a population-based study of 430 first-pregnancy planners. Lancet 352, 1172-1177.

Bungum, M., Humaidan, P., Axmon, A., Spanó, M., Bungum, L., Erenpreiss, J. & Giwercman A. (2007) Sperm DNA integrity assessment in prediction of assisted reproduction technology outcome. Human Reproduction 22:174-179.

Bungum, M., Humaidan, P., Spanó, M., Jepson, K., Bungum, L. & Giwercman, A. (2004).The predictive value of sperm chromatin structure assay (SCSA) parameters for the outcome of intrauterine insemination, IVF and ICSI. Hum Reprod 19:1401-1408.

Bungum, M., Spanó, M., Humaidan, P., Eleuteri, P., Rescia, M. & Giwercman, A. (2008). Sperm Chromatin Structure Assay (SCSA) parameters measured after density gradient centrifugation are not predictive for the outcome of ART. Hum Reprod 23:4-10.

Carrell, D. T. & Liu, L. (2001) Altered protamine 2 expression is uncommon in donors of known fertility, but common among men with poor fertilizing capacity, and may reflect other abnormalities of spermiogenesis. J Androl 22, 604-610.

Carrell, D. T., Liu, L., Peterson, C. M., Jones, K. P., Hatasaka, H. H., Erickson, L. & Campbell, B. (2003). Sperm DNA fragmentation is increased in couples with unexplained recurrent pregnancy loss. Arch Androl 49, 49-55.

Check, J. H., Graziano, V., Cohen, R., Krotec, J. & Check, M. L. (2005). Effect of an abnormal sperm chromatin structural assay (SCSA) on pregnancy outcome following (IVF) with ICSI in previous IVF failures. Arch Androl 51, 121-124.

Chen, Z., Toth, T., Godfrey-Bailey, L., Mercedat, N., Schiff, I. & Hauser, R. (2003) Seasonal variation and age-related changes in human semen parameters. J Androl 24, 226-231.

Chia, S. E., Tay, S. K.& Lim, S. T. (1998) What constitutes a normal seminal analysis? Semen parameters of 243 fertile men. Hum Reprod 13, 3394-3398.

Collins, A.R. (2004). The COMET assay for DNA damage and repair: principles, applications, and limitations. Mol Biotechnol 26(3):249-261.

Collins, J. A., Barnhart, K. T. & Schlegel, P. N. (2008). Do sperm DNA integrity tests predict pregnancy with in vitro fertilization? Fertil Steril 89, 823-831.

Cooper, T. G., Neuwinger, J., Bahrs, S. & Nieschlag, E. (1992). Internal quality control of semen analysis. Fertil Steril 58, 172-178.

Donnelly, E. T., O'Connell, M., McClure, N. & Lewis, S. E. (2000). Differences in nuclear DNA fragmentation and mitochondrial integrity of semen and prepared human spermatozoa. Hum Reprod 15:1552–1561.

Du Plessis SS, Cabler S, McAlister DA, Sabanegh E, Agarwal A. (2010). The effect of obesity on sperm disorders and male infertility. Nat Rev Urol. 7(3):153-161

Duran, E. H., Morshedi, M., Taylor, S. and Oehninger, S. (2002) Sperm DNA quality predicts intrauterine insemination outcome: a prospective cohort study. Hum Reprod 17, 3122-3128.

Enciso, M., Muriel, L., Fernández, J.L., Goyanes, V., Segrelles, E., Marcos, M., Montejo, J.M., Ardoy, M., Pacheco, A. & Gosálvez J. (2006). Infertile men with varicocele show a high relative proportion of sperm cells with intense nuclear damage level, evidenced by the sperm chromatin dispersion test. J Androl 27:106-111.

Erenpreisa, J., Erenpreiss, J., Freivalds, T., Slaidina, M., Krampe, R., Butikova, J., Ivanov, A. & Pjanova, D. (2003). Toluidine blue test for sperm DNA integrity and elaboration of image cytometry algorithm. Cytometry A 52, 19-27.

Erenpreiss, J., Bungum, M., Spano, M., Elzanaty, S., Orbidans, J. & Giwercman , A. (2006). Intra-individual variation in Sperm Chromatin Structure Assay parameters in men from infertile couples: clinical implications. Human Reproduction 21:2061-2064.

Erenpreiss, J., Elzanaty, S. & Giwercman, A. (2008). Sperm DNA damage in men from infertile couples. Asian J Androl 10:786-790.

Erenpreiss, J., Jepson, K., Giwercman, A., Tsarev, I., Erenpreisa, J. & Spanò , M. (2004). Toluidine blue cytometry test for sperm DNA conformation: comparison with the flow cytometric sperm chromatin structure and TUNEL assays. Hum Reprod 19:2277–2282.

Erenpreiss, J., Spano, M., Erenpreisa, J., Bungum, M. & Giwercman, A. (2006a). Sperm chromatin structure and male fertility: biological and clinical aspects. Asian J Androl 8, 11-29.

Evenson, D. & Jost, L. (2000). Sperm chromatin structure assay is useful for fertility assessment. Methods Cell Sci 22, 169-189.

Evenson, D. & Wixon, R. (2006).Meta-analysis of sperm DNA fragmentation using the sperm chromatin structure assay. Reprod Biomed Online 12, 466-472.

Evenson, D. P., Darzynkiewicz, Z. & Melamed, M. R. (1980). Relation of mammalian sperm chromatin heterogeneity to fertility. Science 210, 1131-1133.

Evenson, D. P., Jost, L. K., Baer, R. K., Turner, T. W. &Schrader, S. M. (1991). Individuality of DNA denaturation patterns in human sperm as measured by the sperm chromatin structure assay. Reprod Toxicol 5, 115-125.

Evenson, D. P., Jost, L. K., Marshall, D., Zinaman, M. J., Clegg, E., Purvis, K., de Angelis, P. & Claussen, O. P. (1999). Utility of the sperm chromatin structure assay as a diagnostic and prognostic tool in the human fertility clinic. Hum Reprod 14, 1039-1049.

Evenson, D. P., Larson, K. L. & Jost, L. K. (2002).Sperm chromatin structure assay: its clinical use for detecting sperm DNA fragmentation in male infertility and comparisons with other techniques. J Androl 23, 25-43.

Fairbairn, D.W., Olive, P.L. & O'Neill K.L.. (1995). The COMET assay: a comprehensive review. Mutat Res 339(1):37-59.

Fernández, J.L., Muriel, L., Goyanes, V., Segrelles, E., Gosálvez,J., Enciso, M., LaFromboise, M. & De Jonge C. (2005). Simple determination of human sperm DNA fragmentation with an improved sperm chromatin dispersion test. Fertil Steril;84:833-842.

Fernández, J.L., Muriel, L., Rivero, M.T., Goyanes, V., Vazquez, R., Alvarez, J.G. (2003). The sperm chromatin dispersion test: a simple method for the determination of sperm DNA fragmentation. J Androl 24:59–66.

Fuentes-Mascorro, G., Vergara-Onofre, M., Mercado, E., Hernandez-Perez, O.& Rosado, A. (2000). Participation of DNA structure on sperm chromatin organization. Arch Androl 45, 61-71.

Gandini, L., Lombardo, F., Paoli, D., Caponecchia, L., Familiari, G., Verlengia, C., Dondero, F. & Lenzi, A. (2000). Study of apoptotic DNA fragmentation in human spermatozoa. Hum Reprod 15, 830-839.

Gandini, L., Lombardo, F., Paoli, D., Caruso, F., Eleuteri, P., Leter, G., Ciriminna, R., Culasso, F., Dondero, F.& Lenzi, A. *et al.* (2004). Full-term pregnancies achieved with ICSI despite high levels of sperm chromatin damage. Hum Reprod 19, 1409-1417.

Giwercman, A., Lindstedt, L., Larsson, M., Bungum, M., Spanó, M., Levine, R.J. & Rylander, L. (2010). Sperm chromatin structure assay as an independent predictor of fertility in vivo: a case-control study. Int J Androl 33:221-227.

Giwercman, A., Richthoff, J., Hjollund, H., Bonde, J. P., Jepson, K., Frohm, B. & Spano, M. (2003). Correlation between sperm motility and sperm chromatin structure assay parameters. Fertil Steril 80, 1404-1412.

Giwercman, A., Spano, M., Lahdetie, J. & Bonde, J. P. (1999). Quality assurance of semen analysis in multicenter studies. Asclepios. Scand J Work Environ Health 25 Suppl 1, 23-25; discussion 76-28.

Gorczyca, W., Gong, J & Darzynkiewicz, Z. (1993). Detection of DNA strand breaks in individual apoptotic cells by the in situ terminal deoxynucleotidyl transferase and nick translation assays. Cancer Res 53, 1945-1951.

Govaerts, I., Koenig, I., Van den Bergh, M., Bertrand, E., Revelard, P. & Englert, Y. (1996) Is intracytoplasmic sperm injection (ICSI) a safe procedure? What do we learn from early pregnancy data about ICSI? Hum Reprod 11, 440-443.

Greco, E., Iacobelli, M., Rienzi, L., Ubaldi, F., Ferrero, S. & Tesarik, J. (2005). Reduction of the incidence of sperm DNA fragmentation by oral antioxidant treatment. J Androl 26, 349-353.

Greco, E., Romano, S., Iacobelli, M-, Ferrero, S., Baroni, E., Minasi, M..G., Ubaldi, F., Rienzi L. & Tesarik J. (2005). ICSI in cases of sperm DNA damage: beneficial effect of oral antioxidant treatment. Hum Reprod 20:2590-2594.

Griffin, D. K., Hyland, P., Tempest, H. G. & Homa, S. T. (2003). Safety issues in assisted reproduction technology: Should men undergoing ICSI be screened for chromosome abnormalities in their sperm? Hum Reprod 18, 229-235.

Guzick, D. S., Overstreet, J. W., Factor-Litvak, P., Brazil, C. K., Nakajima, S. T., Coutifaris, C., Carson, S. A., Cisneros, P., Steinkampf, M. P. & Hill, J. A. *et al.* (2001). Sperm morphology, motility, and concentration in fertile and infertile men. N Engl J Med 345, 1388-1393.

Håkonsen, L.B., Thulstrup, A.M., Aggerholm, A.S., Olsen, J., Bonde, J.P., Andersen, C.Y., Bungum, M., Ernst, E.H., Hansen, M.L., Ernst, E.H. & Ramlau-Hansen CH (2011).

Does weight loss improve semen quality and reproductive hormones? results from a cohort of severely obese men. Reprod Health 17;8:24.

Hammadeh, M. E., Zeginiadov, T., Rosenbaum, P., Georg, T., Schmidt, W. & Strehler, E. (2001). Predictive value of sperm chromatin condensation (aniline blue staining) in the assessment of male fertility. Arch Androl 46, 99-104.

Hammadeh, M.E., Stieber, M., Haidl, G. & Schmidt W. (1998). Association between sperm cell chromatin condensation, morphology based on strict criteria, and fertilization, cleavage and pregnancy rates in an IVF program. Andrologia 30(1):29-35.

Henkel, R., Hajimohammad, M., Stalf, T., Hoogendijk, C., Mehnert, C., Menkveld, R., Gips, H., Schill, W. B. & Kruger, T. F. (2004). Influence of deoxyribonucleic acid damage on fertilization and pregnancy. Fertil Steril 81, 965-972.

Hewitson, L., Simerly, C, Sutovsky, P., Dominko, T., Takahashi, D. & Schatten, G. (1999). The fate of sperm components within the egg during fertilization: implications for infertility. In C, G. (ed) *The Male Gamete: From Basic Science to Clinical Applications* Cache River Press, Vienna.

Host, E., Lindenberg, S. & Smidt-Jensen, S. (2000). The role of DNA strand breaks in human spermatozoa used for IVF and ICSI. Acta Obstet Gynecol Scand 79(7):559-563.

Huang, C. C., Lin, D. P., Tsao, H. M., Cheng, T. C., Liu, C. H. & Lee, M. S. (2005). Sperm DNA fragmentation negatively correlates with velocity and fertilization rates but might not affect pregnancy rates. Fertil Steril 84, 130-140.

Huszar, G., Zeyneloglu, H. B. & Vigue, L. (1999). Cellular maturity and fertilizing potential of sperm populations in natural and assisted reproduction. In C, G. (ed) *The Male Gamete: From Basic Science to Clinical Applications*. Cache River Press, Vienna, pp. 385–396.

Irvine, D. S., Twigg, J. P., Gordon, E. L., Fulton, N., Milne, P. A. & Aitken, R. J. (2000). DNA integrity in human spermatozoa: relationships with semen quality. J Androl 21, 33-44.

Jorgensen, N., Andersen, A. G., Eustache, F., Irvine, D. S., Suominen, J., Petersen, J. H., Andersen, A. N., Auger, J., Cawood, E. H. & Horte, A. *et al.* (2001). Regional differences in semen quality in Europe. Hum Reprod 16, 1012-1019.

Jorgensen, N., Asklund, C., Carlsen, E. & Skakkebaek, N. E. (2006). Coordinated European investigations of semen quality: results from studies of Scandinavian young men is a matter of concern. Int J Androl 29, 54-61; discussion 105-108.

Keel, B. A. (2006). Within- and between-subject variation in semen parameters in infertile men and normal semen donors. Fertil Steril 85, 128-134.

Kefer, J.C., Agarwal, A., & Sabanegh, E. (2009), Role of antioxidants in the treatment of male infertility. Int J Urol 2009;16:449-457.

Kurinczuk, J. J. (2003) Safety issues in assisted reproduction technology. From theory to reality--just what are the data telling us about ICSI offspring health and future fertility and should we be concerned? Hum Reprod 18, 925-931.

Larson, K. L., DeJonge, C. J., Barnes, A. M., Jost, L. K.& Evenson, D. P. (2000). Sperm chromatin structure assay parameters as predictors of failed pregnancy following assisted reproductive techniques. Hum Reprod 15, 1717-1722.

Larson-Cook, K. L., Brannian, J. D., Hansen, K. A., Kasperson, K. M., Aamold, E. T. & Evenson, D. P. (2003). Relationship between the outcomes of assisted reproductive techniques and sperm DNA fragmentation as measured by the sperm chromatin structure assay. Fertil Steril 80, 895-902.

Lewis, S. E & Agbaje, I. M. (2008). Using the alkaline comet assay in prognostic tests for male infertility and assisted reproductive technology outcomes. Mutagenesis 23:163-170.

Lewis, S. E. (2007). Is sperm evaluation useful in predicting human fertility? Reproduction 134, 31-40.

Lewis, S. E.& Aitken, R. J. (2005) DNA damage to spermatozoa has impacts on fertilization and pregnancy. Cell Tissue Res 322, 33-41.

Lewis, S. E., O'Connell, M., Stevenson, M., Thompson-Cree, L. & McClure, N. (2004). An algorithm to predict pregnancy in assisted reproduction. Hum Reprod 19, 1385-1394.

Li, Z., Wang, L., Cai, J. & Huang, H. (2006). Correlation of sperm DNA damage with IVF and ICSI outcomes: a systematic review and meta-analysis. J Assist Reprod Genet 23, 367-376.

Lin, M. H., Kuo-Kuang Lee, R., Li, S. H., Lu, C. H., Sun, F. J. and Hwu, Y. M. (2007). Sperm chromatin structure assay parameters are not related to fertilization rates, embryo quality, and pregnancy rates in in vitro fertilization and intracytoplasmic sperm injection, but might be related to spontaneous abortion rates. Fertil Steril 90(2):352-359

Lopes, S., Sun, J. G., Jurisicova, A., Merlano, J. & Casper, R. F. (1998). Sperm deoxyribonucleic acid fragmentation is increased in poor-quality semen samples and correlates with failed fertilization in intracytoplasmic sperm injection. Fertil Steril 69, 528-532.

Makhlouf, A. A. and Niederberger, C. (2006) DNA integrity tests in clinical practice: it is not a simple matter of black and white (or red and green). J Androl 27, 316-323.

Marchetti, F. and Wyrobek, A. J. (2005) Mechanisms and consequences of paternally-transmitted chromosomal abnormalities. Birth Defects Res C Embryo Today 75, 112-129.

Matsuda, Y. and Tobari, I. (1988) Chromosomal analysis in mouse eggs fertilized in vitro with sperm exposed to ultraviolet light (UV) and methyl and ethyl methanesulfonate (MMS and EMS). Mutat Res 198, 131-144.

Matzuk,. MM. & Lamb, D.J. (2008), The biology of infertility: research advances and clinical challenges. Nat Med 14:1197–1213.

Ménézo, Y.J., Hazout, A., Panteix , G., Robert, F., Rollet, J., Cohen-Bacrie, P., Chapuis, F., Clément,. P. & Benkhalifa, M. (2007). Antioxidants to reduce sperm DNA fragmentation: an unexpected adverse effect. Reprod Biomed Online 14:418-421.

Morris, I. D., Ilott, S., Dixon, L. & Brison, D. R. (2002). The spectrum of DNA damage in human sperm assessed by single cell gel electrophoresis (Comet assay) and its relationship to fertilization and embryo development. Hum Reprod 17, 990-998.

Muratori, M., Maggi, M., Spinelli, S., Filimberti, E., Forti, G. and Baldi, E. (2003) Spontaneous DNA fragmentation in swim-up selected human spermatozoa during long term incubation. J Androl 24, 253-262.

Nallella, K. P., Sharma, R. K., Aziz, N. and Agarwal, A. (2006) Significance of sperm characteristics in the evaluation of male infertility. Fertil Steril 85, 629-634.

Nasr-Esfahani, M. H., Salehi, M., Razavi, S., Anjomshoa, M., Rozbahani, S., Moulavi, F. & Mardani, M. (2005). Effect of sperm DNA damage and sperm protamine deficiency on fertilization and embryo development post-ICSI. Reprod Biomed Online 11, 198-205.

Neuwinger, J., Behre, H. M. and Nieschlag, E. (1990) External quality control in the andrology laboratory: an experimental multicenter trial. Fertil Steril 54, 308-314.

O'Flynn O'Brien KL, Varghese AC, Agarwal A. The genetic cause of male factor infertility: a review. Fertil Steril 2010;93:1–12.

Olechuk, K., Giwercman, A & Bungum M. (2011). Intra-individual variation of the sperm chromatin structure assay DNA fragmentation index in men from infertile couples. Hum. Reprod. doi: 10.1093/humrep/der328Palermo, G., Joris, H., Devroey, P. and Van Steirteghem, A. C. (1992) Pregnancies after intracytoplasmic injection of single spermatozoon into an oocyte. Lancet 340, 17.

Payne, J. F., Raburn, D. J., Couchman, G. M., Price, T. M., Jamison, M. G. and Walmer, D. K. (2005) Redefining the relationship between sperm deoxyribonucleic acid fragmentation as measured by the sperm chromatin structure assay and outcomes of assisted reproductive techniques. Fertil Steril 84, 356-364.

Perera,. D, Pizzey, A., Campbell, A., Katz, M., Porter, J., Petrou, M., Irvine, D.S. & Chatterjee R. (2002). Sperm DNA damage in potentially fertile homozygous beta-thalassaemia patients with iron overload. Hum Reprod 17:1820-1825.

Potts, R. J., Newbury, C. J., Smith, G., Notarianni, L. J. & Jefferies, T. M. (1999). Sperm chromatin damage associated with male smoking. Mutat Res 423, 103-111.

Rai, R. & Regan, L. (2006). Recurrent miscarriage. Lancet 368, 601-611

Richthoff, J., Spano, M., Giwercman, Y. L., Frohm, B., Jepson, K., Malm, J., Elzanaty, S., Stridsberg, M. and Giwercman, A. (2002) The impact of testicular and accessory sex gland function on sperm chromatin integrity as assessed by the sperm chromatin structure assay (SCSA). Hum Reprod 17, 3162-3169.

Robbins, W. A., Vine, M. F., Truong, K. Y. & Everson, R. B. (1997). Use of fluorescence in situ hybridization (FISH) to assess effects of smoking, caffeine, and alcohol on aneuploidy load in sperm of healthy men. Environ Mol Mutagen 30, 175-183.

Rubes, J., Lowe, X., Moore, D., 2nd, Perreault, S., Slott, V., Evenson, D., Selevan, S. G. & Wyrobek, A. J. (1998). Smoking cigarettes is associated with increased sperm disomy in teenage men. Fertil Steril 70, 715-723.

Sakkas, D. & Alvarez J.G. (2010). Sperm DNA fragmentation: mechanisms of origin, impact on reproductive outcome, and analysis. Fertil Steril 93:1027-1036.

Sakkas, D. & Alvarez, J.G. (2010). Sperm DNA fragmentation: mechanisms of origin, impact on reproductive outcome, and analysis. Fertil Steril 93:1027-1036.

Sakkas, D., Mariethoz, E., Manicardi, G., Bizzaro, D., Bianchi, P. G. & Bianchi, U. (1999). Origin of DNA damage in ejaculated human spermatozoa. Rev Reprod 4, 31-37.

Sakkas, D., Moffatt, O., Manicardi, G. C., Mariethoz, E., Tarozzi, N. and Bizzaro, D. (2002) Nature of DNA damage in ejaculated human spermatozoa and the possible involvement of apoptosis. Biol Reprod 66, 1061-1067.

Sakkas, D., Urner, F., Bianchi, P. G., Bizzaro, D., Wagner, I., Jaquenoud, N., Manicardi, G. & Campana, A. (1996). Sperm chromatin anomalies can influence decondensation after intracytoplasmic sperm injection. Hum Reprod 11, 837-843.

Saleh, R. A., Agarwal, A., Nada, E. A., El-Tonsy, M. H., Sharma, R. K., Meyer, A., Nelson, D. R. and Thomas, A. J. (2003a) Negative effects of increased sperm DNA damage in relation to seminal oxidative stress in men with idiopathic and male factor infertility. Fertil Steril 79 Suppl 3, 1597-1605.

Saleh, R. A., Agarwal, A., Nelson, D. R., Nada, E. A., El-Tonsy, M. H., Alvarez, J. G., Thomas, A. J., Jr. and Sharma, R. K. (2002b) Increased sperm nuclear DNA damage in normozoospermic infertile men: a prospective study. Fertil Steril 78, 313-318.

Sanchez-Pena, L. C., Reyes, B. E., Lopez-Carrillo, L., Recio, R., Moran-Martinez, J., Cebrian, M. E. & Quintanilla-Vega, B. (2004). Organophosphorous pesticide exposure alters sperm chromatin structure in Mexican agricultural workers. Toxicol Appl Pharmacol 196, 108-113.

Sega, G. A., Sotomayor, R. E. & Owens, J. G. (1978). A study of unscheduled DNA synthesis induced by X-rays in the germ cells of male mice. Mutat Res 49, 239-257.

Selevan, S. G., Borkovec, L., Slott, V. L., Zudova, Z., Rubes, J., Evenson, D. P. & Perreault, S. D. (2000). Semen quality and reproductive health of young Czech men exposed to seasonal air pollution. Environ Health Perspect 108, 887-894.

Seli, E., Gardner, D.K., Schoolcraft, W.B, Moffatt, O. & Sakkas, D. (2004). Extent of nuclear DNA damage in ejaculated spermatozoa impacts on blastocyst development after in vitro fertilization. Fertil Steril 82:378-383.

Silver, E. W., Eskenazi, B., Evenson, D. P., Block, G., Young, S. & Wyrobek, A. J. (2005). Effect of antioxidant intake on sperm chromatin stability in healthy nonsmoking men. J Androl 26, 550-556.

Simon, L., Brunborg, G., Stevenson, M., Lutton, D., McManus, J. & Lewis, S.E. (2010). Clinical significance of sperm DNA damage in assisted reproduction outcome. Hum Reprod 25(7):1594-1608.

Simon, L., Lutton, D., McManus, J. & Lewis SE. (2011). Sperm DNA damage measured by the alkaline Comet assay as an independent predictor of male infertility and in vitro fertilization success. Fertil Steril 95(2):652-657.

Singh, N. P., Muller, C. H. and Berger, R. E. (2003) Effects of age on DNA double-strand breaks and apoptosis in human sperm. Fertil Steril 80, 1420-1430.

Skakkebaek, N. E., Giwercman, A. and de Kretser, D. (1994) Pathogenesis and management of male infertility. Lancet 343, 1473-1479.

Smit, M., Dohle, G. R., Hop, W. C., Wildhagen, M. F., Weber, R. F. and Romijn, J. C. (2007) Clinical correlates of the biological variation of sperm DNA fragmentation in infertile men attending an andrology outpatient clinic. Int J Androl 30, 48-55.

Song, G.J., Norkus, E.P.& Lewis, V. (2006). Relationship between seminal ascorbic acid and sperm DNA integrity in infertile men. Int J Androl. 2006;29:569-75.

Spano, M., Bonde, J. P., Hjollund, H. I., Kolstad, H. A., Cordelli, E. and Leter, G. (2000) Sperm chromatin damage impairs human fertility. The Danish First Pregnancy Planner Study Team. Fertil Steril 73, 43-50.

Spano, M., Kolstad, A. H., Larsen, S. B., Cordelli, E., Leter, G., Giwercman, A. and Bonde, J. P. (1998) The applicability of the flow cytometric sperm chromatin structure assay in epidemiological studies. Asclepios. Hum Reprod 13, 2495-2505..

Sun, J. G., Jurisicova, A. & Casper, R. F. (1997). Detection of deoxyribonucleic acid fragmentation in human sperm: correlation with fertilization in vitro. Biol Reprod 56, 602-607.

Swan, S. H. (2006) Semen quality in fertile US men in relation to geographical area and pesticide exposure. Int J Androl 29, 62-68; discussion 105-108.

The Practice Committee of the American Society of Reproductive Medicine (2008). The clinical utility of sperm DNA integrity testing. Fertil Steril 90 (Suppl 5):178-180.

Tomlinson, M. J., Moffatt, O., Manicardi, G. C., Bizzaro, D., Afnan, M. & Sakkas, D. (2001). Interrelationships between seminal parameters and sperm nuclear DNA damage before and after density gradient centrifugation: implications for assisted conception. Hum Reprod 16, 2160-2165.

Tomsu, M., Sharma, V. & Miller, D. (2002). Embryo quality and IVF treatment outcomes may correlate with different sperm comet assay parameters. Hum Reprod 17, 1856-1862.

Varghese, A. C., Goldberg, E. and Agarwal, A. (2007) Current and future perspectives on intracytoplasmic sperm injection: a critical commentary. Reprod Biomed Online 15, 719-727.

Verpoest, W. & Tournaye, H. (2006). ICSI: hype or hazard? Hum Fertil (Camb) 9, 81-92.

Virro, M. R., Larson-Cook, K. L. & Evenson, D. P. (2004). Sperm chromatin structure assay (SCSA) parameters are related to fertilization, blastocyst development, and ongoing pregnancy in in vitro fertilization and intracytoplasmic sperm injection cycles. Fertil Steril 81, 1289-1295.

Ward, W. S. (1993). Deoxyribonucleic acid loop domain tertiary structure in mammalian spermatozoa. Biol Reprod 48, 1193-1201.

Ward, W. S. and Coffey, D. S. (1991). DNA packaging and organization in mammalian spermatozoa: comparison with somatic cells. Biol Reprod 44, 569 574.

WHO (2000). WHO Manual for the Standardised Investigation, Diagnosis and Management of the Infertile Male. Cambridge University Press, Cambridge.

Wolf, J.P., Bulwa, S., Ducot, B., Rodrigues, D. & Jouannet, P. (1996). Fertilizing ability of sperm with unexplained in vitro fertilization failures, as assessed by the zona-free hamster egg penetration assay: its prognostic value for sperm-oolemma interaction. Fertil Steril 65(6):1196-1201.

World Health Organization. WHO laboratory manual for the examination of human semen and sperm-cervical mucus interaction. Cambridge: Cambridge University Press; 1999.

World Health Organization. WHO laboratory manual for the examination and processing of human semen. 5 ed: World Health Organization; 2010.

Zini A, Sigman M. Are tests of sperm DNA damage clinically useful? Pros and cons. J Androl 2009;30:219-229.

Zini, A., Bielecki, R., Phang, D. and Zenzes, M. T. (2001a) Correlations between two markers of sperm DNA integrity, DNA denaturation and DNA fragmentation, in fertile and infertile men. Fertil Steril 75, 674-677.

Zini, A., Boman, J.M., Belzile, E. & Ciampi, A. (2008). Sperm DNA damage is asociated with an increased risk of pregnancy loss after IVF and ICSI: systematic review and meta-analysis. Hum Reprod 23 (1),4-10.

Zini, A., Fischer, M. A., Sharir, S., Shayegan, B., Phang, D. and Jarvi, K. (2002). Prevalence of abnormal sperm DNA denaturation in fertile and infertile men. Urology 60, 1069-1072.

Zini, A., Kamal, K., Phang, D., Willis, J. and Jarvi, K. (2001b) Biologic variability of sperm DNA denaturation in infertile men. Urology 58, 258-261.

3

The Epididymis: Embryology, Structure, Function and Its Role in Fertilization and Infertility

Kélen Fabiola Arrotéia, Patrick Vianna Garcia, Mainara Ferreira Barbieri, Marilia Lopes Justino and Luís Antonio Violin Pereira
State University of Campinas (UNICAMP)
Brazil

1. Introduction

The epididymis, located between the efferent ducts and the vas deferens, is a male accessory organ characterized by a single coiled tubule duct with an estimated length of 5–7 m in men (Sullivan, 2004; O'Hara et al., 2011).

The anatomic segments of the epididymis include the initial segment, the caput, the corpus and the cauda. Each region consists of a lumen and a polarized epithelium composed mostly of principal and basal cells (Lasserre et al., 2001; Dacheux et al., 2005). Although these four anatomical regions of the epididymis are easily identified in most adult male mammals (Yanagimachi et al., 1985; Smithwick & Young, 2001), histological and ultrastructural segmentation of this organ varies among the different phylogenies of mammals. The rat epididymis is most commonly adopted as an experimental model of study (Figure 1).

Several descriptive anatomical and histological studies of the epididymis appeared at the beginning of the twentieth century. The authors hypothesized that the epididymal secretions played a role in the maintenance of sperm vitality, sperm motility (Benoit, 1926) and the capacity to become fertile (Young, 1929a, 1929b, 1931). Relatively little research was done on the excurrent duct system during the ensuing three decades. However, in 1967, Marie-Claire Orgebin-Crist demonstrated that the key event in sperm maturation was not the passage of time but the exposure of the sperm to the luminal environment of the epididymis (Bedford, 1967; Orgebin-Crist, 1967).

The epididymal duct is now recognized as a channel that transports, concentrates and stores the spermatozoa. It is also known that the spermatozoa leaving the testis are immovable, immature and unable to fertilize an oocyte (Yanagimachi, 1994; Flesch & Gadella, 2000), and that under androgen control, the epididymal epithelium secretes proteins within the intraluminal compartment that create a very complex environment surrounding the spermatozoa (Hermo et al., 1994, 2004; Sullivan, 2004). This luminal compartment stores the spermatozoa until ejaculation and specifically prepares the sperm for fertilization by providing the essentials in terms of temperature, oxygen tension, pH and an available energy substrate (Dacheux et al., 2005). The epididymal duct produces the morphological, biochemical, physiological and functional changes to the structures of the spermatozoa through a process known as epididymal maturation, which converts the spermatozoa into fertilization-competent cells (Toshimori, 2003; Gatti et al., 2004; Flesch & Gadella, 2000).

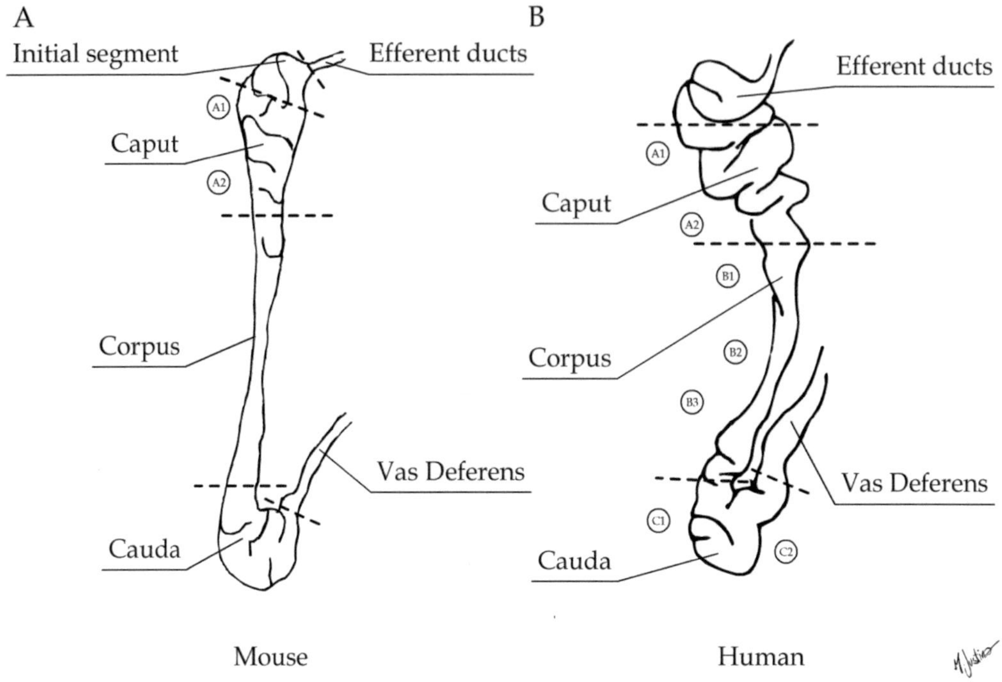

Fig. 1. The regionalization of the epididymis. The illustrations compare the mouse and human epididymides. A) Mouse epididymis (A1: proximal caput, A2: mid-caput). B) Human epididymis (A1: anterior caput, A2: posterior caput, B1: anterior corpus, B2: mid-corpus, B3: posterior corpus, C1: anterior cauda, C2: posterior cauda).

Since 1966, more than 12,000 research articles have been published on the epididymis. These articles, while addressing the various aspects of the epididymis from several points of view, all agree that the epididymis is crucial for the preparation of the spermatozoa prior to fertilization. It is well known that the proteins and small molecules secreted by the epididymal epithelium into the lumen interact with the transiting spermatozoa and directly or indirectly affect the spermatozoa surface; these proteins and small molecules function as signaling molecules to induce activity in the other epididymal proteins (Gatti et al., 2004). A significant number of these molecules will be taken up by the spermatozoa. Several of the acrosome molecules, which were previously formed during spermiogenesis and are involved in the acrosome reaction or in the sperm-zona pellucida interaction and sperm-egg fusion, will be gradually rearranged and compartmentalized in a stage-specific manner during sperm maturation (França et al., 2005).

Through research, we have learned a substantial amount about the function and the biochemical properties of the epididymal structure. However, there are still gaps in nearly all aspects of the research. Studying the complexity of the cellular properties, the spatial and temporal organization of protein syntheses and secretion and the dynamic interactions between the epithelial cells and the contents of the luminal compartment remains challenging but important. A more complete understanding of this organ will allow for the development of male contraceptive agents (Turner et al., 2006). Additionally, up to 40% of infertile men exhibit idiopathic infertility that may reflect sperm maturational disorders (Cornwall, 2009).

A greater understanding of the function and biochemical properties of the epididymis may lead to the development of therapeutic agents to treat certain types of infertility. This chapter provides an overview of the human epididymal embryology, malformation, structure and function.

2. The prenatal development of the epididymis

In mammals, the genetic sex of the embryo is determined mainly by the presence or absence of a single gene on the Y chromosome, the SRY, which is required to initiate male-specific pathways and to repress female-specific pathways of development (Moore & Persaud, 2003). At the onset of sex differentiation, the reproductive primordial organs are indistinguishable in male and female embryos. The gonads (testicles or ovaries), the genital ducts (Wolffian or Müllerian) and the urogenital sinus of both sexes emerge from the morphologically undifferentiated primordia. We can then assume that the genes involved in the establishment of these primordia are the same for both sexes. During development, the first morphological sex difference appear in the gonads of male embryos, followed by the genital ducts, and finally, they appear in the urogenital sinus (Larios & Mendoza, 2001).

In humans, the genital ducts arise from the intermediate mesoderm. Shortly after the formation of the Wolffian ducts, the adjacent Müllerian ducts are formed. In males, sexual differentiation develops through regression of the Müllerian ducts and the development of the Wolffian ducts into multiple reproductive organs, such as the epididymis, the vas deferens ducts and the the seminal vesicles. In females, the Wolffian ducts regress and the Müllerian ducts differentiate to form the uterus, the fallopian tubes, and the proximal vagina (Hannema & Hughes, 2007; Kobayashi & Behringer, 2003).

The development of sexual differentiation involves genetic processes that are primarily controlled by the sexual chromosomes (França et al., 2005).

In the male tract, Müllerian duct regression, which depends on the testicular Sertoli cells, is mediated by the Y chromosome genes and is triggered by a mechanism that is not well understood. The Sertoli cells produce and secrete an anti-Müllerian substance, a non-steroidal hormone. Under the control of human chorionic gonadotrophin (hCG), a placental gonadotrophin, the Leydig cells then differentiate and produce the androgens that will positively regulate the ipsilateral Wolffian duct in a paracrine way. These hormonal combinations induce the cranial Wolffian duct to become highly convoluted and to differentiate into the epididymis.

The major morphogenic event during Wolffian duct/epididymal duct development is the elongation and coiling of the duct (Hannema & Hughes, 2007). The process of elongation is likely a product of potential mechanisms such as cell proliferation coupled with directed cell rearrangements, along with the interactions between the Wolffian duct ephitelium and the surrounding mesenchyme cells (Hinton et al., 2011).

The Wolffian ducts also induce the formation of the mesonephric tubules in the mesonephric mesenchyme, which extend to the epithelial cells of the gonad and subsequently differentiate into efferent ductules. These ductules open into the epididymis (Figure 2). Distal to the epididymis, the Wolffian ducts acquire a thicker investment of smooth muscle and become the ductus deferens (Moore & Persaud, 2003; Hannema & Hughes, 2007).

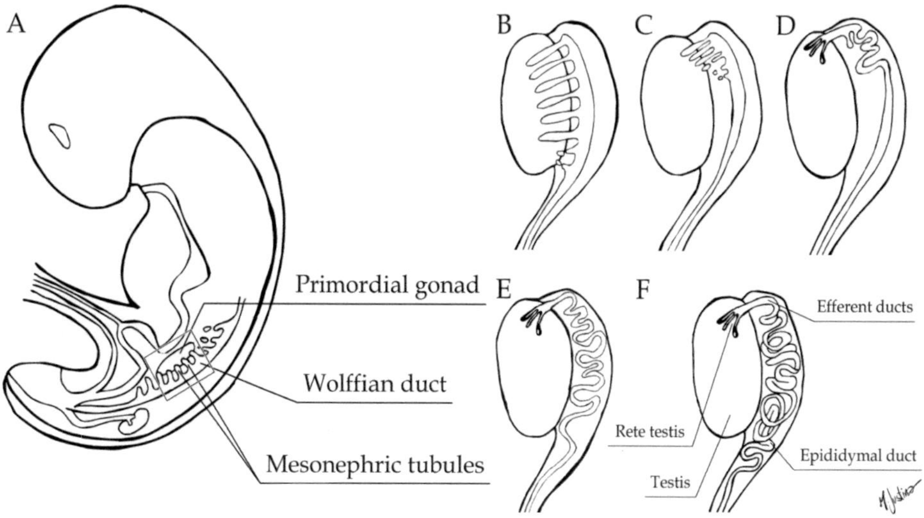

Fig. 2. The intra-uterine development of the epididymis. A) A sagittal section of a mouse embryo showing the relative locations of the developing urogenital structures. B) The testicular production and secretion of testosterone positively regulates the Wolffian ducts. The ducts then induce the formation of the mesonephric tubules in the mesonephric mesenchyme that extend to the epithelial cells of the gonad. C) The cords arising from the apical mesonephric tubules form the efferent ducts and fuse with the adjacent ducts to form the rete testis. The coiling of the initial segment of the efferent duct is not shown here, but it proceeds independently of the coiling of the main epididymal duct shown in D, E and F. D) The coiling shifts from the proximal to the distal duct in a temporal fashion. E) The initial stages of coiling are planar, meaning that the coiling is at the two-dimensional level. F) The three-dimensional coiling proceeds in a caput-to-cauda direction and is completed in the early postnatal period.

3. The postnatal development of the epididymis

While cell proliferation in the Wolffian duct appears to be dependent on the presence of androgens and mesenchymal factors during prenatal development, lumicrine factors produced by the testicles play an additional role during postnatal development (Hinton et al., 2011). Lumicrine factors are molecules (such as androgens, growth factors and enzymes) whose effect on the secretory activity of the epithelial cells of the epididymis and in the epididymis spermatozoa directly participates in the process of epididymal maturation (Lan et al., 1998; Robaire et al., 2006). In general, the postnatal development of the mammalian epididymis can be divided into three periods that take place across two developmental phases (Figure 3).

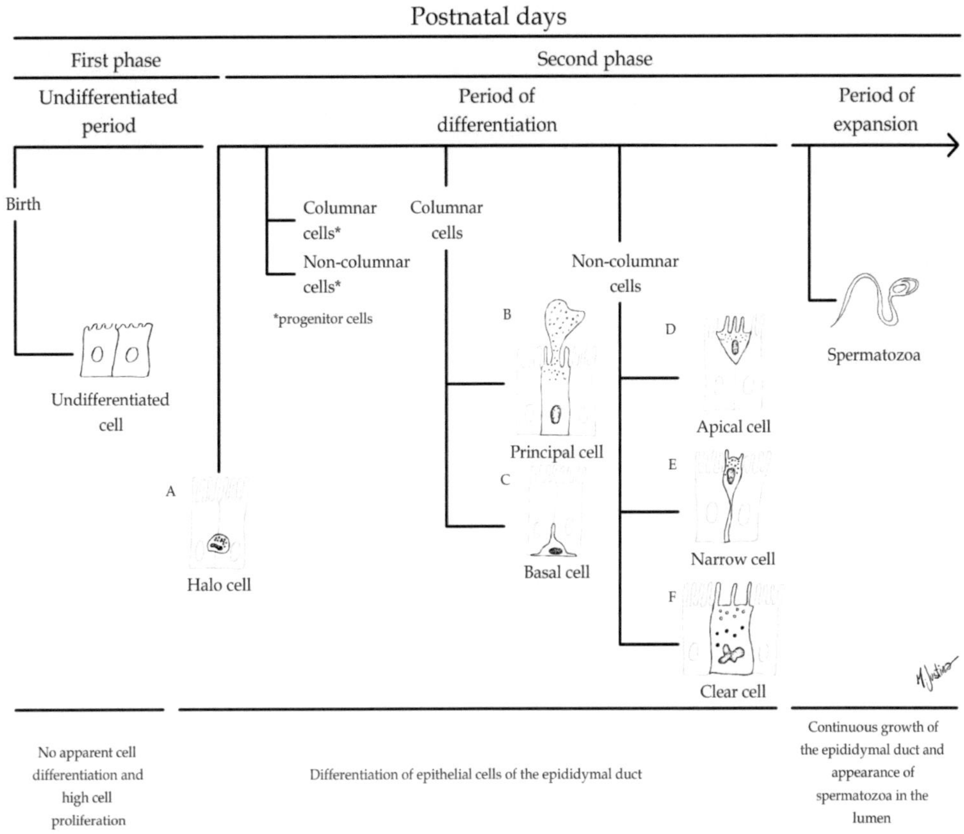

Fig. 3. A schematic diagram that describes the differentiation of the mouse epididymal epithelium from birth until adulthood. A to F indicate the different types of cells in the epithelium of the epididymis.

3.1 The first phase: From birth to early infancy

By birth, the presumptive areas of the initial segment, caput and corpus have experienced considerable coiling compared with the epididymal cauda (Figure 4). Within the first few days after birth, the individual conjunctive septum initiates the anatomical segmentation of epididymis, along with the considerable growth of the duct. The epithelium is still undifferentiated and characterized by columnar cells. The histological changes do not begin until the first stages of early infancy with the appearance of halo cells (Robaire et al., 2000, 2006) (Figure 3). The first phase begins, then, with the undifferentiated period and ends with the first event of the differentiation period, which is defined by the appearance of halo cells.

3.2 The second phase: From early infancy to adulthood

The epididymal epithelium differentiates into the pseudostratified epithelium, which contains principal, basal, apical, narrow and clear cells (Figure 3). Epithelial cell differentiation is completed during puberty when the highest rate of cell division and epididymal expansion occurs. This period of expansion describes the continued growth of the duct and the appearance of spermatozoa in the epididymal lumen. The establishment of the regionalized protein secretion takes place progressively coupled to the consolidation of an anatomical epididymal segmentation (Figure 4). The differentiation and regionalization are related to the different stages of testicular maturation, the steroidogenic activity of the Leydig cells, the androgen dependence of the epididymis itself and the lumicrine factors (De Miguel et al., 1998; Robaire et al., 2000; Dacheux et al., 2005; Cornwall, 2009; Robaire et al., 2006). Then the second phase starts after halo cell differentiation and ends in puberty.

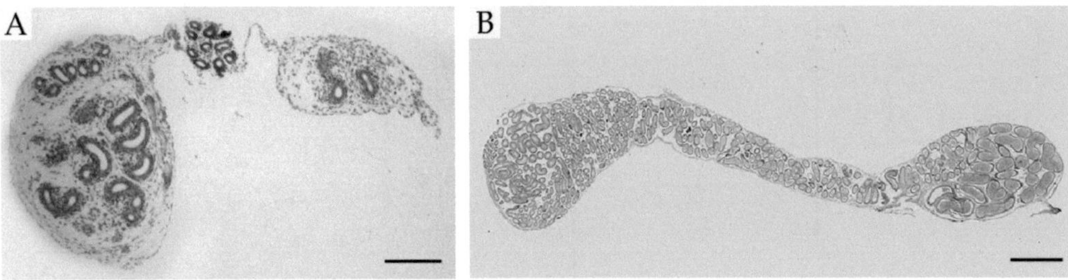

Fig. 4. Photomicrographs of a mouse epididymis. A) At birth, the undifferentiated period shows a poorly coiling duct and little regionalization. B) A well-formed adult epididymis in expansion with a completely coiled and regionalized duct. Scale bar: A) 100μm. B) 1100μm.

4. Structural features

In the testes, the seminiferous tubules converge to form the rete testis, which in turn gives rise to the efferent ducts. The number of tubules contained in the efferent ducts varies depending on the species and on how the tubules converge to form the single coiled duct of the epididymis. The epididymal tubular lumen is continuous with the lumina of the efferent ducts and comes to an end in the vas deferens, which has a thick layer of muscle. The urethra is the final extension of the vas deferens and communicates with the outside of the body.

4.1 Macroscopic features

The mammalian epididymis is an elongated coiled duct suspended within the mesorchium and firmly or loosely bound to the testicular tunica albuginea. The gross aspect of the epididymis allows the identification of the different segments (Figure 1), which are comprised of the proximal region (the initial segment and the caput), the corpus and the distal cauda (Robaire et al., 2006; Turner, 2008). In all mammalian species, each region of the epididymis is further organized into lobules that are separated by the connective tissue septa. These septa not only serve as internal support for the organ but are also thought to provide a functional separation between the lobules that allows for the selective expression of genes and proteins within the individual lobules (Kirchhoff, 1999; Cornwall, 2009).

4.2 Microscopic features

The epididymis also contains region-specific and cell-specific functions specifically located within the epithelium of a given segment. The highly specific regionalization of the epithelium and the luminal protein secretion within the three main epididymal segments can be further subdivided into several regions (França et al., 2005; Robaire et al., 2006; Turner, 2008). In general, the five physiological regions are distinguished as follows: (i) the proximal caput or the initial segment, (ii) the middle caput, (iii) the distal caput and the proximal corpus, (iv) the distal corpus and (v) the cauda (Dacheux et al., 2005; Turner, 2008). Note that the proposed physiological regions are slightly different from the anatomical regions (Figure 1).

The differential response of the segments to androgen withdrawal, stress and aging indicates that each region represents discrete regulatory units (Jervis & Robaire, 2001). The epididymis is more than a uniform channel that transports and stores spermatozoa. The maturation and storage of the spermatozoa depends upon the epithelial tight and adhering junctions. They are important in maintaining the integrity of the epididymal epithelium and in the formation of the blood epididymal barrier. Tight junctions between the adjacent epididymal epithelial cells form the blood-epididymal barrier and restrict the passage of a number of ions, solutes, and macromolecules through the epididymal ephitelium. This barrier also serves as an extension of the blood-testis barrier. The spermatozoa are immunogenic and contain proteins on their surfaces that would be recognized as nonself if they leave the epididymis (Robaire & Hermo, 1988; Dacheux et al., 2005).

Histological characteristics allow for the easy identification of the anterior and posterior extremities of the mammalian epididymis. The thickness of the epididymal epithelium varies with the thickest portion in the proximal caput and the thinnest in the caudal region (Figure 5). Conversely, the luminal diameter and the thickness of the peritubular smooth muscle increases from the proximal to the distal regions (Lasserre et al., 2001; Toshimori, 2003). Few sperm are found in the initial segment, but a large mass of sperm aggregates are located in the cauda (Yanagimachi et al., 1985; Cornwall, 2009). In all of these segments, the epididymal duct is lined with an epithelium composed of principal and basal cells. Other cells, such as apical, narrow, clear and halo cells, are also present in this duct in a segment-specific manner (Figure 6).

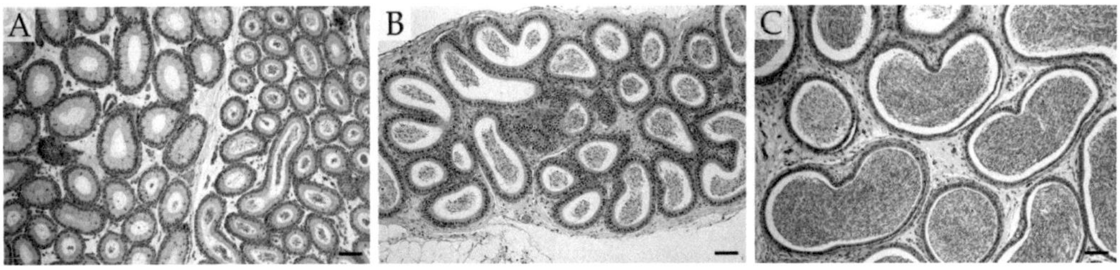

Fig. 5. Photomicrographs showing the histological regionalization of the A) initial segment and caput separated by the connective tissue septa, B) the corpus and C) the cauda of the mouse epididymis. The thickness of the epididymal epithelium varies from the thickest portion in the proximal caput to the thinnest in the caudal region. Conversely, the luminal diameter and the thickness of the peritubular smooth muscle increases from the proximal to the distal regions. Scale bar: 50μm.

Fig. 6. A schematic cross section of epididymis showing the organization of the major cell types in the epithelium of the epididymal duct as observed through a light microscope. A generic epididymal compartment with the relative position and distribution of all cell types found throughout the epithelium is illustrated. The thinnest coat of organized smooth muscle is in the interstitial tissue surrounding the duct and is characteristic of the anterior regions of the epididymis. Note the spermatozoa in the center of the epididymal lumen.

4.2.1 Principal cells

Principal cells (Figure 3B) are the most abundant and extensively studied cell type found in the epididymal epithelium. Principal cells constitute approximately 80% of the total epithelial cell population in the initial segment. The number of principal cells gradually decreases to 65% of the total epithelial cell population in the cauda epididymis (Robaire & Hermo, 1988). These columnar cells present prominent stereocilia and extend into the lumen. Ultrastructurally, the supranuclear region of this cell type contains large stacks of Golgi saccules, mitochondria, multivesicular bodies and apical dilated membranous elements, while the infranuclear region is densely packed with rough endoplasmic reticulum (Robaire et al., 2000, Dacheux et al., 2005). Principal cells are responsible for the bulk of the proteins that are secreted into the lumen and are directly involved in the control of luminal protein concentrations. They frequently exhibit blebs of cytoplasm emanating from their apical cell surface. These cells also form tight junctions with one another, and as such, form the blood-epididymis barrier (Robaire et al., 2006; Cornwall, 2009).

4.2.2 Basal cells

Basal cells (Figure 3C) are the second most abundant cell type found in the epididymal epithelium, constituting 15-20% of the total epithelial cell population of the epididymis. They are triangular and flat cells and they reside in the base of the epithelium. Basal cells cannot access the luminal compartment. They have elongated or round shaped nuclei, and they are in close association with the overlying principal cells or other basal cells through the presence of cytoplasmatic extensions (Robaire et al., 2000; Cornwall, 2009). Because of this contact with the basement membrane, basal cells form an extensive cellular sheet surrounding the epididymal epithelium (Robaire et al., 2006; Cornwall, 2009). Although basal cells are extratubular in origin, some findings have suggested that the cells may have a role within the processes of the epithelial immune system and in the regulation of electrolytes by principal cells. However, the exact functions of these cells are not yet clear (Robaire et al., 2006).

4.2.3 Apical cells

Apical cells (Figure 3D) comprise approximately 10% of the total epithelial population in the initial segment but only approximately 1% of the total epithelial population in the cauda of the epididymis (Adamali & Hermo, 1996). They are clearly defined by the many mitochondria in the apical cytoplasm, the few microvilli at the luminal border and a nucleus that is located in the upper half of the cell cytoplasm (Adamali & Hermo, 1996; Robaire et al., 2006). These cells are related to sperm quiescence and to the regulation of the pH in the lumen through the production of enzymes of the carbonic anhydrase family (Hermo et al., 2005).

4.2.4 Narrow cells

Narrow cells (Figure 3E) are the slender elongated cells. They increase from 3% of the total epithelial population in the initial segment to 6% of the total epithelial population in the corpus. These cells presents numerous C-shaped vesicles and mitochondria with a small flattened nucleus located in the upper half of the cell cytoplasm. The structural features of both apical and narrow cells suggest that these cells are involved in the process of intracellular transport between the lumen and the epithelial cells, in the degradation of specific proteins and carbohydrates within their lysosomes and in protecting spermatozoa from a changing environment of harmful electrophiles (Adamali & Hermo, 1996; Robaire et al., 2006). They also differ dramatically from the neighboring principal cells and display region-specific expression of proteins, such as the glutathione S-transferases and lysosomal enzymes (Adamali & Hermo, 1996).

4.2.5 Clear cells

Clear cells (Figure 3F), along with halo cells, constitute fewer than 5% of the total epithelial cell population. Clear cells are equally distributed through the caput, the corpus and the cauda segments. The dark-stained nucleus of these cells is surrounded by the pale-staining cytoplasm. They are also present in all levels of the epididymal epithelium. Clear cells are also endocytic cells and may be responsible for the clearance of proteins from the epididymal lumen. They normally take up the contents of the cytoplasmic droplets released by the spermatozoa as they transit through the duct (Hermo et al., 1994, 2005; Robaire et al., 2006).

4.2.6 Halo cells

Halo cells (Figure 3A) are usually located in the base of the ephitelium where it does not touch the basement membrane. These cells contain variable numbers of dense core granules.

They develop in the immune system from a combination of B and T lymphocytes and monocytes (Dacheux et al., 2005; Robaire et al., 2006).

4.2.7 Cell interactions in the epididymal duct

The cell types described above are active in the processes of epididymal function, such as protein secretion and absorption (principal cells); endocytosis (clear and apical cells); the secretory activities responsible for the acidification of the luminal fluid (clear cells and narrow cells); immune defense; phagocytosis (halo cells); and the production of antioxidants (basal cells) (Robaire et al., 2000; França et al., 2005). It is important to understand that each cell type may express different proteins within the distinct epididymal regions. This indicates that the cells perform different functions according to their location, and that the specificity of epididymal secretions is progressively established with age (Robaire et al., 2000, 2006). This information confirms the high degree of regionalization involved in the activity of the epididymis. This epididymal regionalization, which is attributed to the diverse patterns of gene expression, is critical to the formation and maintenance of the functions of the epididymal duct (Suzuki et al., 2004).

5. The maintenance of the epididymis

In mammals, the development and maintenance of a fully differentiated epididymal epithelium is dependent on a combination of factors that provide an ideal site where the spermatozoa undergo a series of morphological, biochemical and physiological changes. During this process of epididymal maturation, the spermatozoa move along the ducts in a fluid that dynamically evolves through the processes of absorption and secretion by the epithelial cells, androgens, as well as through lumicrine factors from the testis (Cornwall, 2009).

5.1 Hormones

Androgens play a crucial role in the development of the male reproductive organs, such as the testis, the epididymis, the vas deferens, the seminal vesicle, the prostate and the penis. The role of androgens is an important topic in the study of puberty, male fertility and male sexual function. The effects of androgen withdrawal have been well established through the experimental model of orchiectomy. A decrease in the weight of the epididymis has been commonly observed in animals that have had their testicles removed. In these cases, androgen replacement, even at supraphysiological levels, only partially restored the weight of the epididymis. The removal of the testicles caused the loss of androgens, but it is clear that this approach affected estrogen levels and other testicular factors that may affect the maintenance of epididymis (Robaire et al., 2000).

5.1.1 Androgen: Testosterone and dihydrotestosterone

The formation and function of the epididymis is androgen-dependent. The principal androgen, testosterone (T), is essential for the development of the internal sex organs and is derived from the Wolffian duct system, which consists of the epididymis, the vas deferens, and the seminal vesicle (Umar et al., 2003). Dihydrotestosterone (DHT), the 5α-reduced form of T, is involved in the development of the prostate and the external genitalia. Although T is the predominantly active androgen during the first phase of the postnatal development of the epididymis, it is the effects of DHT that are important in the epididymal fluid of the mature epididymis. DHT can be produced locally in the epididymis by principal cells and is primarily found in the initial segment of the duct (Dacheux et al., 2005; França et al., 2005; Robaire & Henderson, 2006).

The actions of both T and DHT are initiated through the intracellular receptor known as the androgen receptor (AR). DHT is the more potent androgen of the two. The AR is found in all male reproductive organs and can be stimulated by either T or its more potent metabolite, DHT. The binding of either T or DHT to the AR may regulate distinct androgenic effects in target tissues. Clinical syndromes, such as androgen insensitivity (AIS), illustrate the differential actions of T and DHT (Umar et al., 2003). AR expression in the developing male genital tract occurs in a strict temporal pattern. It is first detected in the mesenchymal cells, then in the epithelial cells and then in both the epithelial and stromal compartments of the epididymis (Umar et al., 2003; O'Hara et al., 2011).

The initiation of androgen-dependent differentiation of the Wolffian duct system into epididymis occurs before epithelial cells express a detectable level of the AR protein. In this phase of development, the mesenchymal cells are important androgen targets that elicit androgenic effects in the epithelial cells via paracrine factors and mesenchymal-epithelial interactions (Umar et al., 2003). Genetic mutations in the AR or treatment with AR antagonists during the male embryonic stage results in the regression of the Wolffian ducts and an absence of the epididymis in adult males (O'Hara et al., 2011).

During postnatal development, the luminal secretion of androgens is essential for the maintenance of epithelial cell identity (O´Hara et al., 2011), and for the normal development and function of the stromal cells (Nitta et al., 1993; Robaire et al., 2000; Hess et al., 2001; O'Hara et al., 2011). Both the regionalized differentiation of the epididymis and the variation in the luminal fluid composition take place under the control of androgens (Toshimori, 2003). The production and secretion of at least half of the epididymal proteins, including those later incorporated by the spermatozoa in transit, are under androgenic control. Androgenic control may act positively or negatively, depending on the varying levels of sensitivity (Tezon et al., 1985; Ellerman et al., 1998; Robaire et al., 2000, Dachuex et al. 2005; Robaire & Henderson, 2006).

5.1.2 Estrogen

In addition to testosterone and other androgenic-derived metabolites, estrogen has been reported to target epididymal epithelial cells (Hess et al., 2001, 2011). The presence of two estrogen receptors (ESR) types in the head of the epididymis, as well as in other regions of the epididymis, has been well documented. Their expression appears to be isotype-, species-, and cell-specific (Hess et al., 2001, 2011). Recent studies have shown that the luminal fluid reabsorption that occurs in the efferent ductules and in the initial segment of the epididymis is regulated by estrogen. The estrogen present in the epididymis also regulates the transport of fluid through the duct and is responsible for increasing the concentration of sperm as they enter the caput of the epididymis (França et al., 2005; Hess et al., 2011). Estrogen assists in the maintenance of a differentiated epithelial morphology, which means that it is absolutely necessary for the process of enhancing fertility in the male (Kobayashi & Behringer, 2003).

5.2 Lumicrine factors

Lumicrine factors are molecules produced by an upstream set of cells through a luminal or ductal system in the testes that actively participates through paracrine signaling to the cellular mechanisms of development and maintenance of the epididymis (Lan et al., 1998). In the absence of lumicrine factors, the segments of the epididymal duct regress to a transcriptionally undifferentiated state, which is consistent with a less differentiated histology. The absence of testicular molecules could also stimulate an individual's gene expression in some epididymal segments while suppressing it in others (Turner et al., 2007). Androgen replacement in the experimental models of castrated subjects has shown that apoptotic cell death in the epididymis can be prevented, but the initial segment is dependent on the luminal components coming from the testis and not just on the androgens alone (Robaire et al., 2000). The direct influence of lumicrine factors, in addition to the secretion of androgens by the testicular Leydig cells, is of extreme importance for both cell signaling and the maintenance of the male reproductive tract epithelia. Lumicrine factors regulate the secretion of the molecule-specific androgen-region substances present in the epididymal fluid, which support a peculiar microenvironment that is necessary for the survival and functionality of sperm (Turner et al., 2007; von Horsten et al., 2007).

5.3 Additional molecules

Other substances for which the receptors for or the substance itself have been found in the epididymis include prolactin, retinoic acid, and vitamin D, whose active metabolite is synthesized primarily in the cauda epididymis (Robaire et al., 2000). In a complementary way, another regulator of epididymal function is vitamin E, which plays an important role in maintaining the viability and the functional and structural appearances of the epithelial cells of the epididymal duct (França et al., 2005).

6. Biological function in adulthood

The epididymal functions of transporting, concentrating, maturing, and storing spermatozoa are important processes for male fertility, and their absence or depletion might be a significant factor in male infertility (Turner, 2008). In order to fulfill so many functions, the epididymis is dependent on the establishment of a peculiar microenvironment, which is formed by the highly regionalized secretion of proteins, glycoproteins and other molecules from the testis. Although the composition of the epididymal luminal fluid of several species is known to be the epididymal site where spermatozoa mature and are stored, the manner in which the epididymis contributes to the formation of this specialized milieu is not fully understood (Robaire et al., 2006). Our understanding of the mechanisms behind the regulation of the functions of the epididymal epithelium, and their effects on the spermatozoa, is still fairly limited. The major function of the adult epididymis is to provide the ideal conditions that ensure the progress of the spermatozoa along the duct as they are exposed to a continually changing environment that supports the development and maintenance of fertilization.

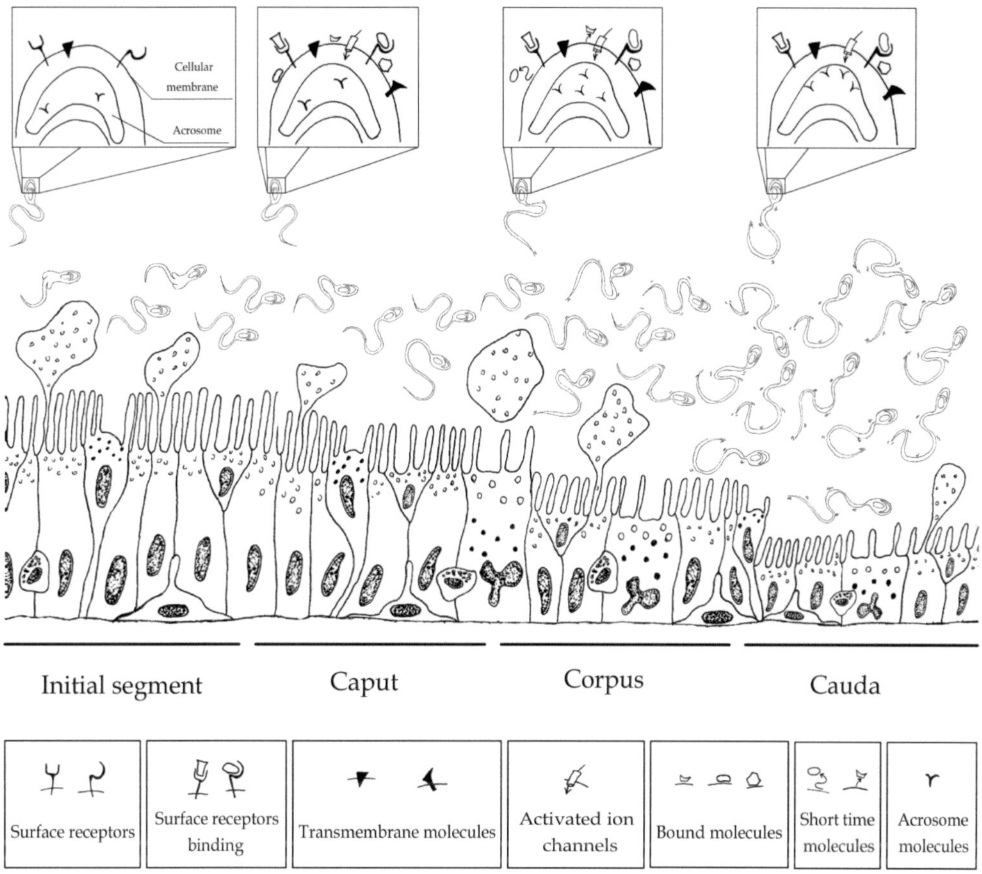

Fig. 7. A schematic illustration representing the distinct changes in the sperm surface throughout the stages of sperm maturation and epididymal cellular distribution throughout the epididymal epithelium. Histological characteristics allow for the identification of the anterior and posterior extremities of the epididymis. The epididymal epithelium varies from its thickest section in the proximal caput to its thinnest section in the caudal region. The relative position and distribution of the main cell types are illustrated in each macroscopic structural region of the epididymis demonstrating that principal and basal cells are the two most abundant cell types throughout the entire epididymis. Detailed illustrations in the upper portion of the figure show the maturation of the spermatozoa, along with the incorporation of the molecules along the passage through the duct. Some molecules are bound, and later, may or may not disappear. Other molecules are incorporated into the cellular membrane of the plasma of the spermatozoa or associate with the surface receptors and ions channels that are activated. Note the molecules that are also incorporated into the acrosomal vesicle. Some of these molecules will persist until the capacitation reaction in the reproductive female tract.
These molecules are indispensable to the specific recognition processes of the egg-spermatozoa interaction.

6.1 The epididymal maturation of the spermatozoa

Sperm maturation involves morphological and biochemical changes in the sperm surface in response to the epididymal secretions of enzymes, proteins and glycoproteins, which are essential in the process of fertilization (Orgebin-Crist, 1967; Robaire et al., 2000, 2006). The knowledge of the interactions between the luminal fluid microenvironment and the dynamics of the epididymal epithelium is indispensable to an understanding of the process of the development and maturation of the spermatozoa in the adult epididymis.

In the process of sperm maturation, molecules are secreted into the luminal fluid by different regions of the epididymis. These molecules interact sequentially with the surface of the spermatozoa or the acrosome, and alter their molecular function (Orgebin-Crist, 1967; Robaire et al., 2000, 2006; Gatti et al., 2004; Dacheux et al., 2005; Sullivan et al., 2005) (Figure 7). In this process, some proteins bind to the sperm and presumably affect sperm function directly (Ellerman et al., 1998; Von Horsten et al., 2007). Some proteins will later bind to the zona pellucida (Ellerman et al., 1998) or the plasma membrane of the oocyte (Cohen et al., 1996; Flesch & Gadella, 2000). Other proteins remain in the lumen throughout the length of the tubule (Dacheux et al., 2005; Fouchécourt et al., 2002). Several of the epididymal proteins that remain in the lumen are either closely bound to each other or they may be integrated into the membrane of the spermatozoa because of their hydrophobic properties (as GPI-anchored proteins) or because of the proteolysis of their carboxy-terminal region (Gatti et al., 2004).

Other proteins are present in the epididymal lumen for only a short time, suggesting that their continued presence may be detrimental to sperm maturation and/or epididymal cell functions. Selective mechanisms are in place for their removal (von Horsten et al., 2007). These short-acting proteins that are loosely bound probably prevent direct interactions with the sperm surface by masking the sites that will be activated during fertilization.

In addition, there are a number of proteins that can be transferred from the epididymal epithelium to the spermatozoa by a very specific and not fully elucidated mechanism mediated by the epididymosomes. Almost all of these proteins are rapidly absorbed or degraded in the first segment of the epididymis. Some of these molecules or aggregates become outer membrane components of the sperm, while others become integral membrane proteins. CRISPs (Cysteine-Rich Secretory Proteins, originally named DE) are epididymal proteins in rats and mice that do not require stable or long-term bounds with the sperm to perform their functions. CRISPs adhere to the sperm surface as epididymosomes during epididymal maturation and are subsequently lost during epididymal transit (Ellerman et al., 1998; Cohen et al., 2011).

Some molecules secreted by the epididymis do not interact directly with the surface of the sperm. Instead, the molecules interact with the sperm acrosome. The sperm acrosome is a highly specialized organelle overlying the anterior part of the sperm nucleus. The acrosome contains a number of hydrolytic enzymes that are believed to be required for fertilization (Yoshinaga & Toshimori, 2003).

The SP-10 and the acrin family (1 and 2) are examples of molecules undergoing intense regulation of the distribution in the movement into and out of the acrosome. These molecules induce changes in the acrosomal proteins of the spermatozoa by maturation-dependent modifications attributed to glycosylation/deglycosylation and/or proteolytic processing during epididymal maturation (Yoshinaga & Toshimori, 2003). The acrosomal human SP-10 protein decreases from 45 kDa to 18-25 kDa as a result of the proteolytic process when the spermatozoa pass into the caput epididymis. Acrin 1 and acrin 2 move from one acrosomal domain to another, which results in a compartmentalization of these molecules during epididymal maturation. Intra-acrosomal proteins initially generated in the testis can change their location and their size simultaneously during epididymal maturation (Yoshinaga & Toshimori, 2003).

There is a range of others proteins that are known to dwell in the epididymal fluid, such as lactoferrin, clusterin, cholesterol transfer protein, glutathione-peroxidase, albuminoidal proteins, prostaglandin, prostaglandin D2 synthase, hexosaminidase and procathepsin D (Fouchecort et al., 2000; Suzuki et al., 2004). It is generally known that these epididymal-secreted proteins exhibit either transport and binding functions or enzymatic activity. They contribute to the fertilization capacity of the spermatozoa by facilitating the exchange of proteins or lipids between the spermatozoa and the surrounding fluid (Dacheux et al., 2005) during epididymal maturation. The proteolysis of the pre-existing proteins, and even the metabolic activity of the spermatozoa, also contributes to the protein profile of the epididymal fluid (Dacheux et al., 2005).

Most of the molecules found in the epididymal luminal fluid participate in the maturation of the spermatozoa by responding to one of the various mechanisms previously described. Although the many of the molecular mechanisms that induce sperm maturation have not been definitively identified, it has been established that the regionalized epididymal fluid microenvironments promote numerous changes in the spermatozoa along the entire length of the epididymal tubule (von Horsten et al., 2007).

6.2 The epididymis and the protection of the mature spermatozoa

In humans, the spermatozoa take approximately ten days to travel from the initial segment to the cauda region of the epididymis (Toshimori, 2003). Once they are fully mature, the spermatozoa may be stored in the terminal region of the cauda epididymis for days or weeks, depending on the species, until ejaculation occurs (Yanagimachi, 1994). The mechanisms of sperm survival in the distal epididymis are poorly understood. The spermatozoa are at risk during transit and during the period of storage within the epididymis. The predominance of polyunsaturated fatty acids (PUFA) in the plasma membrane of the spermatozoa renders them highly susceptible to lipid peroxidation because of attacks by reactive oxygen species (ROS). The development of enzymatic and non-enzymatic strategies for protecting the spermatozoa during this extremely vulnerable period is another role attributed to the secretory activity of the epididymal epithelium. Some of the proteins released by the epididymal epithelium into the lumen seem to be involved in the protection of the spermatozoa from both oxidation reactions and/or bacterial attacks (Robaire et al., 2000; Gatti et al., 2004; Robaire et al., 2006).

A system for the regulated storage of spermatozoa in the distal region of the organ was developed in mammals, ensuring that the stored cells are quiescent and unreactive (Gatti et al., 2004). The exact mechanism of this physiological phenomenon, known as sperm quiescence, is not clearly understood. One hypothesis is that remaining in a quiescent state protects the sperm from the drastic change in ionic composition of the medium in the cauda epididymis. Other factors, such as the inorganic and organic constituents of the luminal fluid, are of secondary importance and might assist in inducing sperm quiescence (Gatti et al., 2004).

Finally, there are a number of epididymal sperm-coating proteins that exert their effects on the male gamete in the female, and not in the male tract. These proteins produced by the epididymis and bind the luminal sperm, but they become functional in the female oviduct. Some of these molecules are considered to be decapacitating factors and might bind to the spermatozoa surface to prevent premature sperm activation (Kobayashi & Behringer, 2003).

6.3 Postepididymal events

After undergoing epididymal maturation, the mature spermatozoa are stored in the cauda epididymal segment until ejaculation. Once in the female reproductive tract, the mature spermatozoa must be capacitated. Capacitation is a complex maturational phenomenon that renders spermatozoa capable of binding and fusing with the oocyte, which is a requirement for mammalian fertilization. Capacitation encompasses plasma membrane reorganization, ion permeability regulation, cholesterol loss and changes in the phosphorylation state of many proteins (Visconti et al., 2011).

To acquire the potential for fusion with the oocyte, the capacitated spermatozoa initially interact in a species-specific manner with the zona pellucida, an extracellular coat that surrounds the mammalian egg. This process, called primary binding, is mediated by the zona pellucida glycoconjugates that recognize the sperm receptors located on the surface of the male gamete. These receptors are incorporated into the spermatozoa surface during epididymal maturation (Bedford, 2008; Robaire et al., 2006). The bound spermatozoa undergo the acrosome reaction and initiate penetration into the zona pellucida. Sperm penetration involves both the digestion of the zona pellucida and vigorous sperm motion. The sperm is also kept bound to the matrix via sperm receptors, an interaction that is called secondary binding (Flesch & Gadella, 2000). Numerous candidates have been postulated as possible sperm receptors for primary and secondary binding, such as human P34H, an epididymal sperm protein secreted predominantly by principal cells in the proximal-distal segment of the corpus epididymis (Boué & Sullivan, 1996).

Acrosome-mediated spermatozoa that have penetrated the zona pellucida enter the perivitelline space, bind and fuse to the egg plasma membrane and release the genetic material that will initiate zygote development. Protein complexes with both binding and fusion functions are present in both the egg and the spermatozoa and interact with their counterparts on the surface of the other gamete. Recent studies have demonstrated that the capacitated spermatozoa contain multiprotein complexes that are present on the sperm surface during capacitation. These complexes display an affinity for the zona pellucida in the event of fertilization (Toshimori, 2003). Several proteins of epididymal origin have also been proposed as participants in the sperm-egg membrane fusion process. In general, fusion proteins may have sequences of hydrophobic residues. They are mostly comprised of an alpha-helical structure and are known as fusion molecules. In a similar manner, the extern proteins that are bound to the sperm surface reveal new functional peptide domains that may trigger the sperm-egg interactions such as protein fertilin β (also known as ADAM2) that initially covers the entire testicular sperm head, is degraded by two successive cleavages during caput transit and is further restricted to the post-acrosomal domain of the spermatozoa (Toshimori, 2003).

7. The aging process of the epididymis

During aging, the male reproductive tract is characterized by testicular dysfunction resulting in the atrophy of the seminiferous epithelium and the Leydig cells. However, in rats, the average epididymal weight is not significantly affected by aging (3 months to 24 months). Aging affects the epididymis in a segment- and cell-dependent manner. Basal cells are primarily affected in the initial segment, clear cells are most modified in the caput segment and principal cells are most damaged in the corpus. In the proximal cauda segment, aging more dramatically affects the appearance of cells. In this segment, clear cells become enlarged and are filled with dense lysosomes, and some principal cells contain large vacuoles (Robaire et al., 2006).

Gene expression significantly decreases with age in the initial segment, the corpus and the cauda of the epididymis. The decrease in total epididymal gene expression from aging can be clearly observed in the corpus and the cauda, where expression of 83% and 62% of genes is respectively reduced to approximately 50%. This is in contrast to the initial segment, in which only 31% of the genes had a decrease in expression of at least 50%. The caput of the epididymis was the only segment in which the expression of large proportions of genes did not drastically change with age (less than 33%) (Robaire et al., 2006).

The function of the epididymis is highly dependent on the presence of androgens, particularly 5α-reduced androgens. Rat models have shown that the ability of the epididymis to produce 5α-reduced androgens becomes compromised with age. However the ability of the organ to respond to androgens is not compromised (Smithwick & Young, 2001).

Finally, the blood-epididymis barrier may be compromised in structure and function with age, resulting in the appearance of, and an increase in the number of, monocytes/macrophages mainly in the initial segment. Several kinematic parameters associated with sperm motility in the cauda of the epididymis may also be decreased. An increased incidence in the rates of pre-implantation loss, lower fetal weight and higher post-natal deaths was observed in the offspring of young female rats mated with older male rats (Robaire et al., 2000, 2006). Whether these effects are due to deficient epididymal maturation or due to a testicular dysfunction requires further study. Dysregulated intracellular trafficking, decreased protein degradation and oxidative stress are a few of the possible hypotheses that may explain the molecular mechanisms behind these changes (Robaire et al., 2000, 2006).

Many social factors have contributed to the increase in the number of men over the age of 35 who wish to become parents over the past several years. In 1970, fewer than 15% of all men fathering children were over the age of 35, but recent percentages have risen to almost 25%. Reproductive function gradually declines with advanced paternal age from multifactorial causes. Pattern quality parameters for fertility significantly decrease from 30-year-old men to 50-year-old men as follows: sperm motility (262 ± 116 ppm to 110 ± 152 ppm), the volume of semen (3.7 ± 8.5 mL to 2.1 ± 1.5 mL) and the concentration of spermatozoa (76 ± 55 million/mL to 59 ± 57 million/mL), respectively. Most evidence suggests that increased aging has negative effects on male fertility and some genetic risk for offspring, but the age at which the risk develops and the magnitude of the risk are poorly defined, as is the role of the epididymis in this process (Stewart & Kim, 2011).

8. Epididymis and infertility

Approximately 13-15% of couples will encounter fertility problems during their reproductive life. Factors involving males are responsible for infertility in approximately 20% of couples and are contributory in another 30–40% of couples. This means that factors involving males and infertility are implicated in more than 50% of the difficulties couples encounter when attempting pregnancy (Hamada et al., 2011). Diagnostic tools and therapies to treat female infertility are relatively well developed. However, many of the causes of male infertility are considered idiopathic (Sullivan, 2004).

The spermogram accurately assesses male fertility through an evaluation of semen quality, sperm concentration, motility and morphology (Hamada et al., 2011). The normal spermogram values were based on multi-centered population studies on fertile men. However, several men presenting normal spermogram values are diagnosed as idiopathically infertile. These men may present with post-testicular defects that result in the ejaculation of spermatozoa with a normal morphology but with a sub-optimal fertilization capacity (Sullivan, 2004). The epididymis could be particularly involved in a number of the pathophysiologies affecting sperm maturation in some of these cases of male infertility (Sullivan, 2004; Hamada et al., 2011).

8.1 The epididymis, xenobiotics and endocrine disruptors

Xenobiotics are substances that are foreign to an organism. Some of these substances can produce adverse effects or damage under specific conditions of use, such as when these substances are present in much higher concentrations than are usual. Through the influence of lifestyle, environmental factors or prenatal exposures to compounds, xenobiotics can act as endocrine disruptors that reduce testosterone synthesis and androgenic signaling (Smithwick & Young, 2001; Marty et al., 2003). As previously described, the epididymis depends on androgens for both the development and maintenance of the organ. The epididymis is therefore a potential target for the toxic effects of xenobiotics, which may then influence male fertility. Experimental models that block the action of androgen produce side effects equivalent to a chemical orchiectomy. The formation of DHT was blocked in the epididymis by inhibiting the expression of genes involved in signal transduction, such as fatty acid and lipid metabolism and the regulation of ion and fluid transport. This inhibition affected sperm quality but did not result in the complete obstruction of fertility. The blocking of the actions of estrogen resulted in a dramatic reduction in the fluid uptake capacity of the epididymal tissue and caused infertility (Robaire et al., 2000).

Epididymal histopathological studies and sperm function tests have not been standard in the fields of drug development or in the assessment of toxicants. The effects of xenobiotics on the epididymis may often have been overlooked (Robaire et al., 2000). In addition to the damaging potential of xenobiotic substances, there are concerns about the effects of new anti-neoplastic drugs and herbal products, chemicals and plastics (cyclophosphamide, bisphenol A, phthalates sulfonates, and plasticizer), cleaning agents (alkylphenols) and pesticides/fungicides (mainly organochlorine compounds).

8.2 Epididymal abnormalities in cryptorchidism

Epididymal anomalies have been more commonly associated with undescended, rather than descended, testes. Testicular maldescent, then, is commonly associated with epididymal anomalies. However, most epididymal abnormalities are not likely to have contributed to testicular maldescent (Elder, 1992; Han & Kang, 2002). Cryptorchidism is a disease in which the testes and the epididymis are retained in the inguinal tract and the seminiferous tubules become atrophic as a result of the increase in temperature, which does not favor spermatogenesis (Garcia et al., 2011). The incidence of cryptorchidism is 1-4% in human male neonates (Toppari et al., 2001). The cause of cryptorchidism is multifactorial, although possible causes and risk factors, such as endocrine disorders, anatomical abnormalities and environmental and genetic factors, can explain the etiology of this phenomenon (Nieschalag et al., 2000).

The damage caused by cryptorchidism is reflected in a reduced diameter of the duct and a reduction in the length of the epithelium within the duct, along with the absence of spermatozoa in the lumen of the duct itself (Arrotéia et al., 2005; Garcia et al., 2011) (Figure 8).

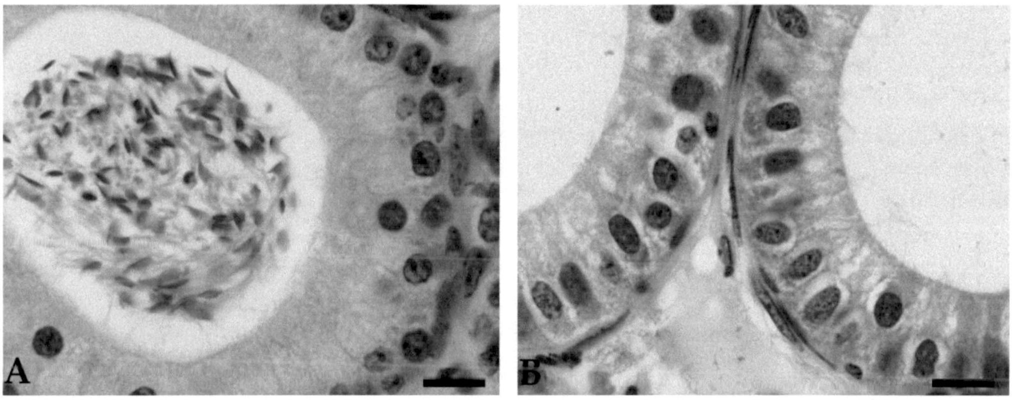

Fig. 8. Photomicrographs of a mouse epididymal caput. A: control photomicrograph of an adult epididymis; B: the epididymis of a cryptorchidic mouse. Sperm is absent because of the failure of production in the cryptorchidical testis. Scale bar: 50μm

Alterations in temperature also affect the ionic and protein composition of the cauda fluid, which effects the cauda epithelium by eliminating the special ability of the cauda to store and prolong the life of spermatozoa through the promotion of rapid epididymal transport (Nieschalag et al., 2000). In a series of studies on the epididymides of cryptorchidic animals and on primary epithelial cell cultures, it was shown that some epididymal gene products are exquisitely sensitive to small changes in temperature (Kirchhoff et al., 2000). Although there is no information on whether epididymal function (and hence sperm maturation) is irreversibly compromised in adult men who were submitted to orchidopexy as children, experimental results in mice indicate that gross and histological alterations caused by cryptorchidism can be restored by orchidopexy (Arrotéia et al., 2005; Garcia et al., 2011). However, the decrease in the transit time of the spermatozoa in the epididymis (i.e., acceleration of sperm passage through the epididymal duct) and the reduction of fertility, fecundity and potency observed in cryptorchidic mice were not fully restored following orchidopexy (Garcia et al., 2011).

8.3 The expresssion of epididymal proteins and infertility

The fate of the proteins involved in spermatozoa maturation may interfere with successful fertilization. The proteins secreted along the epididymis under androgenic control vary from one segment to the other and modify the maturing male gamete in a sequential manner (Sullivan, 2004). The presence or absence of given molecules on the spermatozoa has often been correlated to certain traits of the sperm, such as the recognition and binding to the zona pellucida or oocyte membrane, or to the movement of the spermatozoa (Dacheux et al., 2005).

During the passage through the epididymal duct, molecules are incorporated into the development of the spermatozoa. Some molecules are more relevant than others in the maturation process and these or other molecules may play a relevant role in the acrosomal process later on. In the case of the protein P34H, which is related to male idiopathic infertility, the spermatozoa maturation process is a key event in the process of fertilization. Men diagnosed with idiopathic male infertility have an unexplained reduction in semen quality. These patients have no abnormal findings on physical examination and no laboratory abnormalities associated endocrine function. P34H is a 34 kDa human epididymal sperm protein synthesized and secreted predominantly by principal cells. P34H is physiologically undetectable on the spermatozoa in the caput of the epididymis and progressively accumulates on the sperm surface covering the acrosomal cap from the corpus to the distal cauda of the epididymis. This protein remains inactive in capacitated spermatozoa but apparently is lost during the acrosome reaction (Boué & Sullivan, 1996). This suggests that P34H is a human epididymal protein involved in the sperm interaction with the zona pellucida. P34H is associated with the spermatozoa of semen samples obtained from fertile donors, but it is undetectable in approximately 40% of the semen samples obtained from men presenting with idiopathic infertility. Spermatozoa with undetectable levels of P34H were related to the inability to bind to the zona pellucida through *in vitro* assays. These cases of infertility could be attributed to the inefficient epididymal maturation of male gametes that are then unable to fertilize. The human P34H protein and other epididymal molecules could be considered as markers for epididymal function in sperm maturation and can be used as a diagnostic tool to identify cases of infertility in men that have not been diagnosed through the classic methods of semen analysis (Boué & Sullivan, 1996).

Another possible protein interaction that may affect fertility is a protein complex recognized by the monoclonal antibody (mAb) TRA 54 (a high molecular mass albumin-containing protein complex) located in the acrosome of the spermatozoa that binds to the spermatozoa membrane during passage through the epididymal lumen. Experiments using *in vitro* fertilization have demonstrated that the addition of mAb TRA 54 to the fertilization medium significantly decreases the fertilization rate (Arrotéia et al., submitted for publication).

Similarly, the mAb 4A8 inhibited sperm penetration into the zona pellucida (Batova et al., 1998). Although the mAb used in these studies have the potential to recognize specific molecules that are directly involved in the formation of functional spermatozoa and form the molecular basis of gamete interaction in mammals, the recognition epitopes of these mAb are still not fully characterized (Arrotéia et al., 2004).

In addition to the causes indicated above, any disruption of the epididymal microenvironment through congenital abnormalities, intrinsic alterations in pH, protein composition and concentration, temperature, and other factors may lead to male post-testicular infertility. The study of the fate of molecules involved in sperm maturation and sperm-oocyte recognition represents a key to the etiology of the idiopathic male infertility.

9. Conclusion

The mammalian epididymis promotes the modifications of the spermatozoa that are necessary for the spermatozoa to become fertilization-competent cells and to be stored safely in the male reproductive tract. Since epididymal dysfunctions are related to cases of idiopathic male infertility, a focus on the major epididymal proteins related to the spermatozoa is important. From a clinical point of view, an increase in our understanding of spermatozoa maturation should provide the specific markers that will assist in the development of new criteria for both the prediction of male infertility and for improving the treatment of male infertility. The implementation of research from the fields of genomics and proteomics has assisted in the characterization of some of the already identified proteins, as well as in the description of novel epididymal components. We anticipate that these great advances will be helpful in the elucidation of the sperm maturation process.

10. Acknowledgments

The research projects related to the Biology of the Epididymis and Fertilization conducted at the Department of Histology and Embryology, Institute of Biology at the State University of Campinas has been supported by São Paulo Research Foundation (FAPESP, grants n. 01/12773-7, 01/00016-7; 05/04007-3). K.F.A., P.V.G., and M.L.J. were supported by fellowships from The National Council for Scientific and Technological Development (CNPq).

11. References

Adamali, H.I. & Hermo, L. (1996). Apical and narrow cells are distinct cell types differing in their structure, distribution, and functions in the adult rat epididymis. *Journal of Andrology*, Vol. 17, No. 3, (May-June 1996), pp. 208-222, ISSN 0196-3635

Arrotéia, K.F.; Joazeiro, P.P.; Yamada, A.T.; Tanaka, H.; Nishimune, Y. & Pereira, L.A.V. (2004). Identification and characterization of an antigen recognized by monoclonal antibody TRA 54 in mouse epididymal and vas deferens epithelial cells. *Jounal of Andrology*, Vol. 25, No. 6, (November - December 2004), pp. 914-921, ISSN 0196-3635

Arrotéia, K.F.; Joazeiro, P.P. & Pereira, L.A.V. (2005). Does orchidopexy revert the histological alterations in epididymal and vas deferens caused by cryptorchidism?

Archives of Andrology, Vol. 51, No. 2, (March-April 2005), pp. 109-119, ISSN 0148-5016

Batova, I.N.; Ivanova, M.D.; Mollova, M.V. & Kyurkchiev, S.D. (1998). Human sperm surface glycoprotein involved in sperm-zona pellucida interaction. *International Journal Andrology*, Vol. 21, No. 3, (June 1998), pp. 141-153, ISSN 0105-6263

Bedfort, J.M. (1967). Effect of duct ligation on the fertilizing ability of spermatozoa from different regions of the rabbit epididymis. *Journal of Experimental Zoology*, Vol. 166, (November 1967), pp. 271-281, ISSN 0022-104X

Bedford, J.M. (2008). Puzzles of mammalian fertilization - and beyond. *International Journal of Developmental Biology*, Vol. 52, No. 5-6, (June 2008), pp. 415-426, ISSN 0214-6282

Benoit, J. (1926). Recherches anatomiques, cytologiques et histophysiologiques sur les voies excentrices du testicule chez les mammifères. *Archives d'anatomie, d'histologie et d'embryologie*, Vol. 5, No. 3, (June 1926), pp. 175-412, ISSN 003-9586

Boué, F. & Sullivan, R. (1996). Cases of human infertility are associated with the absence of P34H, an epididymal antigen. *Biology of Reproduction*, Vol. 54, No. 5, (May 1996), pp. 1018-1024, ISSN 0006-3363.

Cohen, D.J.; Maldera, J.A.; Vasen, G.; Ernesto, J.I.; Muñoz, M.W.; Battistone, M.A. & Cuasnicú, P.S. (2011). Epididymal protein CRISP1 plays different roles during the fertilization process. *Journal of Andrology*, Vol. 32, No. 6, (March 2011), ISSN 1939-4640

Cornwall, G.A. (2009). New insights into epididymal biology and function. *Human Reproduction Update*, Vol. 15, No. 2, (January 2009), pp. 213-227, ISSN 1460-2369

Dacheux, J.L., Castella, S.; Gatti, L.J. & Dacheux, F. (2005). Epididymal cell secretory activities and the role of the proteins in boar sperm epididymis. *Theriogenology*, Vol. 63, No. 2, (November 2005), pp. 319-341, ISSN 0093-691X

De Miguel, M.P.; Marino, J.M.; Martinez-Garcia, F.; Nistal, M.; Paniagua, R. & Regadera, J. (1998) Pre- and post-natal growth of the human ductus epididymidis. A morphometric study. *Reproduction, Fertility and Development*, Vol. 10, No. 3, (July 1998), pp. 271-277, ISSN 1031-3613

Ellerman, D.A.; Brantúa, S.; Martínez, S.P.; Cohen, D.J.; Conesa, D. & Cuasnicú, P.S. (1998). Potential contraceptive use of epididymal proteins: immunization of male rats with epididymal protein DE inhibits sperm fusion ability. *Biology of Reproduction*, Vol. 59, No. 5, (November 1998), pp. 1029-1036, ISSN 0006-3363

Flesch, F.M. & Gadella, B.M. (2000). Dynamics of the mammalian sperm plasm membrane in the process of fertilization. *Biochimica et Biophysica Acta*, Vol. 1469, No. 3, (November 2000), pp. 197-235, ISSN 0006-3002

Fouchecourt, S.; Matayer, S.; Locatelli, A.; Dacheux, F. & Dacheux, J.L. (2000). Stallion epididymal fluid proteome: qualitative and quantitative characterization; secretion and dynamic changes of major proteins. *Biology of Reproduction*, Vol. 62, No. 6, (June, 2000), pp. 1790-1803, ISSN 0006-3363

França, L.R.; Avelar, G.F. & Almeida, F.F.L. (2005). Spermatogenesis and sperm transit through the epididymis in mammals with emphasis on pigs. *Theriogenology*, Vol. 63, No. 2, (January 2005), pp. 300-318, ISSN 0093-691X

Garcia, P.V.; Arroteia, K.F.; Joazeiro, P.P.; Mesquita, S.F.P.; Kempinas, W.G. & Pereira, L.A.V. (2011). Orchidopexy restores morphometric-stereologic changes in the caput epididymis and daily sperm production in cryptorchidic mice, although sperm transit time and fertility parameters remain impaired. *Fertility and Sterility*, Vol. 96, No. 3, (July, 2011), pp. 739-744, ISSN 1556-5653

Gatti, J.L.; Castella, S.; Dacheux, F.; Ecroyd, H.; Métayer, S.; Thimon, V. & Dacheux, J.L. (2004). Post-testicular sperm environment and fertility. *Animal Reproduction Science*, Vol. 82-83, (July 2004), pp. 321-339, ISSN 0378-4320

Hamada, A.; Esteves, S.C. & Agarwal, A. (2011). Unexplained male infertility: potential causes and management. Review article. *Human Androloly*, Vol. 1, (March 2011), pp. 2–16, ISSN 1537-744X

Han, C.H. & Kang, S.H. (2002). Epididymal anomalies associated with patent processus vaginalis in hydrocele and cryptorchidism. *Journal of Korean Medical Science*, Vol. 17, No. 5, (October 2002), pp. 660-662, ISSN 1011-8934

Hannema, S.E. & Hughes, I.A. (2007). Regulation of Wolffian duct development. *Hormone Research*, Vol. 67, No. 3, (October 2006), pp. 142-151, ISSN 0301-0163

Hermo, L.; Oko, R. & Morales, C.R. (1994). Secretion and endocytosis in the male reproductive tract: a role in sperm maturation. *Internacional Review of Cytology*, Vol. 154, (May 1994), pp. 106-189, ISSN 0074-7696

Hermo, L.; Chong, D.L.; Moffatt, P.; Sly, W.S.; Waheed, A. & Smith, C.E. (2005). Region - and cell - specific differences in the distribution of carbonic anhydrases II, III, XII, and XIV in the adult rat epididymis. *The Journal of Histochemistry and Cytochemistry*, Vol. 53, No. 6 (June 2005), pp. 699-713, ISSN 0022-1554

Hess, R.A.; Zhou, Q.; Nie, R.; Oliveira, C.; Cho, H.; Nakai, M. & Carnes, K. (2001). Estrogens and epididymal function. *Reproduction, Fertility and Development*, Vol. 13, No. 4 (February 2001), pp. 273-283, ISSN 1031-3613

Hess, R.A.; Fernandes, S.A.F.; Gomes, G.R.O.; Oliveira, C.A.; Lazari, M.F.M. & Porto, C.S. (2011). Estrogen and its Receptors in Efferent Ductules and Epididymis. *Jounal of Andrology*, Vol. 32, No. 6, (March 2011), ISSN 1939-4640

Hinton, B.T.; Galdamez, M.M.; Sutherland, A.; Bomgardner, D.; Xu, B.; Abdel-Fattah, R. & Yang, L. (2011). How Do You Get Six Meters of Epididymis Inside a Human Scrotum? *Journal of Andrology*, In press, (March 2011), ISSN 1939-4640

Jervis, K.M. & Robaire, B. (2001). Dynamic changes in gene expression along the rat epididymis. *Biology of Reproduction*, Vol. 65, No. 3, (September 2001), pp. 696-703, ISSN 0006-3363

Kirchhoff, C. (1999). Gene expression in the epididymis. *International Review of Cytology*, Vol. 188, (May 1999), pp. 133-202, ISSN 0074-7696

Kirchhoff, C.; Carballada, R.; Harms, B; & Kascheike, I. (2000). CD52 mRNA is modulated by androgens and temperature in epididymal cell cultures. *Molecular Reproduction and Development*, Vol. 56, No. 1, (May 2000), pp. 26–33, ISSN 1040-452X

Kobayashi, A. & Behringer, R.R. (2003). Developmental genetics of the female reproductive tract in mammals. *Nature Reviews Genetics*, Vol. 4, No. 12, (December 2003), pp. 969-980, ISSN 0028-0836

Lan, Z.J.; Labus, J.C. &, Hinton, B.T. (1998). Regulation of Gamma-Glutamyl Transpeptidase Catalytic Activity and Protein. Level in the Initial Segment of the Rat Epididymis by Testicular Factors: Role of Basic Fibroblast Growth Factor. *Biology of Reproduction*, Vol. 58, No. 1, (January 1998), pp. 197-206, ISSN 0006-3363

Larios, H.M. & Mendoza, N.M. (2001). Onset of sex differentiation. Dialog between genes and cells. *Archives of Medical Research,* Vol. 32, No. 6, (November-December 2001), pp. 553-558, ISSN 0188-4409

Lasserre, A.; Barrozo, S.; Tezón, J.G.; Miranda, P.V. & Vazquez-Levin, M.H. (2001). Human epididymal proteins and sperm function during fertilization: an update. *Biological Research,* Vol. 34, No. 3-4, (July 2001), pp. 165-178, ISSN 0716-9760

Marty, M.S.; Chapin, R.E.; Parks, L.G. & Thorsrud, B.A. (2003). Development and maturation of the male reproductive system. *Birth Defects Research, Part B, Developmental and Reproduction Toxicoly,* Vol. 68, No. 2, (April 2003), pp. 125-136, ISSN 1542-9733

Moore, K.L. & Persaud, T.V.N. (2003). Urogenital System. In: *The Developing Human: Clinical Oriented Embryology,* Saunders Elsevier, (Ed.), pp. 246-285, Elsevier, ISBN: 978-85-352-2662-1, Philadelphia: Saunders.

Nieschalag, E.; Behre, H.M.; Mesched, D. & Kamischke, A. (2000). Disorders at the testicular level. In: *Andrology: Male reproduction health and dysfunction*. Nieschalag E, Behre HM, eds, Springer, pp. 143-176, ISBN: 3-540-67224-9, Berlin.

Orgebin-Crist, M.C. (1967). Sperm maturation in rabbit epididymis. *Nature,* Vol. 216, No. 5117, (November 1967), pp. 816-818, ISSN 0028-0836

O'Hara, L., Welsh, M., Saunders, P.T.K. & Smith, L.B. (2011). Androgen receptor expression in the caput epididymal epithelium is essential for development of the initial segment and epididymal spermatozoa transit. *Endocrinology,* Vol. 152, No. 2, (November 2010), pp. 718-729, ISSN 1945-7170

Robaire, B. & Hermo, L. (1988). Efferent ducts, epididymis and vas deferens: structure, functions and their regulation. In: *The physiology of reproduction,* Knobil and J. Neil, (Ed.), pp 999-1080, Raven Press, ISBN 0881672815, New York, EUA.

Robaire, B.; Syntin, P. & Jervis, K. (2000). The coming of age of the epididymis. In: *Testis, Epididymis and Technologies in the Year 2000,* Jégou B, Pineau C, Saez J, (Ed.), pp. 229-262, Springer: Hildenberg, ISBN 978-3-540-67345-3, New York, EUA.

Robaire, B.; Hinton, B.T. & Orgebin-Crist, MC. (2006). The Epididymis. In: *The physiology of reproduction,* Knobil and J. Neil, (Ed.), pp. 1071-1148, Elsevier, ISBN 978-0-12-515-400, New York, EUA.

Robaire, B. & Henderson, N.A. (2006). Actions of 5α-reductase inhibitors on the epididymis. *Molecular and Cellular Endocrinology,* Vol. 250, No. 1-2, (2006), pp. 190-195, ISSN 0303-7207

Sullivan, R. (2004). Male fertility markers, myth or reality. *Animal reproduction science,* Vol. 82- 83, (July 2004), pp. 341- 347, ISSN 0378-4320

Suzuki, K.; Drevet, J.; Hinton, B.T.; Huhtaniemi, I.; Lareyere, J.J.; Matusik, R.J.; Pons, E.; Poutanen, M.; Sipila, P. & Orgebin-Christ, M.C. (2004). Epididymis-specific promoter-driven gene targeting: a new approach to control epididymal function?

Molecular and Cellular Endocrinology, Vol. 216, No. 1-2, (March 2004), pp. 15-22, ISSN 303-7207

Smithwick, E.B. & Young, L.G. (2001). Histological effects of androgen deprivation on the adult chimpanzee epididymis. *Tissue & Cell*, Vol. 33, No. 5, (November 2001), pp. 450-461, ISSN 0040-8166

Stewart, A.F. & Kim, E.D. (2011). Fertility concerns for the aging male. Review. *Urology*, Vol. 78, No. 3, (September 2011), pp. 496-499, ISSN 1527-9995

Toshimori, K. (2003). Biology of spermatozoa maturation: an overview with an introduction to this issue. *Microscopy Research and Technique*, Vol. 61, No. 1, (May 2003), pp. 1-6, ISSN 1059-910X

Toppari, J.; Kaleva, M. &, Virtanen, H.E. (2001). Trends in the incidence of cryptorchidism and hypospasdias and methodological limitations or registry based data. Human Reproduction Update, Vol. 7, No. 3, (May-Jun 2001), pp. 282-286, ISSN: 1355-4786

Turner, T.T.; Johnston, D.S. & Jelinsky, S.A. (2006). Epididymal genomics and the search for a male contraceptive. *Molecular and Cellular Endocrinology*, Vol. 250, No. 1-2, (February 2006), pp. 178-183, ISSN 0303-7207

Turner, T.T.; Johnston, D.S.; Jelinsky, S.A.; Tomsig, J.L. & Finger, J.N. (2007). Segment boundaries of the adult rat epididymis limit interstitial signaling by potential paracrine factors and segments lose differential gene expression after efferent duct ligation. *Asian Journal of Andrology*, Vol. 9, No. 4, (July 2007), pp. 565-73, ISSN 1008-682X

Turner, T.T. (2008). De Graaf's thread: the human epididymis. *Journal of Andrology*, Vol. 29, No. 3, (May-June 2008), pp. 237-50, ISSN 1939-4640

Umar, A.; Ooms, M.P.; Luider, T.M.; Grootegoed, J.A. & Brinkmann, A.O. (2003). Proteomic Profiling of Epididymis and Vas Deferens: Identification of Proteins Regulated during Rat Genital Tract Development. *Endocrinology*, Vol. 144, No. 10, (October 2003), pp. 4637–4647, ISSN 0013-7227

Visconti, P.E.; Krapf, D.; de la Vega-Beltrán, J.L.; Acevedo, J.J. & Darszon, A. (2011). Ion channels, phosphorylation and mammalian sperm capacitation. *Asian Journal of Andrology*, Vol. 13, No. 3, (May 2011), pp. 395-405, ISSN 1745-7262

Von Horsten, H.H.; Johnson, S.S.; SanFrancisco, S.K.; Hastert, M.C.; Whelly, S.M. & Cornwall, G.A. (2007). Oligomerization and transglutaminase cross-linking of the cystatin CRES in the mouse epididymal lumen: Potential mechanism of extracellular quality control. *The Journal of Biological Chemistry*, Vol. 282, No. 45, (November 2007), pp. 32912-32923, ISSN 0021-9258

Yanagimachi, R.; Kamiguchi, Y.; Mikamo, K.; Suzuki, F. & Yanagimachi, H. (1985). Maturation of spermatozoa in the epididymis of the Chinese hamster. *American Journal of Anatomy*, Vol. 172, No. 4, (April 1985), pp. 317-330, ISSN 0002-9106

Young, W.C. (1929a). A study of the function of the epididymis: I. Is the attainment of full spermatozoon maturity attributable to some specific action of the epididymal secretion? *Jounal of Morphology and Physiology*, Vol. 47, (June 1929), pp. 479-495, ISSN 0095-9626

Young, W.C. (1929b). A study of the function of the epididymis: II. The importance of an aging process in sperm for the length of the period during wich fertilizating capacity is retained by sperm isolated in the epididymis of the guinea pig. *Journal of Morphology and Physiology*, Vol. 48, (June 1929), pp. 475-491, ISSN 0095-9626

Young, W.C. (1931). A study of the function of the epididymis. III. Functional changes undergone by spermatozoa during their passage through the epididymis and vas deferens in the guinea pig. *Journal of Experimental Biology*, Vol. 47, (September 1931), pp. 479-495, ISSN 0022-0949

Yoshinaga, K. & Toshimori, K. (2003). Organization and modifications of the sperm acrosomal molecules during spermatogenesis and epididymal maturation. *Microscopy Research and Technique*, Vol. 61, No. 1, (May 2003), pp. 39-45, ISSN 1059-910X

Part 2

Implantation, Placentation and Early Development

4

Endometrial Receptivity to Embryo Implantation: Molecular Cues from Functional Genomics

Alejandro A. Tapia
Instituto de Investigaciones Materno Infantil (IDIMI), Universidad de Chile, Santiago Chile

1. Introduction

The endometrium is the mucous lining the uterine cavity comprised of a basal and a functional layer being the latter the one that sheds during menses and regenerates from the basal portion. The main cell populations within the functional stratus are epithelial and stromal cells accompanied by a variable number of leukocytes. Epithelial cells are found covering the luminal surface and tubular glands in basal and functional layers. Endometrial stroma contains reticular connective tissue comprised mainly by uterine fibroblasts that rapidly differentiate into decidualized cells when stimulated by an implanting blastocyst. The stromal compartment contains also abundant lymphocytes, granulocytes and macrophages during luteal phase of the menstrual cycle. These cells along with epithelial and stromal fibroblasts are source and target of paracrine signals of proliferation and differentiation. Both components respond to ovarian steroid hormones and depend on each other for their structure, function and responsiveness to estrogen (E_2) and progesterone (P_4) (Tabibzadeh, 1998, review). During a normal menstrual cycles, human endometrium display unique features for an adult tissue: undergoes cyclic construction and sloughing. The outer layer of the endometrium is loss while the basal layer containing the deep glandular epithelium gets preserved. Later on, stem cells located in this layer will originate the various endometrial cell types in response to the appropriate hormonal stimulus, regenerating the whole endometrium (Padykula, 1991).

The endometrial cycle is driven by the ovarian steroidal hormones and can be divided in three phases: proliferative, secretory and menstrual. Proliferative phase lasts around 10 - 20 days averaging 14 days. During this phase, glands grow and become winding due to the active mitosis of the epithelial cells driven by rising levels of serum E_2 resulting in growing about 10 times the original thickness of the endometrium. Indeed, extensive DNA synthesis in epithelial cells and some in stromal cells is seen during this stage (Padykula, 1991). Once ovulation has taken place, the increase of circulating P_4 triggers the transition to the secretory phase. During this phase, mitotic activity is inhibited and a complex secretory activity is induced beginning with glycogen vesicles polarization in glandular epithelial cells, locating subnuclearly which is further transported by microfilaments to the apical region where glycogen is actively secreted to the lumen of glands. In addition, epithelial cells initiate a complex secretory activity along with the establishment of an adequate environment for embryo implantation that take place only during a restricted time frame called 'window of implantation' (Psychoyos, 1986). During this period, morphological and molecular changes take place leading to a coordinated expression or repression of key molecules that ultimately enable the blastocyst to attach and invade the endometrial tissue. Such changes occur independently of the presence of a blastocyst; however the endometrium undergoes further biochemical and morphological changes induced by signals from the blastocyst and the following trophoblast invasion. With no embryo implantation, the endometrium undergoes a series of processes that end toward late secretory phase with sloughing and menses. When a successful embryo implantation takes place, luteolysis is prevented and the endometrium is not just maintained but differentiates to decidua and undergoes dramatic vascular changes at the implantation site. Therefore, gene expression in the human endometrium is likely to exhibit neat and distinct changes throughout the various stages of the menstrual cycle in accordance with the oscillations in estrogen and progesterone serum levels and their tissue receptor levels. Since these ovarian steroid hormones drive these processes eliciting an array of cellular and biochemical responses, mostly through genomic pathways (O'malley & Tsai, 1992), current thinking suggests that at the onset of receptivity, expression of some genes in given cell types of this tissue, is temporarily turned on or increased while some others are temporarily turned off or decreased (Tabibzadeh, 1998). Some of these changes are essential for establishing and maintaining pregnancy. Likewise, when implantation has occurred, another program of gene expression takes place in the endometrium, not only maintaining it, but also triggering further differentiation to decidua and facilitating and regulating trophoblast invasion and placenta development.

2. Hormonal regulation of the endometrial cycle

The endometrial cycle depends mainly on the steroidal ovarian hormones, acting though cytoplasmic receptors that on its inactive form are found forming a complex with chaperone proteins (O'malley & Tsai, 1992). Upon binding of the steroidal hormone with its cognate receptor, the chaperone-receptor complex dissociates and the new hormone-receptor complex translocates to the nucleus, binding to specific elements of DNA in target genes. As a result from this binding and the recruitment of co-activator and co-repressor proteins, the transcription rate to mRNA is modified. This process ultimately increases or decreases the mRNA transcribed from target genes, which is transported to the cytoplasm where is translated to peptides or proteins. Steroid hormones can also elicit rapid actions on target cells independently of its genomic regulatory effects. Such actions occur in a time scale from seconds to minutes and have been commonly denoted as non-genomic actions so they can be distinguished from their direct actions over nuclear gene expression (Gellersen et al., 2009).

Cytoplasmic expression of receptors for estrogen (ER) and progesterone (PR) in the endometrium is mainly regulated by the own steroidal hormones. ER expression increases in response to rising levels of E_2 during follicular phase of the menstrual cycle, peaking during proliferative phase (Bergeron et al., 1988; Lessey et al., 1988). After ovulation, ER decrease by P_4 influence. The highest expression of PR occur at the time of ovulation driven by circulating E_2 and are more abundant in glandular epithelium than stroma, disappearing almost completely toward mid secretory phase by effect of the own P_4 action. However, stromal cells exhibit moderately high PR expression during proliferative and secretory phases (Lessey et al., 1988).

Although E_2 and P_4 have long been believed to be essential for endometrial development, it is now evident that these effects are further mediated and modulated by peptide hormones and peptide growth factors secreted by a variety of cell types within the uterine endometrium. Cooke et al. (Cooke et al., 1997) using mice model ER deficient showed that the proliferative effects of E_2 on endometrial epithelium was mediated by stromal ER through a paracrine mechanism. The paracrine messenger appears to be insulin growth factor (IGF)-1 (Pierro et al., 2001). Several cytokines have bee also described as part of endometrial signaling networks such as interleukin (IL)-1, transforming growth factor (TGF)-β, vascular-endothelial growth factor (VEGF) and colony stimulating factor (CSF)-1 (Salamonsen et al., 2000, review).

The endometrial basal layer which is adjacent to the myometrium, undergoes few changes during the menstrual cycle; whereas the functional layer is very sensitive to E_2 and P_4. Estrogens induce proliferation and growth of the endometrial tissue during the proliferative phase while post-ovulatory rising levels of circulating P_4 from the *corpus luteum* inhibits proliferation and induces the secretory phenotype. This latter hormone has been shown to be critical for endometrial receptivity (Baulieu, 1989) regulating the expression of several cytokines and growth factors, as well as morphological and molecular changes of the endometrial epithelial cells lining the uterine lumen (Giudice, 1999; Lessey, 2003). In addition induces the influx of distinct immune cells and subsequently triggers the differentiation of the fibroblast from the stromal compartment, a process termed decidualization (Irving & Giudice, 1999) characterized by vascular remodeling and extensive secretion of prolactin, insulin-like growth factor binding protein (IGFBP)-1 and tissue factor (Tseng & Mazella, 1999; Christian et al., 2001).

3. Endometrial receptivity and embryo implantation

Experimental evidence that showed embryo and endometrial development synchronicity as a critical factor for successful pregnancy, has underpinned the importance of determinants for uterine receptivity to further improve implantation rates in couples under assisted reproductive technologies (ART) such as *in vitro* fertilization. The concept of endometrial receptivity is referred to the ability of the endometrium to allow embryo implantation, which is the process whereby the blastocyst gets fixed to the uterine epithelium and penetrates though it. During this process, complexand synchronized interactions between the endometrial and the embryonic cells take place and it has been divided in three consecutive stages. The first one is the apposition or the orientation of the blastocyst embryonic pole toward the uterine epithelium. During the second stage of implantation or adhesion phase, the embryonic throphoectodermal cells attach to the endometrial epithelium a firm adhesion is established. Thereafter the invasion phase occurs where blastocyst braches the endometrial epithelium and invades the entire endometrium reaching the inner third of the myometrium and remodeling the uterine vasculature.

Endometrial receptivity is not permanent, if fact the uterus does not allow embryo implantation during most of the endometrial cycle. This particular feature was first described in the rat and in the mice was described the existence of a 'window of implantation', which is controlled by the ovarian steroidal hormones: a narrow time frame in which the endometrium allows blastocyst implantation (Mclaren, 1956; Psychoyos, 1986). These studies showed that depending on the hormonal stimulus used, the endometrium can be driven to a neutral, receptive or non-receptive (or refractory) state to embryo implantation. Since such window was found to be present in other species (Psychoyos, 1973; Psychoyos & Casimiri, 1980) it was postulated this mechanism could be operating also in humans. In this regard, Hertig *et al.* (Hertig et al., 1956) proposed that human embryo implantation occurs 5-6 days after ovulation by examination of uterine samples in women attempting pregnancy before hysterectomy. They observed free floating embryos within the uterine lumen before days 19-20 of the menstrual cycle, whereas from day 21 blastocysts were found already implanted. These data have been corroborated in oocyte donation cycles in which fertilized oocytes are transferred to the uterus of recipient women during spontaneous and induced cycles with exogenous steroids (Navot et al., 1986; Navot et al., 1991; Bergh & Navot, 1992), leading to the conclusion that the window of implantation in humans lasts for 4-6 days during the secretory phase coinciding with peak P_4 plasmatic levels. It should be noted that unlike the situation in rodents, in humans could not operate the switch from the receptive to the non-receptive state. Insufficient release of human chorionic gonadotrophin to maternal systemic circulation may lead to failure to rescue the *corpus luteum*. As a consequence, serum P_4 will decline leading to menstruation and conceptus loss. Embryo-uterine interactions that allow implantation can only occur when embryo development is synchronized with the endometrial receptivity period since lack of coordination between both events lead to implantation failure (Pope, 1988).

4. Cellular and molecular changes associated to endometrial receptivity

Cowel (Cowell, 1969) found that removing the uterine luminal epithelium in the rat, blastocyst implants regardless of any hormonal control suggesting that endometrial refractoriness lies on the endometrial epithelial cells. Recent data from IVF cycles (Huang et al., 2011) seems to support this fact in humans. In animal experiments and *in vitro* models for human implantation have revealed that the endometrial surface undergoes significant changes in its adhesive properties. In the pre-receptive state, the endometrium displays a structural and functionally polarized epithelium with differentiated basal-lateral and apical domains. During the endometrial receptive state, a reduction of the glycocalyx thickness and electrostatic charge has been seen in the surface of epithelial cells (Murphy & Rogers, 1981; Morris & Potter, 1984). In addition, the long and abundant epithelial microvilli retract, creating multiple flat areas in the surface (Schlafke & Enders, 1975; Murphy, 1993). This process could be related to the destabilization of the actin cytoskeletal network observed in these structures (Luxford & Murphy, 1989; Luxford & Murphy, 1992). On the other hand during the receptivity period it has been reported biosynthesis and expression of a different repertoire of surface proteins in the apical (Aplin, 1997; Lessey, 1998; Kirn-Safran & Carson, 1999) and basal-lateral domains (Rogers & Murphy, 1992; Albers et al., 1995; Murphy, 1995; Nikas, 1999). Considering the above mentioned evidence, the acquisition of adhesive properties by the epithelium may occur by disruption of the polarized apical-basolateral phenotype (Denker, 1983; Denker, 1994). Although it is not well understood yet the relation between the epithelial polarity loss and the initiation of the adhesion stage of implantation, it is speculated that facilitates close apposition between the endometrial epithelium and the blastocyst.

Several molecules contributing to trophoectoderm adhesion to endometrial epithelium have been proposed. During the window of implantation, there is an up-regulation of oligosaccharides ligands for selectin in uterine epithelial cells while human trophoectoderm express L-selectin, establishing a ligand-receptor system since it promotes binding between both cellular types (Genbacev et al., 2003). Other glycoproteins, oligosaccharides chains and their receptors found in the endometrial luminal epithelium have been proposed as mediators of the blastocyst adhesion. Amongst them is heparan sulphate proteoglycan and heparan sulphate binding proteins (Carson et al., 1998; Fukuda & Nozawa, 1999), H type-1 carbohydrate antigen (Fukuda & Nozawa, 1999). In addition the cell surface mucin with antiadhesive properties MUC1 has been involved in endometrial receptivity (Surveyor et al., 1995). MUC1 is expressed at the luminal endometrial surface in the mid-secretory phase (Aplin et al., 1998; Aplin, 1999) and *in vitro* evidence has shown a local cleavage from endometrial epithelial cells at the site of blastocyst attachment (Meseguer et al., 2001). Amongst the most studied adhesion molecules is the integrin family, which act as extracellular matrix elements receptors mediating adhesion events and signal transduction between cells. Some of these glycoproteins display a cycle-dependent endometrial expression (Lessey et al., 1992; Tabibzadeh, 1992; Lessey et al., 1994). At least three integrins seem to be flanking the opening and closure of human window of implantation, which are expressed in glandular epithelium only between days 20-24 of the menstrual cycle (Lessey et al., 1992). These integrins are $\alpha_v\beta_3$, $\alpha_v\beta_5$ y $\alpha_5\beta_1$ and recognize the RGD peptide motif. The best characterized integrin in endometrial receptivity is integrin $\alpha_v\beta_3$ (Lessey & Castelbaum, 2002, review). Intrauterine injection of an antibody against integrin $\alpha_v\beta_3$ before implantation has taken place reduces the number of implantation sites y mice and rabbits (Illera et al., 2000). However the precise role of integrins in the implantation process is not known yet.

The transmembrane protein trophinin mediated the hemophilic adhesion between cells along with the cytoplasmic proteins tastin and bystin forming a complex with cytoskeletal elements (Suzuki et al., 1998). These three proteins have been detected in both trophoblast and decidual cells at the embryo-maternal interphase (Suzuki et al., 1999) suggesting a potential role in the implantation process.

Temporal-spatial expression of Epidermal Growth Factor (EGF) family members and their receptors (ErbBs) in the embryo and endometrium during the peri-implantation period, suggest these growth factors may be mediating the interaction between them (Das et al., 1997). Members of the EGF family expressed in mice uterus at the moment of implantation are the own EGF, the transforming growth factor (TGF)-α, heparin-binding EGF (HB-EGF), amphiregulin (Ar), β-cellulin (BTC), epiregulin (Er) and Herregulin (HRG) (Das et al., 1997). HB-EGF is expressed in humans during the window of implantation (Leach et al., 1999), and also stimulates the development of human embryos generated in IVF cycles (Martin et al., 1998). The relative importance of the other members from the EGF-family in the implantation process has not been determined; however the expression of multiple ligands and receptors of such family may assure an adequate embryo development and further, a successful implantation.

The expression of the leukemia inhibitory factor (LIF) cytokine increases in mice endometrial glands prior to implantation and this regulation is under maternal control (Bhatt et al., 1991). LIF is essential for embryo implantation in mice (Stewart et al., 1992). In human endometrium, LIF is expressed in glandular and luminal epithelium (Cullinan et al.,

1996; Vogiagis et al., 1996). Although its biological functions are not well understood, the intrauterine injection of a monoclonal anti-leukemia inhibitory factor antibody inhibits blastocyst implantation in the rhesus monkey (Sengupta et al., 2006), suggesting a potential role in human embryo implantation.

5. Morphological and molecular assessment of the endometrium

Histomorphological changes of the endometrium throughout the menstrual cycle have been described over half a century ago by Noyes (Noyes et al., 1950) where particular features of the endometrial histology were correlated to specific days of the menstrual cycle allowing the dating of endometrial specimens. Since then, the Noyes criteria remained as the gold standard for endometrial evaluation. However the usefulness of endometrial dating for couples with infertility has been questioned since histological delay in endometrial maturation fails to discriminate between fertile and infertile couples (Coutifaris et al., 2004).In addition, other studies (Murray et al., 2004; Dietrich et al., 2007) have shown that endometrial histological features failed to reliably distinguish specific menstrual cycle days or narrow intervals of days, leading to the conclusion that histological dating has neither the accuracy nor the precision to be useful in clinical management. Another approach used to assess the endometrial status based on its morphological features was the use of scanning electron microscopy. Through the use of this technique, it was revealed the cyclic appearance of bulging structures from the apical pole of luminal epithelial cells during mid-secretory phase termed pinopodes (Nikas et al., 1995) or uterodomes (Murphy, 2000), becoming a candidate for endometrial receptivity marker. Although its involvement in embryo implantation has not been demonstrated, it is speculated that since they extend beyond cilia, they may be the first structure contacting the embryo. The molecular structure of pinopodes remains unknown so an adhesive role has yet to be determined. *In vitro* evidence has shown blastocyst attachment to endometrial epithelial cells displaying pinopode-like structures (Bentin-Ley et al., 1999). However, recent studies have failed to show a reliable pattern for the appearance of these structures in human endometrium (Acosta et al., 2000; Usadi et al., 2003; Quinn & Casper, 2009), rising controversy about its usefulness as an endometrial receptivity marker. In addition, morphological features seldom provides information regarding the molecular mechanisms taking place in the tissue throughout the menstrual cycle, which may allow a better understanding of the physiological status of the endometrium.

Molecular changes associated with the acquisition of the endometrial receptive phenotype in natural spontaneous cycles and pathological and pharmacological models in which endometrial function is compromised rendering it refractory to embryo implantation, have been used in search for molecular markers for endometrial receptivity. A number of candidate molecules have been proposed including members of the integrin family (Lessey et al., 1995; Thomas et al., 2003), glycodelin (Chryssikopoulos et al., 1996), Hb-EGF (Yoo et al., 1997), LIF (Ledee-Bataille et al., 2004) and CSF-1 (Kauma et al., 1991). Although much effort has been put on identifying endometrial receptivity markers to date no single one has been proved to be sensitive and specific enough in predicting pregnancy (Hoozemans et al., 2004; Strowitzki et al., 2006).

6. Wide genomic analysis of human endometrial function

The search for reliable molecular predictors for embryo implantation in the endometrium has been mainly focused on the one-by-one approach. With the development of functional genomics analysis tools more than 10 years ago, was possible to identify endometrial gene expression profiles under different conditions of receptivity or pregnancy, using DNA microarrays technology (Horcajadas et al., 2007). Through this technique it is possible to measure the level of expression in a collection of cells for thousands of genes, allows discovering genes or pathways likely to be involved in a biological process, even when there is no hint regarding their identity (Schena et al., 1995).

The global gene expression assessment has been used to characterize in a broader way the molecular bases of endometrial function in the women, determining the corresponding transcript profile to each endometrial phase during the menstrual cycle (Ponnampalam et al., 2004; Punyadeera et al., 2005; Talbi et al., 2006). In addition, this approach has been used to specifically investigate the particular gene signatures that allow acquisition of endometrial receptivity to embryo implantation during spontaneous cycles (Carson et al., 2002; Kao et al., 2002; Borthwick et al., 2003; Riesewijk et al., 2003; Mirkin et al., 2005). Since acquisition of endometrial receptivity is mainly driven by P_4 (Conneely et al., 2002; Spencer & Bazer, 2002), two strategies based on this feature have been used for gene discovery during spontaneous menstrual cycles: comparing gene expression profiles of the endometrium under peak P_4 circulating levels (days 19-23, window of implantation) and under absent (days 8-11, proliferative phase) (Kao et al., 2002; Borthwick et al., 2003) or low (days 15-17, early secretory phase) (Carson et al., 2002; Riesewijk et al., 2003; Mirkin et al., 2005; Haouzi et al., 2009; Haouzi et al., 2009) serum P_4. Several other strategies have been used to determine the repertoire of genes related to endometrial receptivity using animal, *in vitro*, pharmacological and pathological models which are discussed elsewhere in a comprehensive review (Horcajadas et al., 2007).

We studied the endometrial gene expression signatures from women with implantation failure using the oocyte donation model (Tapia et al., 2008). In an oocyte donation cycle, the endometrium from the embryo recipient woman is prepared with exogenous hormones in order to synchronize conceptus and endometrial development (De Ziegler et al., 1994; Younis et al., 1996), providing a better uterine environment than controlled ovarian hyperstimulation for embryo implantation to take place. In this sense, oocyte donation allows a unique opportunity for investigating endometrial factors involved in human blastocyst nidation (Damario et al., 2001). In our study, three groups of subjects were recruited: women who had previously participated as recipients in oocyte donation cycles and repeatedly exhibited implantation failure (Group A, study group) or had at least one successful cycle (Group B, control group); and spontaneously fertile women (Group C, normal fertility group). All were treated with exogenous E_2 and P_4 to induce an oocyte donation mock cycle as recipients. An endometrial biopsy was taken during the window of implantation (*i.e.* the seventh day of P_4 administration) and RNA from each sample was analyzed by cDNA microarrays to identify differentially expressed genes between groups. We found sixty three transcripts differentially expressed ($\geq$ 2-fold) between Groups A and B, of which 16 were subjected to real time RT-PCR validation. Eleven of these were significantly decreased in Group A with regard to Groups B and C. In addition to those genes whose transcript levels was confirmed by real time RT-PCR, we integrated and cross-validated a less stringent and larger data set that was constructed with other data sets about endometrial gene expression profiles publicly available obtained by other groups. Using this strategy we could increase the confidence in gene discovery for endometrial receptivity for many more genes than is tractable with classical validation (Kemmeren et al., 2002; Rhodes et al., 2002). For that we constructed a database with the reported transcript level changes from non-receptive to receptive endometrial phenotype at the time the study was made (Carson et al., 2002; Kao et al., 2002; Borthwick et al., 2003; Riesewijk et al., 2003; Mirkin et al., 2005) and 14 coincident genes were identified. Interestingly, five genes out of the 14 coincident genes were also dysregulated in eutopic endometrium from women with endometriosis. These genes are: Complement component 4 binding protein, alpha (C4BPA), Glycodelin (PAEP, glycodelin), RAP1 GTPase activating protein 1 (RAP1GA1), Endothelin receptor type B (EDNRB) and Ankyrin 3, node of Ranvier [ankyrin G] (ANK3). Interestingly, a detailed analysis of the functions associated to the 14 genes whose transcripts were significantly decreased in endometria without manifest abnormalities showed that 4 of them were related to the regulation of the immune function. This suggest that implantation failure in women from group A could be related to molecules from the immune system, whose function in the endometrium is to destroy infectious agents and foreign bodies, display an exaggerated response in presence of an implanting embryo (Damario et al., 2001).

Other strategy we have used is the integration and cross-validation of all available data sets about endometrial gene expression profiles produced by different groups (Tapia et al., 2011)to determine the up- and down-regulated genes that together orchestrate the acquisition of the receptive phenotype of the endometrium for embryo implantation. We considered studies that had used microarrays technology to determine the gene expression profiles that identify different phases of the endometrial cycle in spontaneous menstrual cycles (Ponnampalam et al., 2004; Punyadeera et al., 2005; Talbi et al., 2006). In addition we included those that also had used this technology during the acquisition of endometrial receptivity to embryo implantation (Carson et al., 2002; Kao et al., 2002; Borthwick et al., 2003; Riesewijk et al., 2003; Mirkin et al., 2005). In two studies the proliferative phase was compared with the 'window of implantation' time (Kao et al., 2002; Borthwick et al., 2003) and in another three studies gene expression differences between the early secretory phase (2–4 days after the luteinizing hormone (LH) surge) and the receptive phase (7–9 days after the LH surge) were included (Carson et al., 2002; Riesewijk et al., 2003; Mirkin et al., 2005). The intersection of lists with regulated genes reported in these studies showed a rather small number of coincident transcripts. We identified 40 up-regulated genes in at least four of seven reports and 21 down-regulated genes present in at least three of six studies considered. We denominated this set of coincident genes the consensus endometrial receptivity transcript list (CERTL) (Tapia et al., 2011). The most consistent up-regulated genes were C4BPA, SPP1, APOD, CD55, CFD, CLDN4, DKK1, ID4, IL15 and MAP3K5; whereas OLFM1, CCNB1, CRABP2, EDN3, FGFR1, MSX1 and MSX2 were the most consistently down-regulated in endometrial tissue for the acquisition of receptivity to embryo implantation.

7. Future perspectives in the clinic

One of the main objectives in reproductive medicine especially in the context of IVF has been the search for markers predictive of endometrial receptivity. Even though great efforts have been made to predict embryo implantation for improving live-births, no successful endometrial evaluation has been clinically validated so far. Moreover, attempts to improve IVF pregnancy rates treating infertile patients with factors thought to be essential for implantation process have turned out to achieve the opposite (Brinsden et al., 2009). Nevertheless gene expression profiling of endometrial biopsies during the window of implantation is one of the most promising strategies for gene discovery related to uterine receptivity. In fact, a genomic tool composed of a customized microarray and a bioinformatic predictor for endometrial dating and detection of endometrial pathologies has been recently described (Diaz-Gimeno et al., 2011). This tool denominated Endometrial Receptivity Array (ERA) assesses the transcriptomic signature defined by 134 genes related to endometrial receptivity, becoming specific for uterine function evaluation. Other study recently published (Tseng et al., 2010), analyzing gene expression profiles of endometrial biopsies and using hierarchical cluster analysis described a 123-gene model for endometrial function with transcripts up-regulated at mid-secretory phase, moderately expressed at late-secretary phase, and down-regulated at late-secretory phase.

The role of the proteins encoded by the transcripts contained in CERTL, ERA and the '123-gene model' in the acquisition of endometrial receptivity and embryo implantation; as well as the prognostic value for each transcript profiling as a marker for endometrial receptivity has yet to be determined. Although is highly possible that a combination of these three approaches may allow defining the actual transcriptomic signature of human endometrial receptivity.

8. Acknowledgment

The author is grateful to the staff from the Molecular Endocrinology lab of IDIMI and to the funding support FONDECYT 11100443, PBCT-PSD51(IDIMI) and FONDAP 15010006.

9. References

Acosta, A. A.; Elberger, L.; Borghi, M.; Calamera, J. C.; Chemes, H.; Doncel, G. F.; Kliman, H.; Lema, B.; Lustig, L. & Papier, S. (2000). Endometrial dating and determination of the window of implantation in healthy fertile women. Fertil Steril Vol.73, No.4, pp. 788-798, ISSN 0015-0282

Albers, A.; Thie, M.; Hohn, H. P. & Denker, H. W. (1995). Differential expression and localization of integrins and CD44 in the membrane domains of human uterine epithelial cells during the menstrual cycle. Acta Anat (Basel) Vol.153, No.1, pp. 12-19, ISSN 0001-5180

Aplin, J. D. (1997). Adhesion molecules in implantation. Rev Reprod Vol.2, No.2, pp. 84-93, ISSN 1359-6004

Aplin, J. D. (1999). MUC-1 glycosylation in endometrium: possible roles of the apical glycocalyx at implantation. Hum Reprod Vol.14 Suppl 2, pp. 17-25, ISSN 0268-1161

Aplin, J. D.; Hey, N. A. & Graham, R. A. (1998). Human endometrial MUC1 carries keratan sulfate: characteristic glycoforms in the luminal epithelium at receptivity. Glycobiology Vol.8, No.3, pp. 269-276, ISSN 0959-6658

Baulieu, E. E. (1989). Contragestion and other clinical applications of RU 486, an antiprogesterone at the receptor. Science Vol.245, No.4924, pp. 1351-1357, ISSN 0036-8075

Bentin-Ley, U.; Sjogren, A.; Nilsson, L.; Hamberger, L.; Larsen, J. F. & Horn, T. (1999). Presence of uterine pinopodes at the embryo-endometrial interface during human implantation in vitro. Hum Reprod Vol.14, No.2, pp. 515-520, ISSN 0268-1161

Bergeron, C.; Ferenczy, A. & Shyamala, G. (1988). Distribution of estrogen receptors in various cell types of normal, hyperplastic, and neoplastic human endometrial tissues. Lab Invest Vol.58, No.3, pp. 338-345, ISSN 0023-6837

Bergh, P. A. & Navot, D. (1992). The impact of embryonic development and endometrial maturity on the timing of implantation. Fertil Steril Vol.58, No.3, pp. 537-542, ISSN 0015-0282

Bhatt, H.; Brunet, L. J. & Stewart, C. L. (1991). Uterine expression of leukemia inhibitory factor coincides with the onset of blastocyst implantation. Proc Natl Acad Sci U S A Vol.88, No.24, pp. 11408-11412, ISSN 0027-8424

Borthwick, J. M.; Charnock-Jones, D. S.; Tom, B. D.; Hull, M. L.; Teirney, R.; Phillips, S. C. & Smith, S. K. (2003). Determination of the transcript profile of human endometrium. Mol Hum Reprod Vol.9, No.1, pp. 19-33, ISSN 1360-9947

Brinsden, P. R.; Alam, V.; De Moustier, B. & Engrand, P. (2009). Recombinant human leukemia inhibitory factor does not improve implantation and pregnancy outcomes after assisted reproductive techniques in women with recurrent unexplained implantation failure. Fertil Steril Vol.91, No.4 Suppl, pp. 1445-1447, ISSN 1556-5653

Carson, D. D.; Desouza, M. M. & Regisford, E. G. (1998). Mucin and proteoglycan functions in embryo implantation. Bioessays Vol.20, No.7, pp. 577-583, ISSN 0265-9247

Carson, D. D.; Lagow, E.; Thathiah, A.; Al-Shami, R.; Farach-Carson, M. C.; Vernon, M.; Yuan, L.; Fritz, M. A. & Lessey, B. (2002). Changes in gene expression during the early to mid-luteal (receptive phase) transition in human endometrium detected by high-density microarray screening. Mol Hum Reprod Vol.8, No.9, pp. 871-879, ISSN 1360-9947

Conneely, O. M.; Mulac-Jericevic, B.; Demayo, F.; Lydon, J. P. & O'malley, B. W. (2002). Reproductive functions of progesterone receptors. Recent Prog Horm Res Vol.57, pp. 339-355, ISSN 0079-9963

Cooke, P. S.; Buchanan, D. L.; Young, P.; Setiawan, T.; Brody, J.; Korach, K. S.; Taylor, J.; Lubahn, D. B. & Cunha, G. R. (1997). Stromal estrogen receptors mediate mitogenic effects of estradiol on uterine epithelium. Proc Natl Acad Sci U S A Vol.94, No.12, pp. 6535-6540, ISSN 0027-8424

Coutifaris, C.; Myers, E. R.; Guzick, D. S.; Diamond, M. P.; Carson, S. A.; Legro, R. S.; Mcgovern, P. G.; Schlaff, W. D.; Carr, B. R.; Steinkampf, M. P.; Silva, S.; Vogel, D. L. & Leppert, P. C. (2004). Histological dating of timed endometrial biopsy tissue is not related to fertility status. Fertil Steril Vol.82, No.5, pp. 1264-1272, ISSN 0015-0282

Cowell, T. P. (1969). Implantation and development of mouse eggs transferred to the uteri of non-progestational mice. J Reprod Fertil Vol.19, No.2, pp. 239-245, ISSN 0022-4251

Cullinan, E. B.; Abbondanzo, S. J.; Anderson, P. S.; Pollard, J. W.; Lessey, B. A. & Stewart, C. L. (1996). Leukemia inhibitory factor (LIF) and LIF receptor expression in human endometrium suggests a potential autocrine/paracrine function in regulating embryo implantation. Proc Natl Acad Sci U S A Vol.93, No.7, pp. 3115-3120, ISSN 0027-8424

Christian, M.; Marangos, P.; Mak, I.; Mcvey, J.; Barker, F.; White, J. & Brosens, J. J. (2001). Interferon-gamma modulates prolactin and tissue factor expression in differentiating human endometrial stromal cells. Endocrinology Vol.142, No.7, pp. 3142-3151, ISSN 0013-7227

Chryssikopoulos, A.; Mantzavinos, T.; Kanakas, N.; Karagouni, E.; Dotsika, E. & Zourlas, P. A. (1996). Correlation of serum and follicular fluid concentrations of placental protein 14 and CA-125 in in vitro fertilization-embryo transfer patients. Fertil Steril Vol.66, No.4, pp. 599-603, ISSN 0015-0282

Damario, M. A.; Lesnick, T. G.; Lessey, B. A.; Kowalik, A.; Mandelin, E.; Seppala, M. & Rosenwaks, Z. (2001). Endometrial markers of uterine receptivity utilizing the donor oocyte model. Hum Reprod Vol.16, No.9, pp. 1893-1899, ISSN 0268-1161

Das, S. K.; Yano, S.; Wang, J.; Edwards, D. R.; Nagase, H. & Dey, S. K. (1997). Expression of matrix metalloproteinases and tissue inhibitors of metalloproteinases in the mouse uterus during the peri-implantation period. Dev Genet Vol.21, No.1, pp. 44-54, ISSN 0192-253X

De Ziegler, D.; Fanchin, R.; Massonneau, M.; Bergeron, C.; Frydman, R. & Bouchard, P. (1994). Hormonal control of endometrial receptivity. The egg donation model and controlled ovarian hyperstimulation. Ann N Y Acad Sci Vol.734, pp. 209-220, ISSN 0077-8923

Denker, H. W. (1983). Basic aspects of ovoimplantation. Obstet Gynecol Annu Vol.12, pp. 15-42, ISSN 0091-3332

Denker, H. W. (1994). Endometrial receptivity: cell biological aspects of an unusual epithelium. A review. Anat Anz Vol.176, No.1, pp. 53-60, ISSN 0940-9602

Diaz-Gimeno, P.; Horcajadas, J. A.; Martinez-Conejero, J. A.; Esteban, F. J.; Alama, P.; Pellicer, A. & Simon, C. (2011). A genomic diagnostic tool for human endometrial receptivity based on the transcriptomic signature. Fertil Steril Vol.95, No.1, pp. 50-60, 60 e51-15, ISSN 1556-5653

Diedrich, K.; Fauser, B. C.; Devroey, P. & Griesinger, G. (2007). The role of the endometrium and embryo in human implantation. Hum Reprod Update Vol.13, No.4, pp. 365-377, ISSN 1355-4786

Fukuda, M. N. & Nozawa, S. (1999). Trophinin, tastin, and bystin: a complex mediating unique attachment between trophoblastic and endometrial epithelial cells at their respective apical cell membranes. Semin Reprod Endocrinol Vol.17, No.3, pp. 229-234, ISSN 0734-8630

Gellersen, B.; Fernandes, M. S. & Brosens, J. J. (2009). Non-genomic progesterone actions in female reproduction. Hum Reprod Update Vol.15, No.1, pp. 119-138, ISSN 1460-2369

Genbacev, O. D.; Prakobphol, A.; Foulk, R. A.; Krtolica, A. R.; Ilic, D.; Singer, M. S.; Yang, Z. Q.; Kiessling, L. L.; Rosen, S. D. & Fisher, S. J. (2003). Trophoblast L-selectin-mediated adhesion at the maternal-fetal interface. Science Vol.299, No.5605, pp. 405-408, ISSN 1095-9203

Giudice, L. C. (1999). Potential biochemical markers of uterine receptivity. Hum Reprod Vol.14 Suppl 2, pp. 3-16, ISSN 0268-1161

Haouzi, D.; Assou, S.; Mahmoud, K.; Tondeur, S.; Reme, T.; Hedon, B.; De Vos, J. & Hamamah, S. (2009). Gene expression profile of human endometrial receptivity: comparison between natural and stimulated cycles for the same patients. Hum Reprod Vol.24, No.6, pp. 1436-1445, ISSN 1460-2350

Haouzi, D.; Mahmoud, K.; Fourar, M.; Bendhaou, K.; Dechaud, H.; De Vos, J.; Reme, T.; Dewailly, D. & Hamamah, S. (2009). Identification of new biomarkers of human endometrial receptivity in the natural cycle. Hum Reprod Vol.24, No.1, pp. 198-205, ISSN 1460-2350

Hertig, A. T.; Rock, J. & Adams, E. C. (1956). A description of 34 human ova within the first 17 days of development. Am J Anat Vol.98, No.3, pp. 435-493, ISSN 0002-9106

Hoozemans, D. A.; Schats, R.; Lambalk, C. B.; Homburg, R. & Hompes, P. G. (2004). Human embryo implantation: current knowledge and clinical implications in assisted reproductive technology. Reprod Biomed Online Vol.9, No.6, pp. 692-715, ISSN 1472-6483

Horcajadas, J. A.; Pellicer, A. & Simon, C. (2007). Wide genomic analysis of human endometrial receptivity: new times, new opportunities. Hum Reprod Update Vol.13, No.1, pp. 77-86, ISSN 1355-4786

Huang, S. Y.; Wang, C. J.; Soong, Y. K.; Wang, H. S.; Wang, M. L.; Lin, C. Y. & Chang, C. L. (2011). Site-specific endometrial injury improves implantation and pregnancy in patients with repeated implantation failures. Reprod Biol Endocrinol Vol.9, No.1, pp. 140, ISSN 1477-7827

Illera, M. J.; Cullinan, E.; Gui, Y.; Yuan, L.; Beyler, S. A. & Lessey, B. A. (2000). Blockade of the alpha(v)beta(3) integrin adversely affects implantation in the mouse. Biol Reprod Vol.62, No.5, pp. 1285-1290, ISSN 0006-3363

Irving, J. C. & Giudice, L. C. (1999). Decidua. Encyclopedia of Reproduction. Knobil, E. & Niaal, J. New York, Academic Press. Vol.1, pp. 823-835, ISBN 978-0-12-515401-7

Kao, L. C.; Tulac, S.; Lobo, S.; Imani, B.; Yang, J. P.; Germeyer, A.; Osteen, K.; Taylor, R. N.; Lessey, B. A. & Giudice, L. C. (2002). Global gene profiling in human endometrium during the window of implantation. Endocrinology Vol.143, No.6, pp. 2119-2138, ISSN 0013-7227

Kauma, S. W.; Aukerman, S. L.; Eierman, D. & Turner, T. (1991). Colony-stimulating factor-1 and c-fms expression in human endometrial tissues and placenta during the menstrual cycle and early pregnancy. J Clin Endocrinol Metab Vol.73, No.4, pp. 746-751, ISSN 0021-972X

Kemmeren, P.; Van Berkum, N. L.; Vilo, J.; Bijma, T.; Donders, R.; Brazma, A. & Holstege, F. C. (2002). Protein interaction verification and functional annotation by integrated analysis of genome-scale data. Mol Cell Vol.9, No.5, pp. 1133-1143, ISSN 1097-2765

Kirn-Safran, C. B. & Carson, D. D. (1999). Dynamics of uterine glycoconjugate expression and function. Semin Reprod Endocrinol Vol.17, No.3, pp. 217-227, ISSN 0734-8630

Leach, R. E.; Khalifa, R.; Ramirez, N. D.; Das, S. K.; Wang, J.; Dey, S. K.; Romero, R. & Armant, D. R. (1999). Multiple roles for heparin-binding epidermal growth factor-like growth factor are suggested by its cell-specific expression during the human endometrial cycle and early placentation. J Clin Endocrinol Metab Vol.84, No.9, pp. 3355-3363, ISSN 0021-972X

Ledee-Bataille, N.; Olivennes, F.; Kadoch, J.; Dubanchet, S.; Frydman, N.; Chaouat, G. & Frydman, R. (2004). Detectable levels of interleukin-18 in uterine luminal secretions at oocyte retrieval predict failure of the embryo transfer. Hum Reprod Vol.19, No.9, pp. 1968-1973, ISSN 0268-1161

Lessey, B. A. (1998). Endometrial integrins and the establishment of uterine receptivity. Hum Reprod Vol.13 Suppl 3, pp. 247-258; discussion 259-261, ISSN.

Lessey, B. A. (2003). Two pathways of progesterone action in the human endometrium: implications for implantation and contraception. Steroids Vol.68, No.10-13, pp. 809-815, ISSN 0039-128X

Lessey, B. A. & Castelbaum, A. J. (2002). Integrins and implantation in the human. Rev Endocr Metab Disord Vol.3, No.2, pp. 107-117, ISSN 1389-9155

Lessey, B. A.; Castelbaum, A. J.; Buck, C. A.; Lei, Y.; Yowell, C. W. & Sun, J. (1994). Further characterization of endometrial integrins during the menstrual cycle and in pregnancy. Fertil Steril Vol.62, No.3, pp. 497-506, ISSN 0015-0282 (Print) 0015-0282 (Linking).

Lessey, B. A.; Castelbaum, A. J.; Sawin, S. W. & Sun, J. (1995). Integrins as markers of uterine receptivity in women with primary unexplained infertility. Fertil Steril Vol.63, No.3, pp. 535-542, ISSN 0015-0282

Lessey, B. A.; Damjanovich, L.; Coutifaris, C.; Castelbaum, A.; Albelda, S. M. & Buck, C. A. (1992). Integrin adhesion molecules in the human endometrium. Correlation with the normal and abnormal menstrual cycle. J Clin Invest Vol.90, No.1, pp. 188-195, ISSN 0021-9738

Lessey, B. A.; Killam, A. P.; Metzger, D. A.; Haney, A. F.; Greene, G. L. & Mccarty, K. S., Jr. (1988). Immunohistochemical analysis of human uterine estrogen and progesterone receptors throughout the menstrual cycle. J Clin Endocrinol Metab Vol.67, No.2, pp. 334-340, ISSN 0021-972X

Luxford, K. A. & Murphy, C. R. (1989). Cytoskeletal alterations in the microvilli of uterine epithelial cells during early pregnancy. Acta Histochem Vol.87, No.2, pp. 131-136, ISSN 0065-1281

Luxford, K. A. & Murphy, C. R. (1992). Reorganization of the apical cytoskeleton of uterine epithelial cells during early pregnancy in the rat: a study with myosin subfragment 1. Biol Cell Vol.74, No.2, pp. 195-202, ISSN 0248-4900

Martin, K. L.; Barlow, D. H. & Sargent, I. L. (1998). Heparin-binding epidermal growth factor significantly improves human blastocyst development and hatching in serum-free medium. Hum Reprod Vol.13, No.6, pp. 1645-1652, ISSN 0268-1161

Mclaren A, M. D. (1956). Studies in the transfer of fertilised mouse eggs to uterine foster mothers I: factors affecting the implantation and survival of native and transferred eggs. Journal of Experimental Biology Vol.33, pp. 394-416

Meseguer, M.; Aplin, J. D.; Caballero-Campo, P.; O'connor, J. E.; Martin, J. C.; Remohi, J.; Pellicer, A. & Simon, C. (2001). Human endometrial mucin MUC1 is up-regulated by progesterone and down-regulated in vitro by the human blastocyst. Biol Reprod Vol.64, No.2, pp. 590-601, ISSN 0006-3363

Mirkin, S.; Arslan, M.; Churikov, D.; Corica, A.; Diaz, J. I.; Williams, S.; Bocca, S. & Oehninger, S. (2005). In search of candidate genes critically expressed in the human endometrium during the window of implantation. Hum Reprod Vol.20, No.8, pp. 2104-2117, ISSN 0268-1161

Morris, J. E. & Potter, S. W. (1984). A comparison of developmental changes in surface charge in mouse blastocysts and uterine epithelium using DEAE beads and dextran sulfate in vitro. Dev Biol Vol.103, No.1, pp. 190-199, ISSN 0012-1606

Murphy, C. R. (1993). The plasma membrane of uterine epithelial cells: structure and histochemistry. Prog Histochem Cytochem Vol.27, No.3, pp. 1-66, ISSN 0079-6336

Murphy, C. R. (1995). The cytoskeleton of uterine epithelial cells: a new player in uterine receptivity and the plasma membrane transformation. Hum Reprod Update Vol.1, No.6, pp. 567-580, ISSN 1355-4786

Murphy, C. R. (2000). Understanding the apical surface markers of uterine receptivity: pinopods-or uterodomes? Hum Reprod Vol.15, No.12, pp. 2451-2454, ISSN 0268-1161

Murphy, C. R. & Rogers, A. W. (1981). Effects of ovarian hormones on cell membranes in the rat uterus. III. The surface carbohydrates at the apex of the luminal epithelium. Cell Biophys Vol.3, No.4, pp. 305-320, ISSN 0163-4992

Murray, M. J.; Meyer, W. R.; Zaino, R. J.; Lessey, B. A.; Novotny, D. B.; Ireland, K.; Zeng, D. & Fritz, M. A. (2004). A critical analysis of the accuracy, reproducibility, and clinical utility of histologic endometrial dating in fertile women. Fertil Steril Vol.81, No.5, pp. 1333-1343, ISSN 0015-0282 (Print) 0015-0282 (Linking).

Navot, D.; Laufer, N.; Kopolovic, J.; Rabinowitz, R.; Birkenfeld, A.; Lewin, A.; Granat, M.; Margalioth, E. J. & Schenker, J. G. (1986). Artificially induced endometrial cycles and establishment of pregnancies in the absence of ovaries. N Engl J Med Vol.314, No.13, pp. 806-811, ISSN 0028-4793

Navot, D.; Scott, R. T.; Droesch, K.; Veeck, L. L.; Liu, H. C. & Rosenwaks, Z. (1991). The window of embryo transfer and the efficiency of human conception in vitro. Fertil Steril Vol.55, No.1, pp. 114-118, ISSN 0015-0282

Nikas, G. (1999). Cell-surface morphological events relevant to human implantation. Hum Reprod Vol.14 Suppl 2, pp. 37-44, ISSN 0268-1161

Nikas, G.; Drakakis, P.; Loutradis, D.; Mara-Skoufari, C.; Koumantakis, E.; Michalas, S. & Psychoyos, A. (1995). Uterine pinopodes as markers of the 'nidation window' in cycling women receiving exogenous oestradiol and progesterone. Hum Reprod Vol.10, No.5, pp. 1208-1213, ISSN 0268-1161

Noyes, R. W.; Hertig, A. T. & Rock, J. (1950). Dating the endometrial biopsy. Feril Steril Vol.1, No.1, pp. 3-25

O'malley, B. W. & Tsai, M. J. (1992). Molecular pathways of steroid receptor action. Biol Reprod Vol.46, No.2, pp. 163-167, ISSN 0006-3363

Padykula, H. A. (1991). Regeneration in the primate uterus: the role of stem cells. Ann N Y Acad Sci Vol.622, pp. 47-56, ISSN 0077-8923

Pierro, E.; Minici, F.; Alesiani, O.; Miceli, F.; Proto, C.; Screpanti, I.; Mancuso, S. & Lanzone, A. (2001). Stromal-epithelial interactions modulate estrogen responsiveness in normal human endometrium. Biol Reprod Vol.64, No.3, pp. 831-838, ISSN 0006-3363

Ponnampalam, A. P.; Weston, G. C.; Trajstman, A. C.; Susil, B. & Rogers, P. A. (2004). Molecular classification of human endometrial cycle stages by transcriptional profiling. Mol Hum Reprod Vol.10, No.12, pp. 879-893, ISSN 1360-9947

Pope, W. F. (1988). Uterine asynchrony: a cause of embryonic loss. Biol Reprod Vol.39, No.5, pp. 999-1003, ISSN 0006-3363

Psychoyos, A. (1973). Endocrine control of egg implantation. Handbook of physiology. Greep, R.O., Astwood, E.G. & Geiger, S.R. Washington, DC, American Physiological Society, pp. 187-215

Psychoyos, A. (1986). Uterine receptivity for nidation. Ann N Y Acad Sci Vol.476, pp. 36-42, ISSN 0077-8923

Psychoyos, A. & Casimiri, V. (1980). Uterine state of non-receptivity: demonstration of an ovotoxic substance (blastocidine) in rats. C R Seances Acad Sci D Vol.291, No.12, pp. 973-976, ISSN 0567-655X

Punyadeera, C.; Dassen, H.; Klomp, J.; Dunselman, G.; Kamps, R.; Dijcks, F.; Ederveen, A.; De Goeij, A. & Groothuis, P. (2005). Oestrogen-modulated gene expression in the human endometrium. Cell Mol Life Sci Vol.62, No.2, pp. 239-250, ISSN 1420-682X

Quinn, C. E. & Casper, R. F. (2009). Pinopodes: a questionable role in endometrial receptivity. Hum Reprod Update Vol.15, No.2, pp. 229-236, ISSN 1460-2369

Rhodes, D. R.; Barrette, T. R.; Rubin, M. A.; Ghosh, D. & Chinnaiyan, A. M. (2002). Meta-analysis of microarrays: interstudy validation of gene expression profiles reveals pathway dysregulation in prostate cancer. Cancer Res Vol.62, No.15, pp. 4427-4433, ISSN 0008-5472

Riesewijk, A.; Martin, J.; Van Os, R.; Horcajadas, J. A.; Polman, J.; Pellicer, A.; Mosselman, S. & Simon, C. (2003). Gene expression profiling of human endometrial receptivity on days LH+2 versus LH+7 by microarray technology. Mol Hum Reprod Vol.9, No.5, pp. 253-264, ISSN 1360-9947

Rogers, P. A. & Murphy, C. R. (1992). Morphometric and freeze fracture studies of human endometrium during the peri-implantation period. Reprod Fertil Dev Vol.4, No.3, pp. 265-269, ISSN 1031-3613

Salamonsen, L. A.; Dimitriadis, E. & Robb, L. (2000). Cytokines in implantation. Semin Reprod Med Vol.18, No.3, pp. 299-310, ISSN 1526-8004

Schena, M.; Shalon, D.; Davis, R. W. & Brown, P. O. (1995). Quantitative monitoring of gene expression patterns with a complementary DNA microarray. Science Vol.270, No.5235, pp. 467-470, ISSN 0036-8075

Schlafke, S. & Enders, A. C. (1975). Cellular basis of interaction between trophoblast and uterus at implantation. Biol Reprod Vol.12, No.1, pp. 41-65, ISSN 0006-3363

Sengupta, J.; Lalitkumar, P. G.; Najwa, A. R. & Ghosh, D. (2006). Monoclonal anti-leukemia inhibitory factor antibody inhibits blastocyst implantation in the rhesus monkey. Contraception Vol.74, No.5, pp. 419-425, ISSN 0010-7824

Spencer, T. E. & Bazer, F. W. (2002). Biology of progesterone action during pregnancy recognition and maintenance of pregnancy. Front Biosci Vol.7, pp. d1879-1898, ISSN 1093-4715

Stewart, C. L.; Kaspar, P.; Brunet, L. J.; Bhatt, H.; Gadi, I.; Kontgen, F. & Abbondanzo, S. J. (1992). Blastocyst implantation depends on maternal expression of leukaemia inhibitory factor. Nature Vol.359, No.6390, pp. 76-79, ISSN 0028-0836

Strowitzki, T.; Germeyer, A.; Popovici, R. & Von Wolff, M. (2006). The human endometrium as a fertility-determining factor. Hum Reprod Update Vol.12, No.5, pp. 617-630, ISSN 1355-4786

Surveyor, G. A.; Gendler, S. J.; Pemberton, L.; Das, S. K.; Chakraborty, I.; Julian, J.; Pimental, R. A.; Wegner, C. C.; Dey, S. K. & Carson, D. D. (1995). Expression and steroid hormonal control of Muc-1 in the mouse uterus. Endocrinology Vol.136, No.8, pp. 3639-3647, ISSN 0013-7227

Suzuki, N.; Nakayama, J.; Shih, I. M.; Aoki, D.; Nozawa, S. & Fukuda, M. N. (1999). Expression of trophinin, tastin, and bystin by trophoblast and endometrial cells in human placenta. Biol Reprod Vol.60, No.3, pp. 621-627, ISSN 0006-3363

Suzuki, N.; Zara, J.; Sato, T.; Ong, E.; Bakhiet, N.; Oshima, R. G.; Watson, K. L. & Fukuda, M. N. (1998). A cytoplasmic protein, bystin, interacts with trophinin, tastin, and cytokeratin and may be involved in trophinin-mediated cell adhesion between trophoblast and endometrial epithelial cells. Proc Natl Acad Sci U S A Vol.95, No.9, pp. 5027-5032, ISSN 0027-8424

Tabibzadeh, S. (1992). Patterns of expression of integrin molecules in human endometrium throughout the menstrual cycle. Hum Reprod Vol.7, No.6, pp. 876-882, ISSN 0268-1161

Tabibzadeh, S. (1998). Molecular control of the implantation window. Hum Reprod Update Vol.4, No.5, pp. 465-471, ISSN 1355-4786

Talbi, S.; Hamilton, A. E.; Vo, K. C.; Tulac, S.; Overgaard, M. T.; Dosiou, C.; Le Shay, N.; Nezhat, C. N.; Kempson, R.; Lessey, B. A.; Nayak, N. R. & Giudice, L. C. (2006). Molecular phenotyping of human endometrium distinguishes menstrual cycle phases and underlying biological processes in normo-ovulatory women. Endocrinology Vol.147, No.3, pp. 1097-1121, ISSN 0013-7227

Tapia, A.; Gangi, L. M.; Zegers-Hochschild, F.; Balmaceda, J.; Pommer, R.; Trejo, L.; Pacheco, I. M.; Salvatierra, A. M.; Henriquez, S.; Quezada, M.; Vargas, M.; Rios, M.; Munroe, D. J.; Croxatto, H. B. & Velasquez, L. (2008). Differences in the endometrial transcript profile during the receptive period between women who were refractory to implantation and those who achieved pregnancy. Hum Reprod Vol.23, No.2, pp. 340-351, ISSN 1460-2350

Tapia, A.; Vilos, C.; Marin, J. C.; Croxatto, H. B. & Devoto, L. (2011). Bioinformatic detection of E47, E2F1 and SREBP1 transcription factors as potential regulators of genes associated to acquisition of endometrial receptivity. Reprod Biol Endocrinol Vol.9, pp. 14, ISSN 1477-7827

Thomas, K.; Thomson, A.; Wood, S.; Kingsland, C.; Vince, G. & Lewis-Jones, I. (2003). Endometrial integrin expression in women undergoing in vitro fertilization and the association with subsequent treatment outcome. Fertil Steril Vol.80, No.3, pp. 502-507, ISSN 0015-0282

Tseng, L. & Mazella, J. (1999). Prolactin and its receptor in human endometrium. Semin Reprod Endocrinol Vol.17, No.1, pp. 23-27, ISSN 0734-8630

Tseng, L. H.; Chen, I.; Chen, M. Y.; Yan, H.; Wang, C. N. & Lee, C. L. (2010). Genome-based expression profiling as a single standardized microarray platform for the diagnosis of endometrial disorder: an array of 126-gene model. Fertil Steril Vol.94, No.1, pp. 114-119, ISSN 1556-5653

Usadi, R. S.; Murray, M. J.; Bagnell, R. C.; Fritz, M. A.; Kowalik, A. I.; Meyer, W. R. & Lessey, B. A. (2003). Temporal and morphologic characteristics of pinopod expression across the secretory phase of the endometrial cycle in normally cycling women with proven fertility. Fertil Steril Vol.79, No.4, pp. 970-974, ISSN 0015-0282

Vogiagis, D.; Marsh, M. M.; Fry, R. C. & Salamonsen, L. A. (1996). Leukaemia inhibitory factor in human endometrium throughout the menstrual cycle. J Endocrinol Vol.148, No.1, pp. 95-102, ISSN 0022-0795

Yoo, H. J.; Barlow, D. H. & Mardon, H. J. (1997). Temporal and spatial regulation of expression of heparin-binding epidermal growth factor-like growth factor in the human endometrium: a possible role in blastocyst implantation. Dev Genet Vol.21, No.1, pp. 102-108, ISSN 0192-253X

Younis, J. S.; Simon, A. & Laufer, N. (1996). Endometrial preparation: lessons from oocyte donation. Fertil Steril Vol.66, No.6, pp. 873-884, ISSN 0015-0282

5

The Actors of Human Implantation: Gametes, Embryo, Endometrium

Virginie Gridelet et al.*
University of Liège (ULg)
Belgium

1. Introduction

The success of pregnancy depends on a receptive endometrium, a normal blastocyst, a synchronized cross-talk at the maternal–fetal interface at the time of implantation, and finally a successful placentation and remodeling of uterine vasculature. In routine, less than 5% of oocytes collected in *in vitro* fertilization (IVF) cycles and only 20 to 25% of embryos transferred lead to a birth. Implantation and placentation processes remain the black box of fertility, involving following steps: fertilization, endometrial receptivity, embryo implantation (apposition-adhesion-invasion), trophoblastic differentiation and invasion (Cartwright et al., 2010).

At the time of fertilization, a cascade of cytokines mediates the oocyte-sperm dialogue long time before embryo implantation in the endometrium. At the oocyte side, genomic and proteomic profiles of good follicle are under investigation. Markers of oocytes with high subsequent embryo implantation potential are one of the main goals of the actual research. For example, granulocyte colony stimulating factor (G-CSF or CSF-3) in individual follicular fluids (FF) appears to correlate with the birth potential of the corresponding embryo in two opposite models of ovarian monitoring, standard ovarian hyperstimulation and modified natural IVF/ICSI cycles (Ledee et al., 2008b; Ledee et al., 2011a). To achieve the fertilization step, a good quality oocyte must meet a normal sperm with low DNA damages, leading to the development of a functionally normal blastocyst able to dialogue with maternal endometrium. Sperm DNA damages include fragmentation, trouble of condensation and epigenetic modification that could impair implantation process and methylation of imprinted genes (Boitrelle et al., 2011; Tavalaee et al., 2009).

Endometrial receptivity is established during a limited period of time called the *implantation window*. During this 4-days window (d20-d24), endometrium is highly receptive to the different signals and ligands produced by embryo throughout apposition, adhesion and invasion steps of implantation. The complex signaling networks that regulate this tightly coordinated maternal-fetal crosstalk are clearer as studies on endometrial receptivity and early pregnancy are performed. Many target molecules have been identified in the receptive

* Olivier Gaspard, Barbara Polese, Philippe Ruggeri, Stephanie Ravet, Carine Munaut, Vincent Geenen, Jean-Michel Foidart, Nathalie Lédée and Sophie Perrier d'Hauterive
University of Liège (ULg)
Belgium

endometrium such as specific cytokines equilibrium, growth factors and angiogenic factors but also some specific immunological target cells. Despite research and technique progresses, despite a lot of publications defining profiles – normal or pathological- at the genomic and proteomic level, the molecular fingerprint of the receptive endometrium remains unknown (Berlanga et al., 2011; Rashid et al., 2011; Singh et al., 2011).

The progression of implantation and then pregnancy requires immunological tolerance which allows conceptus survival. It has been proposed that uterine natural killer cells (uNK) could exert, directly or indirectly, either positive or negative control over these early steps (Dosiou and Giudice, 2005). These cells secrete an array of cytokines important for adequate local immune regulation, angiogenesis, placental development, and establishment of pregnancy. Successful subsequent placentation and remodeling of the uterine vasculature is a fundamental step for a healthy pregnancy that requires also a highly orchestrated reciprocal signaling process. Deficiencies in this process are implicated in a number of dangerous pregnancy complications with excess (percreta/accreta placentation) or defective implantation (preeclampsia, intra uterine growth restriction). Implantation failure, recurrent miscarriage and preeclampsia have several recognized causes in common, but in most cases, the precise etiology remains obscure.

The most limiting and difficult issue to evaluate during the implantation process is the dialogue at the materno-fetal interface. Embryo is able to cross-talk with the endometrium through different molecules, cytokines and hormones. It is able to actively participate to its own implantation and to influence endometrial gene expression (Kashiwagi et al., 2007). Inversely, endometrium is competent to differently answer to the implanting embryo, to favor or reject implantation (Baucroachs et al., 2009). Moreover, the cytokine network acting in the female reproductive tract around implantation integrates environmental information to program the embryo and fine-tune the maternal immune response and endometrial remodeling to determine implantation success (Robertson et al., 2011). All these interactions are not accessible to the researcher for obvious ethical reasons that let understand why implantation remains the black box of reproduction, even in 2012.

Among the factors produced by the embryo, its specific signal chorionic gonadotropin hormone (hCG) and its hyperglycosylated form H-hCG are another example of target molecules at this crossroads of immune tolerance, angiogenesis, and invasive process at the maternal-fetal interface.

This chapter will overview the recent literature and personal data concerning impact of gametes, endometrium and embryo during implantation process.

2. The oocyte

Maternal factors play a predominant role during early embryo development. Oocyte supports indeed by itself the early cleavage of the zygote until the 4-8 cells stage. Unfortunately, oocyte morphology is poorly discriminative and allow mainly a negative selection (Balaban and Urman, 2006; Rienzi et al.) and oocyte quality remains one of the main limiting factors of success of Assisted Reproductive Technology (ART) in human. This is due not only to the prime impact during early embryo growth but also to the fragility of oocytes along the life time. Less than 5% of oocytes collected after ovarian hyperstimulation in IVF program lead to birth. The selection of oocytes able to give rise to implanting embryo is crucial to avoid such wastage. Actual research focuses on the oocyte physiology and aim to evidence functional markers of good quality oocyte and competence to complete morphological observation of embryos which are insufficient to highly predict subsequent successful implantation.

Throughout the process of folliculogenesis and ovulation, the oocyte maturation from resting primary oocyte to secondary oocyte includes a complex sequence of nuclear and cytoplasmic events that prepare oocyte to fertilization and initiation of embryo development. During all their maturation, oocytes grow and develop in a highly coordinated and mutually dependent manner with the cumulus cells surrounding them. The formation of meiotic spindles during the two phases of the meiosis is essential for accurate chromosome segregation and for the highly asymmetric cell divisions necessary for the formation of small polar bodies and a large polarized oocyte (Schatten and Sun, 2011). The first polar body is extruded from the oocyte just before fertilization while the second one is expelled at the end of the second meiotic division which occurs after the fertilization. Chromosomal segregation errors occurs in approximately 15-20% of oocytes (Pellestor et al., 2005) and 5% of all pregnancies are aneuploidy (Hassold and Hunt, 2001).

During mostly IVF procedures, high doses of gonadotropins are administered to hyperstimulate the development of multiple oocytes to obtain a maximum of matures eggs in a single cycle. The aim of such an ovarian stimulation is the production of more than one embryo allowing the embryologist to select the best to transfer (while the others are cryopreserved for a possible later transfer). The assumption that ovarian stimulation could impair oogenesis, embryo quality and endometrial receptivity becomes more and more evident. The underlying mechanisms of these detrimental effects are still poorly understood, and further knowledge is needed in order to increase the safety of ovarian stimulation and to reduce potential effects on embryo development and implantation (Santos et al., 2010). A better understanding of oocyte oocyte-cumulus physiology/interactions, as well as, the improvement of oocyte selection based on its competence of giving rise to a healthy embryo, will ultimately increase the safety of ovarian stimulation and then pregnancy rates in IVF.

2.1 Oocyte quality and physiology

A "high quality" oocyte is an oocyte that is able of maturing, being fertilized, realizing a good implantation and giving rise to a healthy baby.

The poor quality of an oocyte seems to be characterized by a range of morphological defects or abnormalities, although there is still no precise quantification of the relative importance of each different anomaly. There is still not a comprehensive morphological oocyte grading scheme able to enough optimize the selection of normal oocytes (Lasiene et al., 2009; Patrizio et al., 2007; Rienzi et al., 2011) for fertilization in IVF or for oocyte cryopreservation. Criteria usually used to determine the grade of quality of oocytes morphology include the evaluation of the structure of oocyte: cumulus complex, oocyte cytoplasm, polar body, perivitelline space, zona pellucida and meiotic spindle (Lasiene et al., 2009). An oocyte with a large potential to give a competent embryo is describe with different characteristics:

- The oocyte is surrounded by a compact cumulus (at least five layers of cells)(Mayes and Sirard, 2001; Nagano et al., 2006; Warriach and Chohan, 2004).

- Ooplasm is almost transparent and homogeneous or a dark ring is seen around the cytoplasm. It is not granular (Balaban et al., 2008).
- There is none extracytoplasmic abnormalities (like dark zona pellucida, granular, vacuoles or cytoplasmic fragments) (Balaban et al., 2008; Rienzi et al., 2008).
- The perivitelline space is not too large and without grains (Hassan-Ali et al., 1998; Xia, 1997).
- It is estimated that oocytes observed with a polarization light microscope that shown birefringent spindle had higher developmental potential after fertilization than oocytes without birefringent spindle (Fang et al., 2007; Moon et al., 2003; Rienzi et al., 2004; Shen et al., 2006; Wang et al., 2001).

The morphology of the first polar body expulsed can also be studied to determine oocyte quality. Some morphological criteria such as the shape, the size, the surface and the integrity of cytoplasm of the polar body can be used to determine oocyte quality (Fancsovits et al., 2006; Navarro et al., 2009; Rienzi et al., 2008).

The contribution of each of these characteristics for the selection of a good quality embryo are largely discuss in the literature but they are still unclear (Lasiene et al., 2009; Rienzi et al., 2011; Setti et al., 2011). The selection of good quality oocyte is primordial in order to help during the management of ART but also to cryopreserve them to realize a bank of oocytes able to support thawing. Recently pregnancies were obtained with oocyte that has been cryopreserved or vitrified (Cobo and Diaz, 2011). Cryopreservation of oocytes is a desired tool for the possibility of extending the reproductive capability of young women with malignant diseases in cases where the treatment may compromise the ovarian reserve (Herrero et al., 2011; Saragusty and Arav, 2011).

Chromosomally abnormal oocyte is morphologically impossible to distinguish with traditional observation by the biologist, while it is an important cause of development or implantation failure of human embryo produced *in vitro*. Beside the necessity to optimize oocytes observation in order to fertilize them, there is a great need to better characterize a good quality oocytes with the purpose of cryopreserve them during fertility saving and egg donation programmes.

The transcriptomic profiling of human oocytes has been studied in normal oocytes and in aneuploidy oocyte to find a non-invasive biomarker of aneuploidy (Fragouli et al., 2010). Furthermore, the dialogue between the oocyte and its cumulus could be a key factor for the development of a competent embryo (Royere et al., 2009). Individual oocytes and their associated cumulus cells has been analysed by microarray technologies to provide potential viability markers related to oocyte competence. Increasing number of papers have shown correlation between cumulus gene expression and oocyte maturation, fertilization rate and pregnancy outcome. (Assou et al., 2010; Feuerstein et al., 2007; Hamel et al., 2008; van Montfoort et al., 2008; Yerushalmi et al., 2011). The study of Assou et al.(Assou et al., 2010), have demonstrated a good correlation between the expression profile of 45 genes from cumulus cells analysed by microarray and pregnancy outcome without relationship to the morphological grade of the embryo. The transcription of genes of cumulus cells has also been studied by Feuerstein et al., the genes were chosen because their expression was induced by the LH peak. Expression levels of all genes investigated, except one, were increased after resumption of meiosis (Feuerstein et al., 2007).

Furthermore, the transcriptome of the first polar body (extruded from the oocyte before fertilization) was analysis with the assumption that the polar body transcriptome is representative of that of the oocyte. The results show that the transcriptome of the human polar body reflect the one of the oocyte. This study could lead at the first molecular diagnostic for gene expression in oocyte ready for the fertilization, using mRNA detection and quantification (Reich et al., 2011).

Currently, the selection of the embryo to transfer is based on a morphological base such as kinetic of growth, number/size/form of blastomeres, early cleavage, fragmentation rate, aspect of ooplasm, aspect of pronuclei, and absence of multinuclear blastomeres. These usual morphological criteria are judged not sufficient for selecting the ideal embryo able to develop until the blastocyst stage and to transfer and to cryopreserve (Guerif et al., 2007; Guerif et al., 2009). Identification of non-invasive test of oocyte competence would undoubtedly improve the efficiency of assisted reproductive technology in selecting competent embryos.

The fluid that is in the follicular antrum, accumulated from the early stage of follicle development is rich of components and ease of access for studies that may contribute to a better understanding of the mechanisms underlying follicular development, oocyte quality or even ovarian hyperstimulation (Fahiminiya et al., 2011; Jarkovska et al., 2011). The aim of the study realize by Jarkovska *et al.* (Jarkovska et al., 2011) was to identify candidate proteins in follicular fluid (FF) which may help to identify patient at risk of ovarian hyperstimulation syndrome of women undergoing *in vitro* fertilization. Three proteins were found as potential markers among which kininogen-1 was highlighted by computer modeling as a potential key factor for mediated inflammation and angiogenesis. In the study realize Fahiminiya *et al.* (Fahiminiya et al., 2011), FF were collected from ovaries at three different stages of follicle development (early dominant, late dominant and preovulatory) and were analyzed by 2D-PAGE, 1D-Page and mass spectrometry to observe the proteomic expression in crude, depleted and enriched FF. They demonstrate that the enrichment method could be used to visualize and further identify the low-abundance proteins in FF, which could reflect the physiological status of the follicle.

A new marker measurable in FF has been proposed to select embryo with a high potential of implantation: the granular colony stimulating factor (G-CSF) (Ledee et al., 2008b).

2.2 Biomarker of the competence of the oocyte: G-CSF in follicular fluid

Research performed by Lédée team has extensively explored the follicular Granulocyte – Colony Stimulating Factor (G-CSF or CSF-3) properties as non-invasive immune biomarker of the oocyte competence able to predict subsequent birth (Ledee et al., 2008b). In a first study, FF were collected individually and the traceability of each fluid was ensured until birth or failure of the attempt was known. Twenty seven cytokines and chemokines were simultaneously measured in each FF collected from 132 individual follicles of oocyte subsequently fertilized and transferred after conventional ovarian hyperstimulation. The conclusion of this study was that the level of G-CSF in individual FF samples correlates with the implantation potential of the corresponding embryo. These data were reproduced in a cohort of 200 embryos while detailing the adequacy of distinct methods for measuring follicular fluid G-CSF (Ledee et al., 2010).

A third study was subsequently conducted including 83 patients undergoing a modified natural IVF/ICSI cycle to measured 26 soluble factors in the FF of these patients. The aim of this study was to provide an experimental model where the traceability was complete: only one oocyte was recovered and therefore only one embryo was transferred. Each of the 26 factors was evaluated as a potential biomarker of subsequent birth and G-CSF was found to be the best predictor of birth in this study. The combination of FF G-CSF and morphological embryo scoring on day 2 has been suggested as a possible prognostic value before starting the embryo transfer (Ledee et al., 2011a). Through these 3 studies, selection of follicular fluids over thousand fluids was performed to select the ones corresponding to an embryo successfully transferred with the traceability of each sample until birth. Each fluid was analysed through multiplex bead based technology. In all experiments, FF G-CSF appears as an excellent non-invasive biomarker of oocyte competence in regard to its significant strong power of discrimination, independently so adding value, to our daily embryo morphology based-selection.

We postulate that G-CSF factor in individual FF appears to correlate with the birth potential of the corresponding embryo in two opposite models of ovarian monitoring: standard ovarian hyperstimulation and modified natural IVF/ICSI cycles. Prospective randomized studies are needed to confirm the hypothesis and evaluation of adding value when correlated with the embryologist morphological choice is currently studied by our team.

The presence of G-CSF in the female genital tract and its possible role in reproduction have already been studied in the past. G-CSF has been shown to be secreted by granulosa cells at ovulation (Salmassi et al., 2004), then during the luteal phase within the endometrium and finally during gestation in the placenta (Duan, 1990). FF G-CSF may promote local maternofetal tolerance (Rutella et al., 2005) or influence the oocyte's own mRNA levels or its potential for self-repair (Yannaki et al., 2005). It might also interact with environmental cells to produce cytokines and growth factors which are necessary for the embryo's development and implantation.

It has been suggested key interactions within the follicle involving immune cells such as dendritic cells and regulatory T (Treg) cells)(Ledee et al., 2011b). Local maternal-foetal immunotolerance could potentially be promoted by G-CSF: almost all the miscarriages observed in our cohort were found in the group with a low follicular fluid G-CSF level and a recent study reported that G-CSF administration significantly increases live-birth rates in patients with unexplained recurrent miscarriages (Scarpellini and Sbracia, 2009).

The measure of the FF G-CSF is a non-invasive method for the assessment of the potentiality of the embryo to implant, and is cover by a patent. This method could be quickly implemented with a short time response compatible with the selection of the embryo before transfer. FF G-CSF may also help to promote the single embryo policy by helping to distinguish even before the fertilization of the oocyte with a good or a bad potential of implantation. It may also be helpful to evaluate individually the oocyte (for women with a low ovarian reserve), to identify the best protocol of ovarian stimulation to apply (specific indication of minimal stimulation) and to choose in a more powerful way the embryo to cryopreserve.

2.3 Oocyte influence on embryo development

First days of embryo development are characterized by a sequence of cells divisions of the fertilized oocyte into smaller and smaller cells. These divisions will transform the zygote into an implantation-competent blastocyst. The size of the whole embryo doesn't change until the blastocyst formation, capsuled into the zona pellucida.

The preimplantation embryo is able to realize a form of autonomous development firstly fueled by products provided by the oocyte and then from products coming from the activation of its own genome (Schultz, 2005). At the beginning of its life, right after the fertilization, the developmental program is initially directed by the maternally inherited protein and transcripts. Thereafter the embryonic cells will need to become totipotents. This totipotency is acquired by the reset of the epigenetic states of the DNA of the germ cells (the oocyte and the spermatozoa) and then by the activation of the embryo's genome and its own epigenetics arrangements. This activation will influence gene expression patterning to produce two specific cell types necessary for the survival of the embryo: the trophoblast and the internal cellular mass. It will exhibit correct spatiotemporal activity of the sequence of events necessary for the embryo development achievement (Corry et al., 2009). The next step will be the progressive replacement of maternal derived transcripts located in the cytoplasm with the transcripts specifics newly formed from the embryo (Schultz, 2002). A variety of epigenetic mechanisms underlies each step of the development to allow the specific identity of each cell type. These epigenetic mechanisms may change depending on environmental and temporal conditions (Corry et al., 2009).

Compared to spermatozoa, the DNA methylation status in oocyte prior to fertilization is little known. It is difficult to obtain mature, ready-to fertilize oocyte in large enough quantities (compared to spermatozoa).

About the embryo, the organization and the introduction of different epigenetic marks are done early during the development of the embryo and are essential for the development of the new organism. These epigenetic marks, DNA methylation, histone modifications and noncoding RNAs, have a critical role in the cell memory during the development. During gametogenesis, spermatozoa and oocyte are produced with distinctive chromatin resulting from a epigenetic reprogramming from the original cells (Hales et al., 2011).

2.4 Environmental influences on oocyte

- Oocyte aging

Females who show a progressive decline in fecundity can have oocyte aging. It is known that after a certain age the ovarian reserve diminish more quickly. Treatments have been proposed to increase the maintenance of the ovarian reserve. Dehydroepiandrosterone (DHEA) has been reported to improve pregnancy chances in case of diminished ovarian reserve. Currently best available evidence suggests that DHEA improves ovarian function, increases pregnancy chances and, by reducing aneuploidy, lowers miscarriage rates. DHEA over time also appears to improve ovarian reserve (Gleicher and Barad, 2011; Gleicher et al., 2010a, b; Sonmezer et al., 2009). Although levels of DHEA produced locally in FF have been measured and levels correlate negatively with *in vitro* fertilization outcomes (Li et al., 2011). But age has also an influence on the oocyte itself. When oocyte is getting older, cellular and molecular abnormalities have a greater probability to occur. A lot of functional changes associated with oocyte aging have been observed, including furthermore: decreased fertilization rate, polyspermy, parthenogenesis, chromosomal anomalies, apoptosis and abnormal and/or retarded development of embryo (Miao et al., 2009). A study realized by

Wilcox et al., showed that oocyte aging increase significantly the risk of early pregnancy loss (Wilcox et al., 1998). It has moreover been showed that embryos resulting from aged oocyte fertilized with ICSI showed low implantation rates and low developmental potential after transfer (Esfandiari et al., 2005; Javed et al., 2010; Liu et al., 1995; Nagy et al., 1993; Van Steirteghem et al., 1993). In a healthy body, reactive oxygen species (ROS) and antioxidants remain in balance. When there is an unbalanced towards an overabundance of ROS, oxidative stress occurs and can do damages. It has been suggested that oxidative stress modulates the age-related decline in fertility (Agarwal et al., 2008; Agarwal et al., 2005) . Antioxidants are actually used to enhance female and male fertility, although benefits on fertility are still controverted (Ruder et al., 2008; Ruder et al., 2009; Visioli and Hagen, 2011).

- Obesity

Obesity is known as a cause of increased risk of infertility, mostly because of ovulatory dysfunction, but also because obese women have increased risks for miscarriage and stillbirth. In case of ART, implantation and pregnancy rates are lower in obese subjects versus subjects with normal weight (Maheshwari et al., 2007). A retrospective study using oocyte donation model analyse 97-first cycles recipients of oocyte donation under conditions of controlled hormonal stimulation and embryos quality was evaluated. The conclusion of this paper said that they found no correlation between the BMI of the women receiving the embryo (given by another woman with a high Body Mass Index) in their uterus and the implantation rates for the same grade of embryo morphology quality. These results could mean that uterine receptivity is not negatively influenced by obesity and that the oocyte or the embryo is the cause of decreased pregnancy rate in obese women (Wattanakumtornkul et al., 2003). Another retrospective study on 536 first-cycle recipients with donor oocytes had confirmed these results (Styne-Gross et al., 2005). Metabolic syndrome and polycystic ovary syndrome are two diseases related to obesity; they can also have a negative impact on the women fertility. In mice, it has been recently shown that the negatives effects of obesity and insulin resistance persist beyond the pre- and the peri-conception period, affecting embryonic development and reproductive outcomes (Cardozo et al., 2011). The mechanisms leading to an altered oocyte and embryo quality in obese women could result from an altered maternal metabolic environment in FF. A study has explored different molecules present in FF from obese women compared to women with a moderate BMI and have found that obese women have an increased level of C-reactive protein in FF. This molecule might indicate increased oxidative stress in the oocyte's microenvironment, which impairs its development (Robker et al., 2009).

- Smoking

In a study realize by Gruber et al., in 2008, smokers presented a higher number of non-fertilized oocytes than nonsmokers (20.1% vs. 10.8% of fertilization failure)(Gruber et al., 2008). Another study realize in 2006 in a cohort of twenty-seven patients undergoing IVF classified as smokers and 32 as non-smokers showed that smokers had decreased number of retrieved oocytes compared with non-smokers ($p < 0.05$). This study demonstrates that active cigarette smoking increases the zona pellucida thickness of oocytes and decreases the quality of oocytes (Depa-Martynow et al., 2006). Tobacco compounds exert a deleterious effect on the process of ovarian follicle maturation. This effect is expressed by decreased pregnancy rate, increased early spontaneous abortions and altered ovarian reserve. (Sepaniak et al., 2006; Soares and Melo, 2008).

3. The sperm

The role of the sperm has long been reduced to its progression through the female genital tract, and the penetration of the oocyte in order to provide the paternal genetic material. However, the quality of this genetic inherited material is of great importance for embryo development and early pregnancy. In fact, it has little influence on fertilization and early embryo cleavage until the passage of 4 to 8 cells-stage, because the embryo uses almost exclusively the maternal RNA and proteins produced and stored during oocyte growth and maturation. By the third day of development, the embryonic genome begins to be expressed and embryonic RNAs and proteins gradually replace maternal ones. In other words, even if ART allows *in vitro* production of embryos, these embryos can be unable to implant and give rise to an ongoing pregnancy and/or a healthy offspring. Their development will depend on the activation and quality of its genome, half of which being of paternal origin.

3.1 The role of spermatozoa and the classical evaluation of semen

The different roles of the sperm during reproductive process are classified according to the stage at which embryo development fails (Tesarik et al., 2004):

- Early paternal effects includes inability to go through the female genital tract (problems of concentration, mobility, classic morphology), to spontaneously or in an assisted way (IVF, ICSI) fertilize the oocyte (inability to fix the zona pellucida glycoproteins ZP 1, 2 or 3, or to perform the acrosome reaction, ...), to activate the oocyte (deficiency in phospholipase zeta (Saunders et al., 2002) ...)
- Late paternal effects will become visible during (or after) the activation of the embryonic genome, particularly the paternal genome, resulting in the development of apparently normal early embryos (until the third day of development), but which are unable to implant or to continue a long-term development after implantation.
- Very late paternal effects are suspected when faced with recurrent miscarriages or with pathologies of imprinted genes (Lucifero et al., 2004; Zini, 2011b) leading potentially to inherited diseases.

Since the beginning of ART, the evaluation of male fertility has often been restricted to a conventional semen analysis as recommended by World Health Organization (WHO) (World Health Organization., 2010). In the last edition of the WHO manual for the semen analysis and processing, limit thresholds have been modified according to the results published by Cooper (Cooper et al., 2010). Actually, 1953 men from 5 studies in 8 countries have been included in this review. The studies have analyzed semen from men who recently become parent with a known TTP (time-to-pregnancy: time between the beginning of unprotected intercourse and occurrence of a pregnancy) of maximum 12 months. Percentiles 95 for the different results were then defined as the new limit thresholds. But in daily ART practice, it appears that these tests are not sufficient to permit the diagnostic of possible late or very late paternal effects. Some men first judged as infertile succeed to obtain a pregnancy with their partner (Haugen et al., 2006), while other men judged above limit thresholds as defined by WHO failed to obtain a pregnancy (Bonde et al., 1998) or achieved it with miscarriage. Despite these limitations, conventional sperm analysis has permit to propose adapted ART treatments in a majority of patients, and the Intracytoplasmic Sperm Injection (ICSI) technique (Palermo et al., 1992) used since the early nineties has given the chance for a part of these subfertile or infertile men to become parents.

3.2 Sperm DNA particular organization

The first studies on the genetic material carried by sperm have been published at the end of the 70s with the establishment of the first human sperm karyotypes by Rudak (Rudak et al., 1978). In the 90s, the technique of Fluorescent in situ hybridization (FISH) permits to study aneuploidy, abnormalities such as translocations, and recombinations during meiosis (Shi et al., 2001). Over the past decade, a growing number of papers have been published on the organization of sperm chromatin and the integrity of the DNA double helix. It appears that different problems in the sperm genetic material - called DNA damages - can affect male fertility (Shen et al., 1999; Spano et al., 2000). These DNA damages include abnormalities in chromatin condensation and organization, in fragmentation of the double helix (single and/or double strand), in DNA bases modifications, in epigenetic damages, etc... (Aitken et al., 2009).

During spermiogenesis, the final phase of spermatogenesis after meiosis, chromatin is radically reorganized and undergoes an extreme condensation resulting in a shift from a nucleosome-based genome organization to the sperm-specific, highly compacted nucleoprotamine structure. The DNA double helix of somatic cells is indeed organized into chains of nucleosomes by a family of nuclear proteins, the histones. At spermiogenesis, histones are mostly replaced by transition proteins, which are themselves replaced by two types of specific nuclear proteins of the male gametes, protamines 1 and 2. These proteins are particularly rich in arginines (which help to neutralize negative charges of DNA phosphate groups), in cysteines (which allow the formation of disulfide bonds intra and inter -protamines), and in histidines (which, in combination with cysteines, permit the establishment of zinc bridges). Because of this particular composition, the sperm chromatin fibers are strongly compacted, and have almost a crystalline organization (Bjorndahl and Kvist, 2010). In human sperm, about 85% of histones are replaced by protamines, while the other 15% remain. So sperm chromatin is organized predominantly in toroids (by protamines), the rest being organized in nucleosomes (by histones) or attached to the nuclear matrix (MARs: Matrix attached regions) (Govin et al., 2011; Jonge and Barratt, 2006; Ward, 2010). Recent data support the idea that region-specific programming of the haploid male genome is of high importance for the post-fertilization events and for successful embryo development. The molecular basis of post-meiotic male genome reorganization and compaction constitutes one of the last black boxes in modern biology of reproduction. Although the successive transitions in DNA packaging have been well described, the molecular factors driving these near genome-wide reorganizations remain obscure (Rousseaux et al., 2011).

This particular organization has several roles. Firstly sperm DNA occupies about 10 times less space than somatic cell DNA, which reduces the volume of the head and facilitates the movement of sperm. Secondly, this extreme condensation makes the DNA inaccessible to molecules such as free radicals (ROS produced by sperm mitochondria and by leukocytes), providing a protection against oxidative attacks. Finally, the distribution of remaining nucleosomes is not random but concerns gene regions involved in the early embryonic development. Moreover it seems increasingly clear that this organization and the inheritance of paternal histones have a very important role in epigenetic control (Miller et al., 2010; Tavalaee et al., 2009; Ward, 2010). Firstly remaining histones preferentially bind to DNA regions involved in early embryo development. As they are less compacted than if organized by protamines, these regions can be easily decondensed and transcripted. Secondly, sperm histones undergo different modifications (methylation, acetylation,...) that impact expression of the linked DNA domains (Rousseaux et al., 2008), and that can be perturbed in in infertile patients. Similarly, the ratio of protamines 1/2, which should normally be close to 1, can have a significant negative impact on fertility when disturbed (Hammoud et al., 2011). So it is clear that bad condensation of sperm DNA have an impact on male fertility (Venkatesh et al., 2011), and a recent study demonstrates the great differences between a fertile control group and a patient group presenting repeated spontaneous abortions in terms of chromatin condensation and stability (Talebi et al., 2011).

3.3 Sperm DNA damages and oxidative stress

Damages to the DNA double helix are quite varied (Bennetts et al., 2008). The most common is the fragmentation of the DNA double helix, which involves the breaking of a bond between a phosphate group and the neighboring deoxyribose on one or both strands: this is called single- or double-strand fragmentation. Another damage consists of base modifications, becoming oxidized or alkylated. Covalent bonds between a strand and another molecule or the other strand may also appear. DNA adducts are formed from xenobiotics and their metabolites, or from the oxidation of membrane lipids of the sperm, such as ethenonucleosides. These adducts can be either stable or result in the loss of a base. All these changes interfere with DNA replication or transcription. Similarly, an aberrant base repair may cause the appearance of a mutation that can be transmitted to the offspring (Aitken and Roman, 2008).

Another important group of damages to sperm genetic material concerns the epigenetic control. One of the most studied epigenetic marks is DNA methylation. Knowing that the epigenetic environment of the sperm, especially histones and their modifications, has an important role in the establishment and maintenance of epigenetic marks in the embryo, it appears that aberrant epigenetic regulation during spermatogenesis has a major impact on male fertility and embryonic development (Rajender et al., 2011). For example, poor sperm quality has been shown to be associated with hypermethylation of sperm genome (Houshdaran et al., 2007). This is probably due to a failure in epigenetic marks erasure that normally occurs during spermatogenesis. Promoter methylation aberrations of the gene encoding the enzyme methylenetetrahydrofolate reductase (MTHFR, which plays a key role in maintaining the bioavailability of methyl groups) are associated with secretory azoospermia (Khazamipour et al., 2009). Moreover, abnormal methylation can affect gene imprinting, and have an impact on embryonic development. Similarly, methylation aberrations were detected in imprinted genes in men with idiopathic infertility (Poplinski et al., 2010). Finally, many syndromes (Prader-Willi, Angelman, Silver-Russell and Beckwith-Wiedemann syndromes) are known to be in a number of cases due to imprinting defects (Rajender et al., 2011). All these methylation abnormalities observed in patients are fortunately not directly transmitted to the embryo: the question is how far they can be hereditary? In the epigenetic reprogramming of sperm DNA, these marks are erased and then re-established differentially in men and women. Chromatin organization plays a key role in the establishment and maintenance of the methylation profile.

An important matter about the DNA damages concerns their origin. A theory has been proposed by Aitken (Aitken et al., 2009): DNA damages can be considered as a process taking place in two stages. First of all, spermatogenesis, especially spermiogenesis, leads to inadequate sperm production, with an improperly compacted DNA. In a second step, these badly compacted spermatozoa are vulnerable to outside attacks, mainly of oxidative type (ROS). There is indeed a significant correlation between DNA fragmentation and the level of 8-hydroxy-2'-deoxyguanosine [8OHdG], a DNA adduct derived from the oxidation of DNA (Aitken and De Iuliis, 2010). Apoptosis also appears to play an important role since the sperm cell is unable to perform a complete process of programmed cell death (Aitken and De Iuliis, 2010). The compacted DNA, mainly silent, contained in a cytosol-depleted cell, and physically separated from mitochondria limit the action of classical apoptosis effectors. The problem is that these defective sperm cells, normally silently eliminated in the female genital tract, remain capable of fertilization, especially in ARTs (Barratt et al., 2010; Kurosaka et al., 2003; Sakkas et al., 2004; Weng et al., 2002).

In a normal sperm, there is a balance between antioxidant processes and ROS production, the latter being normal below a certain level, since it plays a role in sperm capacitation, acrosome reaction, and fertilization (Griveau and Le Lannou, 1997). However, oxidative stress remains controlled, and the presence of many antioxidants in seminal fluid (taurin, vitamin C, vitamin E, glutathione, uric acid, thioredoxin, glutathione peroxidase) limits the damaging effects (Tremellen, 2008). However, the balance between ROS and antioxidants may be disturbed by various external factors or diseases, which reduce the production of antioxidants, increase production of ROS, and / or influence the senescence of spermatozoa. All this may lead, especially in cases of inadequate chromatin compaction, to DNA damages (Koppers et al., 2008).

3.4 Impact of sperm DNA damages on fertility and ART

DNA damages have been studied in different populations of patients or sperm donors. In patients with an elevated production of seminal ROS, DNA fragmentation is also increased (Mahfouz et al., 2009). Avendano *et al.* (Avendano et al., 2009) analyzed simultaneously sperm morphology and DNA fragmentation, and showed an increased fragmentation in the infertile patient group, despite the selection of morphologically normal spermatozoa for the DNA analysis. Correlations exist between fragmentation and classical sperm features, but not always those expected: in a population of 1633 patients, an inverse correlation between sperm count, fast progressive motility and DNA fragmentation has been found (Cohen-Bacrie et al., 2009). Altogether, these results allow us to conclude that the study of fragmentation gives further additional information about male fertility to those provided by conventional analysis of semen.

In a great part of all the published data concerning fragmentation, authors try to assess the predictive value of different tests in terms of chances for achieving pregnancy, spontaneous or by different ARTs. Zini showed a strong association between DNA damage and failure to obtain a pregnancy naturally or by intrauterine insemination (Zini, 2011a; Zini and Sigman, 2009). This association, although weaker, exists also with the clinical pregnancy rate in IVF, and in a lesser extent for ICSI. On the other hand, fragmentation increases the risk of miscarriage for IVF and ICSI. The impact of DNA damages is also more significant for the IVF and ICSI in terms of take-home baby rate than clinical pregnancy rate (Avendano et al., 2010; Benchaib et al., 2007; Boe-Hansen et al., 2006; Borini et al., 2006; Bungum et al., 2007; Check et al., 2005; Duran et al., 2002; Evenson et al., 1999; Evenson and Wixon, 2008;

Frydman et al., 2008; Gandini et al., 2004; Henkel et al., 2003; Host et al., 2000; Huang et al., 2005; Lin et al., 2008; Loft et al., 2003; Micinski et al., 2009; Muriel et al., 2006; Simon et al., 2010; Spano et al., 2000; Speyer et al., 2010; Tarozzi et al., 2009; Zini et al., 2005).

However, the sensitivity of fragmentation tests remains low. The explanation for this observation is that the predictive value of DNA fragmentation tests depends on many factors (Sakkas and Alvarez, 2010): the type of DNA damage (single or double strand), the method of analysis, the site of injury (introns or exons), the percentage of fragmented cells, the extent of damage per cell, or the presence of other damage such as DNA adducts. On the other hand, it is clear that the sperm DNA quality is not the only explanation for the failure to obtain a pregnancy: oocyte quality and in particular its ability to repair sperm DNA (Meseguer et al., 2011), the number of oocytes available in the case of IVF, and the quality of the endometrium have a direct impact on the chances of pregnancy.

Another important point is to avoid a negative impact of ARTs - and in particular of the semen processing - on oxidative stress and fragmentation. It is therefore important to separate safe sperm from leucocytes and dead cells. Gradient centrifugation or swim-up allows the separation of different cell types, while the centrifugation of semen without cell separation should be rejected (Jackson et al., 2010; Marchesi et al., 2010; Monqaut et al., 2011). Anyway, it seems important to limit the number of centrifugations, which stimulate the ROS production by spermatozoa (Aitken and Clarkson, 1988; Aitken et al., 2010). The elimination of apoptotic sperm by Magnetic Activated Cell Sorting (MACS) could also improve pregnancy rates (Dirican et al., 2008; Lee et al., 2010; Polak de Fried and Denaday, 2010; Rawe et al., 2010). Another issue is to avoid transition metals (which may increase the damages caused by ROS) and to add antioxidants to the sperm preparation medium (Aitken et al., 2010). Finally, the addition of antioxidants in the cryopreservation medium seems to decrease the oxidative stress caused by the cryopreservation itself (Thomson et al., 2009).

Concerning the technique to fertilize the oocyte when sperm DNA damage is present, it seems that ICSI is the least influenced by the DNA fragmentation, although the impact of the latter remains significant on the risk of miscarriage (Zini and Sigman, 2009). Several techniques derived to improve ICSI aim to add new sperm selection criteria in relation to the quality of their genetic material and / or their fertilizing ability.

The use of polarized light highlights the well-organized structures such as the oocyte meiotic spindle, but also the well-organized sperm chromatin which then appears bright, in contrast to the improperly condensed one. Used in ICSI, this technique seems to improve implantation and clinical pregnancy rates (Gianaroli et al., 2008; Gianaroli et al., 2010). Sperm selection based on its ability to recognize and bind hyaluronic acid (HA), a major component of cumulus-oocyte complexes extracellular matrix, has been also proposed. Using HA-coated Petri dishes, this technique, called PICSI or HA-ICSI, allows the selection of sperm with a less fragmented DNA and a better head morphology assessed by high magnification (X6500-X10000) living sperm observation (Motile Sperm Morphology Examination organelle or MSOME) (Parmegiani et al., 2010a). It also appears to improve embryo quality and implantation rate (Parmegiani et al., 2010b).

Finally, the technique of IMSI (Intracytoplasmic Morphologically Selected Sperm Injection), which has been developed 10 years ago, looks promising in particular for the selection of sperm without problems of chromatin organization. In 2001 Bartoov and his team have associated their MSOME technique with ICSI to give birth to IMSI. Sperm is then observed and selected on a morphology basis at a 10000 times magnification, as compared to a conventional ICSI magnification of 200 to 400 times. In IMSI, a Nomarski differential interference contrast is used and allows a more precise evaluation of the classical strict criteria of sperm morphology (shape of the head, midpiece...), and of the presence at the head of surface abnormalities called vacuoles. These vacuoles vary in size and number between sperm an also between patients. Several teams have shown a correlation between a higher rate of vacuolated sperm and higher DNA damages (Franco et al., 2008; Garolla et al., 2008; Oliveira et al., 2010), and IMSI was presented as a potential solution for the selection of non-fragmented sperm in order to increase the implantation rate and to reduce the miscarriage rate (Antinori et al., 2008; Bartoov et al., 2003; Bartoov et al., 2002; Berkovitz et al., 2006a; Berkovitz et al., 2006b; Berkovitz et al., 2005). These date are however controversial and the question about the origin - fragmentation or decondensation - of the vacuoles remain open. Recently, different teams aimed to answer this question. They studied a specific population of sperm with large vacuoles (more than 13% of the head surface)(Boitrelle et al., 2011; Franco Jr et al., 2011; Perdrix et al., 2011) and came to the conclusion that their presence is correlated with abnormal DNA condensation. In this particular population, the surface depressions observed in MSOME reflects the presence of a nuclear vacuole containing a badly condensed chromatin, as confirmed by aniline blue staining (Perdrix et al., 2011). On the other side, no positive correlation was demonstrated between these large vacuoles and DNA fragmentation (Boitrelle et al., 2011; Perdrix et al., 2011). So it seems now clear that the presence of large vacuoles reflects a disruption of sperm chromatin condensation. Other types of vacuoles also exist (particularly small vacuoles, sometimes numerous, with an area <4% of the surface of the head). Further studies are still needed to understand if these vacuoles also reflect DNA damages.

3.5 Impact of environment on sperm DNA damages

Numerous factors affect semen oxidative balance, sperm quality and DNA damages. Tobacco has an impact on oxidative stress, by increasing the presence of leukocytes in semen (Saleh et al., 2002; Soares and Melo, 2008). Pesticides provoke sperm DNA damages, especially in cases of occupational exposure, as well as pollutants such as heavy metals or polycyclic aromatic hydrocarbons in the air (Delbes et al., 2010; Perry, 2008; Somers and Cooper, 2009). Occupational exposure of testicles to heat also has deleterious effects on male fertility (Mieusset et al., 1987; Paul et al., 2008) as well as sporadic exposure (Rockett et al., 2001). The effects of obesity on semen quality are marked not only on the concentration of sperm, but also on DNA fragmentation (Kort et al., 2006), probably as a result of a high scrotal temperature and of an important hormonal disruption (Du Plessis et al., 2010). Mobile phone port to the belt, especially when using a hands-free kit (the emission of electromagnetic waves is more important during a call) could have an adverse effect on male fertility (Desai et al., 2009). Age has also an effect on sperm quality and in particular on the quality of genetic material (Sartorius and Nieschlag, 2010). It has been demonstrated by Singh (Singh et al., 2003) that the double-strand fragmentation increases with age. Finally, other factors such as high consumption of alcohol, dietary exposure to plasticizers, stress, poor diet, are potentially impacting the maintenance of the balance ROS / antioxidants in the semen (Tremellen, 2008).

Infection of the genitourinary tract is an important medical cause of DNA damages, as the influx of activated leukocytes provokes an increased production of free radicals. Systemic infections may also increase oxidative stress in sperm through an increase in the concentration of leukocytes in semen (HIV) (Umapathy et al., 2001), or via a systemic oxidative stress (Tremellen, 2008). Patients who underwent surgery to treat cryptorchidism still have a production of ROS and a sperm DNA fragmentation higher than those of fertile men (Smith et al., 2007). The presence of varicoceles also increases oxidative stress and DNA damage (Chen et al., 2004; Smith et al., 2006). Finally, there is a variety of iatrogenic origins of DNA damages. Cancer treatments (chemotherapy, pelvic radiotherapy) are generally detrimental to the integrity of sperm DNA (Tremellen, 2008). Concerning ARTs, various semen manipulations may affect sperm DNA. Centrifugation and cryopreservation increased oxidative stress, and consequently the DNA damage (Aitken et al., 2010; Thomson et al., 2009).

Several therapeutic strategies have been proposed (Hazout et al., 2008). The most effective is of course to treat infections of the urogenital tract. The patient must also be aware of the harmful effects of lifestyle (smoking, obesity, exposure to heat ...) on the quality of his sperm. If possible, it may be beneficial to limit exposure to pollutants. In the presence of varicoceles, embolization or surgery improve sperm quality, not really in term of numeration or morphology but clearly in terms of fragmentation of DNA (Smit et al., 2010). Finally there are several studies that have examined the effect of antioxidant treatment semen quality and sperm DNA. While it is necessary to have new controlled-randomized studies, many preliminary publications suggest beneficial effects (Agarwal and Sekhon, 2010).

In conclusion, the study of sperm DNA has highlighted abnormalities in chromatin organization, in DNA double helix and in epigenetic marks that impact male fertility and sperm role during reproduction. Sperm DNA analyses enable new diagnosis in infertilities previously classified as of idiopathic origin. These problems of DNA damages appear to influence fertility in early pregnancy by causing failure of implantation or repeated miscarriages. New treatments begin to be developed, but need to be well evaluated in prospective controlled-randomized studies. Improved lifestyle, environment and antioxidant treatments could improve semen and sperm DNA quality. However the newly developed sperm selection techniques remain limited to the quality of semen used: if all the sperm have an abnormal DNA, it becomes impossible to make a selection.

Finally, in accordance with the conclusions of the Position Report published by the Working Group of the ESHRE Andrology (Barratt et al., 2010), several recommendations can be made. First, it is necessary to continue to develop basic research on the chromatin organization to better understand the causes and the nature of DNA damages. It is also important to standardize analysis techniques to allow a correct interpretation of results, and also to be able to compare the different studies. The development of animal models allowing studies on the long-term effects of sperm DNA damages and ARTs, and well conducted studies paralleling DNA damages and results in ARTs (with neonatal data) will help the clinician in the guidance of patients in treatment options. Finally, the long-term monitoring of children from ARTs should be systematized.

4. Embryo

Successful implantation requires a competent blastocyst able to cross-talk with the receptive endometrium. This process includes dynamic and coordinated changes at the intricate crossroad between endocrinology, immunity and angiogenesis. Implantation process is not easily accessible *in vivo* to the research for obvious ethical and technical reasons. The only way to evaluate it is to extend data derived from animal studies, from *in vivo* studies about endometrium and embryo separately or from *in vitro* endometrium-early blastocyst co-culture. Despite extensive research in this field, the implantation process remains the black box of the reproduction, even in 2012. An important wastage of embryos is observed since the majority of blastocysts will perish before or around implantation (Macklon et al., 2002; Robertson et al., 2011; Teklenburg et al., 2010a). Implantation failure would come from an inadequate gamete/embryo quality, an erroneous cross-talk at the implantation site or an inadequate peri-conceptual environment. As recently demonstrated by Robertson et al., the peri-conceptual period and environment are critical for pregnancy success: the local cytokine network of the reproductive tract integrates environmental information and provides a signaling system programming both maternal receptivity and embryo development (Robertson et al., 2011). On the other hand, data from Kashiwagi (Kashiwagi et al., 2007) and from Bauersachs (Bauersachs et al., 2009) evidence that both embryo and endometrium are closely related and are actively influenced by each other, in order to favor pregnancy or, at contrary, to reject it.

Implantation is a 3-steps process including free-floating apposition of the blastocyst then adhesion between two epitheliums: endometrium and trophoblast, and finally, the fine-tuned process of trophoblastic invasion and differentiation. This materno-fetal dialogue is mediated through a broad array of molecules released at the implantation site both by trophoblast and/or endometrium. During implantation process, embryo and maternal tissues talk about immunity and angiogenesis, on an autocrine, paracrine and/or juxtacrine mood (Singh et al., 2011). In definitive, implantation process is the best example of a successfully tolerated graft and controlled tumor invasion.

4.1 Trophoblast differentiation

Placentation in Humans is characterized by the formation of a highly invasive hemochorial placenta accompanied with dramatical changes in the vasculature of the uterus. Differentiation of cytotrophoblast includes villous trophoblast ensuring exchanges at the maternal-fetal interface, and extravillous cytotrophoblast (EVCT) anchoring the placenta and participating to the vascular remodeling of the wall of spiral arteries. This deep trophoblastic invasion through the entire depth of the endometrium and the inner third of the myometrium is considered the hallmark of human pregnancy. Trophoblast deriving from trophectoderm, clothes the terminal villi (the outermost branches of the villous trees) that descend from the chorionic plate and fix to the basal plate by anchoring villi (Carter, 2011). The chorionic villi are then the structural and functional unit of the placenta, floating within the maternal blood present in the intervillous chambers. Villous cytotrophoblast that mantles the chorionic villi, aggregates and fuses to form the syncytiotrophoblast (ST), allowing efficient communication and signal exchanges. It ensures the endocrine function of placenta, releasing hormones involved in the homeostasis of pregnancy such as chorionic gonadotropin and placental lactogen. In the other hand, EVCT refers to the invading trophoblast that exits the villi and colonizes the uterine wall and spiral arteries, the lumen of which is plugged by trophoblastic cells during the first 8 weeks of gestation (Fournier et al., 2011).

Subsequent transformation of spiral arteries is a 2-steps process: the first seems to be mediated to factors secreted by uNK and covers endothelial swelling and vascular smooth muscle cells (VSMC) loss of coherence (Harris, 2011); the second is linked to the loss of endothelium, the beak down of VSMC/elastic fibers and finally the incorporation into the vessel wall of extravillous trophoblast cells, embed in a fibrinoid-rich matrix (Carter and Pijnenborg, 2011). The consequences of these changes are the widening of the vessel lumen, highly increasing the supplying blood to the intervillous space. In the placenta formation as during implantation process, immunity and angiogenesis are closely related.

Successful invasion is mediated through both maternal and embryo factors. Differentiation of cytotrophoblast to ST or EVCT is accurately controlled by transcription factors, hormones, growth factors, cytokines, and O2 level. Implantation, early embryo development and placentation processes take indeed place under a low oxygen environment during the first trimester of gestation. Many pregnancy disorders such as fetal growth restriction or pre-eclampsia are associated to the loss of invasiveness, with the extravillous trophoblast that does not reach the myometrium and the corresponding arterial segments retaining therefore their endothelium; whereas an excessive trophoblast invasion is associated with invasive mola, placenta accrete or choriocarcinoma (Lunghi et al., 2007). Pre-eclampsia is a pregnancy-specific syndrome characterized by hypertension, proteinuria and edema. It resolves on placenta delivery. Placental hypoxia is likely to be responsible for the maternal vascular dysfunction, through the increased placental release of anti-angiogenic factors such as soluble receptors flt1 and endoglin, both binding vascular endothelium growth factor (VEGF), placental growth factor (PLGF) and transforming growth factor (TGF)beta1/3 in the maternal circulation and causing endothelial dysfunctions (Lorquet et al., 2010).

4.2 Embryonic signals: Example of the embryo-specific hCG

The first known human embryo specific signal is hCG which is produced by the embryo before implantation, since it has been detected in embryo supernatants as soon as day 2 post-fertilization (Ramu et al., 2011). Human chorionic gonadotropin is the major pregnancy glycoprotein hormone, from the cystine knot cytokines superfamily, and is specifically secreted by trophoblast. It is classically well known from it action as corpus luteum progesterone production rescuer but many recent studies has evidenced more and more extra-gonadal actions. HCG is a non covalently linked heterodimer composed of 2 subunits α (produced by cytotrophoblast) and β (produced by syncytium). Maternal concentration and glycosylation of hCG change throughout the pregnancy since 3 majors isoforms are described, all of which sharing the hCG β amino acide sequence (Banerjee and Fazleabas, 2011):

- **native hCG**: produced by syncytium, its concentration reach a peak around week 10 then drops. Its main gonadal action is to rescue corpus luteum and support progesterone production but many extragonadal actions on uterine receptivity, trophoblast functions, immunity and angiogenesis are described (see below). It signals through the LH/hCG receptor (LHR), which has been described on different extra-gonadal human tissues (Berndt et al., 2006; Rao, 2006). Recent data concerning impact of hCG on uNK cells suggest that hCG may also bind to the mannose receptor (Kane et al., 2009).
- **hyperglycosylated hCG (H-hCG):** produced by EVCT, it accounts for 90% of total hCG production with a peak in the week following implantation. It is an autocrine factor

modulating its own production and highly promoting invasiveness of trophoblast. H-hCG 2D structure is different from native hCG, appears to be independent in numerous biological functions, overall promoting placental implantation/invasiveness and is likely to signal through a different receptor than LHR (TGF-beta Receptor?). H-hCG is also critical for choriocarcinoma growth and malignancy (Cole, 2010; Fournier et al., 2011; Guibourdenche et al., 2010).

- **free beta subunit of H-hCG (hCG free β):** produced by all non-trophoblastic malignancies, it inhibits apoptosis and promotes malignant transformation of cancer cells.

Data from literature and from our own work show that through native hCG (and H-hCG concerning invasiveness of EVCT), embryo profoundly intervenes in its own implantation and favors immunological tolerance and active angiogenesis that are crucial for successful implantation (Banerjee and Fazleabas, 2011; Tsampalas et al., 2010).

At the first steps of implantation, hCG is able to prolong the WOI, by inhibiting decidual Insulin like Growth Factor (IGF)-Binding Protein and IGFI (both markers of complete decidualization) (Fluhr et al., 2008b), and in association with interleukin 1 (IL-1), to favor adhesion via the increase in trophinin expression on endometrial epithelium (Sugihara et al., 2008). Concerning trophoblast functions hCG (and its hyperglycosylated variant) has been demonstrated to promote trophoblastic cells migration, and to reduce endometrial barrier, through its action on matrix metalloprotease protein 9 (MMP9, increased), TIMP-1, -2, -3 (reduced), GM-CSF, IL-11 (Chen et al., 2011; Fluhr et al., 2008a; Paiva et al., 2011). We also observed that hCG is able to increase leukemia inhibitory factor (LIF) production by endometrial epithelial cells *in vitro*, a cytokine known to be crucial for implantation in mice (Perrier d'Hauterive et al., 2004; Stewart, 1994). These data has been confirmed by different teams (Licht et al., 2002; Sherwin et al., 2007).

HCG levels coincide with the development of trophoblast tolerance. Indeed, it offers many immunological properties. For example, hCG increases the number of uterine natural killer cells that play a key role in the establishment of pregnancy (Kane et al., 2009). HCG also intervenes in the development of local immune tolerance through apoptosis via Fas/Fas-Ligand (Kayisli et al., 2003). It also modulates the Th1/Th2 balance toward the Th2 pathway (Koldehoff et al., 2011). During pregnancy, the balance of Th1 (cell-mediated immunity) and Th2 (humoral immunity) cytokines is characterized by an initial prevalence (but not an exclusivity) of Th2 cytokines, followed by a progressive shift towards Th1 predominance late in gestation, that when is abnormal, may initiate and intensify the cascade of inflammatory cytokine production involved in adverse pregnancy outcomes (Challis et al., 2009). In a more general way, Khan *et al.* showed that the administration of hCG to nonobese diabetic mice (NOD) before the beginning of the clinical symptoms reduced the increase in glycaemia, reversed establishment of insulitis, and inhibited the development of Th1 autoimmune diabetes (Khan et al., 2001). A recent large proteomic study demonstrated the influence of several molecules produced by the trophoblast that regulate the mother's immune tolerance. Among these molecules, hCG inhibits T lymphocytes (Dong et al., 2008).

The transient tolerance, evident during gestation is at least partially achieved via the presence of regulatory T cells that are expended during pregnancy (Dimova et al., 2011; Ernerudh et al., 2011; Fainboim and Arruvito, 2011; Nevers et al., 2011; Xiong et al., 2010) at the maternal-fetal interface, showing a suppressive phenotype, whereas Treg cells are not increase in the circulation of pregnant women. In part, they are attracted by hCG at the fetal–maternal interface during early pregnancy, via a LHR signalling (Leber et al., 2010; Schumacher et al., 2009). Finally, hCG treatment of activated dendritic cells results in an up-regulation of MHC class II, IL-10 and IDO expression, reducing the ability to stimulate T cell proliferation. Impact of dendritic cells during implantation has recently been review by Blois (Blois et al., 2011).

Interestingly, immunologic properties of hCG are likely to differ as far as urinary versus recombinant molecules are concerned. These data are particularly important to keep in mind if immunomodulation of hCG would apply in clinical practice (Carbone et al., 2010; Kajihara et al., 2011).

Beside immune tolerance, a successful implantation requires also an extensive vascular remodeling of maternal arteries at the placental bed. Recent data demonstrate angiogenic effects of hCG via its interaction with endometrial epithelial cells which are able to produce angiogenic VEGF under hCG-LHR binding. Moreover, hCG highly increases angiogenesis of new mature vessels, via the stimulation endothelial cells proliferation as well as the migration of smooth muscle cells leading to the maturation of vessels, an important step for placentation (Berndt et al., 2009; Berndt et al., 2006).

Taken together the immune and vascular roles of hCG during early pregnancy, it is not surprising that this hormone has been studied during pregnancy disorders such as pre-eclampsia (Kalinderis et al., 2011; Norris et al., 2011).

5. The endometrium

Implantation of the embryo into the maternal endometrium represents a unique biological process example of an immunological (tolerance of an allograft) and biological (adhesion of two epithelia) paradox (Perrier D'hauterive et al., 2002). The success of implantation depends on a receptive endometrium, a functionally normal blastocyst and a synchronized cross-talk between embryonic and maternal tissues. Though sexual steroids control the process (Paulson, 2011), a cascade of cytokines, growth factors and adhesion molecules are the private paracrine mediators of the uterine receptivity. Particularly, progesterone is crucial to establishment of endometrial receptivity, modulating subsequently the appearance or disappearance of a wide network of molecules responsible of the uterine receptivity.

5.1 The window of implantation and the decidua

Endometrium undergoes cyclic morphological and functional changes, including growth, differentiation and desquamation. Altogether, these cyclic physiological modifications lead to the preparation of a receptive endometrium able to tolerate the foetal allograft and to control invasiveness of trophoblastic cells, allowing implantation process during the mid-secretory phase. Control of gene expression is crucial to this process, and inappropriate epigenetic modifications occasioning an altered chromatin structure and transcriptional activity may result in aberrant expression of receptive endometrial pattern. Epigenetics is likely to be associated with the initial regeneration and then proliferation of endometrium, angiogenesis, decidual reaction and angiogenesis (Munro et al., 2010).

Whereas the implantation can occur in any human tissue, the endometrium is one of rare in which the embryo cannot implant, except during one limited period called the *window of implantation* (WOI), at the time when progesterone reaches peak serum concentrations (day 20-day24), allowing embryo to implant only under optimal circumstances. During this period, it offers a high receptivity to the embryo, resulting in a successful pregnancy (Lessey, 2011). This physiological state of the endometrium at the mid-secretory phase allows blastocyst 3-steps implantation process: apposition, then adhesion followed by trophoblast invasion and then induction of localized changes in the stroma called decidualization (Rashid et al., 2011).

WOI ends with the decidual transformation, where endometrium becomes a well vascularized tissue and stromal cell differentiate into specialized decidual cells (Teklenburg et al., 2010b). Factors secreted by decidual cells (such as LEFTY-A) compromise endometrial receptivity, ending the WOI. Oedema, increased vascular permeability, proliferation and transformation of fibroblast into cuboid secreting cells, invasion of leucocytes and angiogenesis are the principal mechanisms enabling decidua to resist to oxidative stress, to allow interaction within immune cells and to restrain the invasiveness of trophoblasts.

Apprehending molecular mechanisms of endometrial receptivity and then decidual transformation would provide a better understanding implantation process and its pathological implications. Failure to express adequate decidualization pattern is likely to induce early pregnancy lost or predispose to obstetrical complications such as implantation failure, recurrent miscarriages, pre-eclampsia, fetal growth restriction... (Blois et al., 2011; Brosens, 2011; Plaisier, 2011).

Actual research aims to dissect the molecular basis for the changes occurring during the WOI at the genomic and proteomic level (Berlanga et al., 2011; Diaz-Gimeno et al., 2011; Horcajadas et al., 2007). Despite a lot of experiments aiming to evidence relevant and selective biomarkers of uterine receptivity, there is to date no molecular fingerprint of the WOI. This is due partly to the discovery of an increasingly higher number of potential new markers according to the evolution of research, partly to the structural composition of endometrium, where the specialized epithelial cells that encounter the embryo are distinct from the glandular part. Finally, it is important to note the absence of consensus to which biomarkers to use for endometrial receptivity, and none of studied mediators (Mucine 1, L-selectin ligand, integrins, heparin-binding epidermal growth factor-like growth factor) has been explored in sufficient detail to validate its usefulness in clinical practice (Lessey, 2011).

The modification occurring in the endometrium during the WOI is closely related to angiogenesis and to immune system, allowing the tolerance of the fetal allograft and an adequate vascular remodelling during plancentation for a successful pregnancy until birth. Angiogenesis refers to the formation of new vessels from existing ones, by elongation, intussusception or sprouting of endothelial cells. One of the key local adaptations to pregnancy is the stimulation of maternal vessel network at the embryo implantation site. Normal fetal development requires extensive angiogenesis and important vascular remodelings allowing adequate supply of nutrients as well as gas and metabolite exchanges. Abnormal uterine blood supply is associated with higher perinatal morbidity and mortality caused by preterm delivery, preeclampsia, or intrauterine growth restriction. Stimulation of angiogenesis in many organs (including uterus) is mediated through enhanced expression of VEGF, an angiogenic cytokine produced by epithelial endometrial and stromal cells (Berndt et al., 2006). As demonstrated at the next section, some other molecules from a trophoblastic origin, are able to actively stimulate maternal vascular remodelling as well.

In parallel, immune cells increasingly infiltrate endometrium from post-ovulation to menstruation in absence of pregnancy. In case of pregnancy, an higher increased number is observed after fertilisation up to mid gestation, since 30% of decidual cells are leukocytes, including 75% of uterine natural killers (uNK) (Plaisier, 2011; Zhang et al., 2011). It has been proposed that uNK could exert, directly or indirectly, either positive or negative control over implantation process. These cells secrete an array of cytokines important for adequate local immune regulation, angiogenesis, placental development, and establishment of pregnancy. Other immune cells or systems such as dendritic cells, T regulatory cells, macrophages, Th1/Th2 equilibrium orchestrate immune tolerance, trophoblast invasion and angiogenesis associated with embryo implantation (Blois et al., 2011; Denney et al., 2011).

ART programs, and particularly egg/embryo donation protocols, has clearly contributed to our knowledge of endometrial physiology. They gave us the opportunity to separately evaluate oocyte and endometrium, and to discover that endometrial receptivity can be controlled artificially with exogenous estrogen and progesterone, with a certain success (Paulson, 2011). Although it ensures obtaining a sufficient number of oocytes and thus embryos, ovarian hyper stimulation with exogenous gonadotropins is likely to be associated with modified endometrial development that may impact endometrial receptivity and success of implantation. These phenomenons would result from the inducing of an embryo-endometrium asynchrony (histologic endometrial advancement), the up-regulation of P-receptor expression, and a negative correlation between implantation and premature progesterone elevation, impacting fresh implantation rates in normal and high responders (Shapiro et al., 2011a, b), despite some other controversial results (Levi et al., 2001).

5.2 Clinical relevance of endometrial immunity

A better understanding of the uterine–embryo interaction and of the "seed and soil" regulations during embryo implantation is mandatory to increase the efficacy of ART. Implantation failure, recurrent miscarriage and preeclampsia have several recognized causes in common, but in most cases, the precise etiology remains obscure. Recent data in reproductive immunology identify the importance of the local immune environment and suggest to the clinician the need to develop tools to explore these endometrial perturbations (Tuckerman et al., 2010).

Indeed the semantic distinction between implantation failure, abortion and preeclampsia might in fact be more quantitative than qualitative (Chaouat, 2008). An important subset of implantation defects is the consequence of a deregulation of the interleukin systems, tumor necrosis factor (TNFα), interferon system as well as the uNK cell-mediated networks. Both deficient and excess expression of cytokines and immune cells number and activation play indeed detrimental key roles in implantation, since these actors can have both positive and negative effects.

For example, in human reproduction, a proper balance in the IL-12, IL-18, IL-15 and a correct uNK activation stat result in successful implantation and pregnancy. Conversely, imbalances in these parameters correlate with implantation failure or early pregnancy loss (Ledee, 2005). Their expression controls the local uNK recruitment and subendometrial angiogenesis as reflected by the vascular flow index (VFI) determined by three-dimensional ultrasound (Ledee et al., 2008a; Ledee et al., 2011b).

The excellent correlation between IL-15 mRNA expression and the sub endometrial vascular flow index (VFI) suggests that this cytokine and the uNK cells that produce it participate to the local control of angiogenesis. Patients with a low sub-endometrial VFI and low IL-15 mRNA are patients with insufficient uNK recruitment and/or inadequate uNK-derived angiogenic-related proteins. In contrast, some patients with implantation failure exhibit very high VFI and, at the same time, high IL-15 and IL-18 mRNAs and CD^{56+} cell count. A Th-1 excess could possibly be involved in implantation failure (Kwak-Kim et al., 2005).

Abnormal subendometrial vascularization assessed by ultrasound may be the consequence of distinct cytokine dysregulation patterns. These may cause implantation failure, abortion or preeclampsia, through abnormal (insufficient or excessive) recruitment of uNK cells or through inadequate endothelial vascular remodeling before implantation. 3-D ultrasonography with vocal analysis may inform on the uterine preparation to a constructive dialogue with the conceptus through an adequate trophicity and angiogenesis. This point on echographic evaluation of endometrium opens up new horizons for the evaluation of the endometrium, well beyond the simple measurement of thickness that improperly correlate with the outcome of IVF, since without absolute cutoff (Noyes et al., 1995; Paulson, 2011). Nevertheless, focusing on vascularization at the time of implantation is absolutely mandatory from a physiological point of view (Ledee, 2005). In daily clinical practice, echographic evaluation of subendometrial VFI and measurement of the IL-15 and IL-18 mRNAs together with CD^{56+} counts may be useful to identify those women at risk of implantation failure (Ledee et al., 2011b).

Moreover, Tumor necrosis factor like WEAK inducer of apoptosis (TWEAK) is a transmembrane protein which, when cleaved, functions as a soluble cytokine. It is highly expressed by different immune cells and triggers multiple roles, including control of angiogenesis. TWEAK and IL-18 mRNA expression are correlated in patients with implantation failures. Basic TWEAK expression influences the IL-18 related uNK recruitment and local cytotoxicity. Actually, as demonstrated by Petitbarat et al., TWEAK doesn't act on IL-18 expression but is likely to control IL-18 related cytotoxicity on uNK cells when IL-18 is over-expressed (Petitbarat et al., 2010).

Using large-scale microarray analysis, endometrial gene expression at the time of the implantation window was compared in fertile control patients (FC) and women displaying previous IVF/ICSI repeated implantation failure (IF) or recurrent unexplained miscarriages (RM). Biological functions and gene networks were explored using the Ingenuity Pathways Analysis software. The number of differentially expressed genes revealed the extent of changes within the preconceptional endometrium as a function of either fertility or infertility. The main similarities of differentially expressed genes between IF and RM relate to immune and hematological system development abnormalities, especially deregulation of the differentiation and development of T lymphocytes and blood cells. As the endometrium is now thought to be a biosensor for the quality of the embryo, such differential expression may have direct consequences on the initial embryo-uterus dialogue (Chaouat et al., 2011; Ledee et al., 2011c).

6. Conclusion: Implantation process

Implantation process and placenta formation are closely associated to angiogenesis and immune tolerance, at the crossroad of endocrinology. Whereas embryo could implant in every human tissues and particularly in Fallopian tubes with sometimes a very huge morbidity (Shaw and Horne, 2011), uterus is tailored to tolerate the fetal allograft and to allow controlled invasiveness of trophoblast only during the WOI, ensuring the success of the intricate cascade of implantation. Blastocyst implantation requires a competent embryo- supported by the fertilization of a good quality oocyte by a top spermatozoa - able to cross-talk with a receptive endometrium, and the dialogue must talk about immunity and angiogenesis. Implantation failure and some pathologies of pregnancy are associated to defect of this close dialogue at one or many steps of the cascade.

Implantation rate in IVF treatment certainly benefits from research about all these fields in reproductive medicine, able to increase selection, diagnostic and treatment of the different actors of implantation: gametes, endometrium, embryo and the most difficult issue: the dialogue between each other.

7. References

Agarwal, A., Gupta, S., and Sharma, R. K., 2005, Role of oxidative stress in female reproduction: Reprod Biol Endocrinol, v. 3, p. 28.

Agarwal, A., Gupta, S., Sekhon, L., and Shah, R., 2008, Redox considerations in female reproductive function and assisted reproduction: from molecular mechanisms to health implications: Antioxid Redox Signal, v. 10, no. 8, p. 1375-1403.

Agarwal, A., and Sekhon, L. H., 2010, The role of antioxidant therapy in the treatment of male infertility: Hum Fertil (Camb), v. 13, no. 4, p. 217-225.

Aitken, R. J., and Clarkson, J. S., 1988, Significance of reactive oxygen species and antioxidants in defining the efficacy of sperm preparation techniques: J Androl, v. 9, no. 6, p. 367-376.

Aitken, R. J., and Roman, S. D., 2008, Antioxidant systems and oxidative stress in the testes: Adv Exp Med Biol, v. 636, p. 154-171.

Aitken, R. J., De Iuliis, G. N., and McLachlan, R. I., 2009, Biological and clinical significance of DNA damage in the male germ line: Int J Androl, v. 32, no. 1, p. 46-56.

Aitken, R. J., and De Iuliis, G. N., 2010, On the possible origins of DNA damage in human spermatozoa: Mol Hum Reprod, v. 16, no. 1, p. 3-13.

Aitken, R. J., De Iuliis, G. N., Finnie, J. M., Hedges, A., and McLachlan, R. I., 2010, Analysis of the relationships between oxidative stress, DNA damage and sperm vitality in a patient population: development of diagnostic criteria: Hum Reprod, v. 25, no. 10, p. 2415-2426.

Antinori, M., Licata, E., Dani, G., Cerusico, F., Versaci, C., d'Angelo, D., and Antinori, S., 2008, Intracytoplasmic morphologically selected sperm injection: a prospective randomized trial: Reprod Biomed Online, v. 16, no. 6, p. 835-841.

Assou, S., Haouzi, D., De Vos, J., and Hamamah, S., 2010, Human cumulus cells as biomarkers for embryo and pregnancy outcomes: Mol Hum Reprod, v. 16, no. 8, p. 531-538.

Avendano, C., Franchi, A., Taylor, S., Morshedi, M., Bocca, S., and Oehninger, S., 2009, Fragmentation of DNA in morphologically normal human spermatozoa: Fertil Steril, v. 91, no. 4, p. 1077-1084.

Avendano, C., Franchi, A., Duran, H., and Oehninger, S., 2010, DNA fragmentation of normal spermatozoa negatively impacts embryo quality and intracytoplasmic sperm injection outcome: Fertil Steril, v. 94, no. 2, p. 549-557.

Balaban, B., and Urman, B., 2006, Effect of oocyte morphology on embryo development and implantation: Reprod Biomed Online, v. 12, no. 5, p. 608-615.

Balaban, B., Ata, B., Isiklar, A., Yakin, K., and Urman, B., 2008, Severe cytoplasmic abnormalities of the oocyte decrease cryosurvival and subsequent embryonic development of cryopreserved embryos: Hum Reprod, v. 23, no. 8, p. 1778-1785.

Banerjee, P., and Fazleabas, A. T., 2011, Extragonadal actions of chorionic gonadotropin: Rev Endocr Metab Disord.

Barratt, C. L., Aitken, R. J., Bjorndahl, L., Carrell, D. T., de Boer, P., Kvist, U., Lewis, S. E., Perreault, S. D., Perry, M. J., Ramos, L., Robaire, B., Ward, S., and Zini, A., 2010, Sperm DNA: organization, protection and vulnerability: from basic science to clinical applications--a position report: Hum Reprod, v. 25, no. 4, p. 824-838.

Balaban, B., and Urman, B., 2006, Effect of oocyte morphology on embryo development and implantation: Reprod Biomed Online, v. 12, no. 5, p. 608-615.

Bartoov, B., Berkovitz, A., Eltes, F., Kogosowski, A., Menezo, Y., and Barak, Y., 2002, Real-time fine morphology of motile human sperm cells is associated with IVF-ICSI outcome: J Androl, v. 23, no. 1, p. 1-8.

Bartoov, B., Berkovitz, A., Eltes, F., Kogosovsky, A., Yagoda, A., Lederman, H., Artzi, S., Gross, M., and Barak, Y., 2003, Pregnancy rates are higher with intracytoplasmic morphologically selected sperm injection than with conventional intracytoplasmic injection: Fertil Steril, v. 80, no. 6, p. 1413-1419.

Bauersachs, S., Ulbrich, S. E., Zakhartchenko, V., Minten, M., Reichenbach, M., Reichenbach, H. D., Blum, H., Spencer, T. E., and Wolf, E., 2009, The endometrium responds differently to cloned versus fertilized embryos: Proc Natl Acad Sci U S A, v. 106, no. 14, p. 5681-5686.

Benchaib, M., Lornage, J., Mazoyer, C., Lejeune, H., Salle, B., and Francois Guerin, J., 2007, Sperm deoxyribonucleic acid fragmentation as a prognostic indicator of assisted reproductive technology outcome: Fertil Steril, v. 87, no. 1, p. 93-100.

Bennetts, L. E., De Iuliis, G. N., Nixon, B., Kime, M., Zelski, K., McVicar, C. M., Lewis, S. E., and Aitken, R. J., 2008, Impact of estrogenic compounds on DNA integrity in human spermatozoa: evidence for cross-linking and redox cycling activities: Mutat Res, v. 641, no. 1-2, p. 1-11.

Berkovitz, A., Eltes, F., Yaari, S., Katz, N., Barr, I., Fishman, A., and Bartoov, B., 2005, The morphological normalcy of the sperm nucleus and pregnancy rate of intracytoplasmic injection with morphologically selected sperm: Hum Reprod, v. 20, no. 1, p. 185-190.

Berkovitz, A., Eltes, F., Ellenbogen, A., Peer, S., Feldberg, D., and Bartoov, B., 2006a, Does the presence of nuclear vacuoles in human sperm selected for ICSI affect pregnancy outcome?: Hum Reprod, v. 21, no. 7, p. 1787-1790.

Berkovitz, A., Eltes, F., Lederman, H., Peer, S., Ellenbogen, A., Feldberg, B., and Bartoov, B., 2006b, How to improve IVF-ICSI outcome by sperm selection: Reprod Biomed Online, v. 12, no. 5, p. 634-638.

Berlanga, O., Bradshaw, H. B., Vilella-Mitjana, F., Garrido-Gomez, T., and Simon, C., 2011, How endometrial secretomics can help in predicting implantation: Placenta, v. 32 Suppl 3, p. S271-275.

Berndt, S., Perrier d'Hauterive, S., Blacher, S., Pequeux, C., Lorquet, S., Munaut, C., Applanat, M., Herve, M. A., Lamande, N., Corvol, P., van den Brule, F., Frankenne, F., Poutanen, M., Huhtaniemi, I., Geenen, V., Noel, A., and Foidart, J. M., 2006, Angiogenic activity of human chorionic gonadotropin through LH receptor activation on endothelial and epithelial cells of the endometrium: FASEB J, v. 20, no. 14, p. 2630-2632.

Berndt, S., Blacher, S., Perrier d'Hauterive, S., Thiry, M., Tsampalas, M., Cruz, A., Pequeux, C., Lorquet, S., Munaut, C., Noel, A., and Foidart, J. M., 2009, Chorionic gonadotropin stimulation of angiogenesis and pericyte recruitment: J Clin Endocrinol Metab, v. 94, no. 11, p. 4567-4574.

Bjorndahl, L., and Kvist, U., 2010, Human sperm chromatin stabilization: a proposed model including zinc bridges: Mol Hum Reprod, v. 16, no. 1, p. 23-29.

Blois, S. M., Klapp, B. F., and Barrientos, G., 2011, Decidualization and angiogenesis in early pregnancy: unravelling the functions of DC and NK cells: J Reprod Immunol, v. 88, no. 2, p. 86-92.

Boe-Hansen, G. B., Fedder, J., Ersboll, A. K., and Christensen, P., 2006, The sperm chromatin structure assay as a diagnostic tool in the human fertility clinic: Hum Reprod, v. 21, no. 6, p. 1576-1582.

Boitrelle, F., Ferfouri, F., Petit, J. M., Segretain, D., Tourain, C., Bergere, M., Bailly, M., Vialard, F., Albert, M., and Selva, J., 2011, Large human sperm vacuoles observed in motile spermatozoa under high magnification: nuclear thumbprints linked to failure of chromatin condensation: Hum Reprod, v. 26, no. 7, p. 1650-1658.

Bonde, J. P., Ernst, E., Jensen, T. K., Hjollund, N. H., Kolstad, H., Henriksen, T. B., Scheike, T., Giwercman, A., Olsen, J., and Skakkebaek, N. E., 1998, Relation between semen quality and fertility: a population-based study of 430 first-pregnancy planners: Lancet, v. 352, no. 9135, p. 1172-1177.

Borini, A., Tarozzi, N., Bizzaro, D., Bonu, M. A., Fava, L., Flamigni, C., and Coticchio, G., 2006, Sperm DNA fragmentation: paternal effect on early post-implantation embryo development in ART: Hum Reprod, v. 21, no. 11, p. 2876-2881.

Brosens, I., 2011, Placental bed & maternal - fetal disorders. Preface: Best Pract Res Clin Obstet Gynaecol, v. 25, no. 3, p. 247-248.

Bungum, M., Humaidan, P., Axmon, A., Spano, M., Bungum, L., Erenpreiss, J., and Giwercman, A., 2007, Sperm DNA integrity assessment in prediction of assisted reproduction technology outcome: Hum Reprod, v. 22, no. 1, p. 174-179.

Carbone, F., Procaccini, C., De Rosa, V., Alviggi, C., De Placido, G., Kramer, D., Longobardi, S., and Matarese, G., 2010, Divergent immunomodulatory effects of recombinant and urinary-derived FSH, LH, and hCG on human CD4+ T cells: J Reprod Immunol, v. 85, no. 2, p. 172-179.

Cardozo, E., Pavone, M. E., and Hirshfeld-Cytron, J. E., 2011, Metabolic syndrome and oocyte quality: Trends Endocrinol Metab, v. 22, no. 3, p. 103-109.

Carter, A. M., 2011, Comparative studies of placentation and immunology in non-human primates suggest a scenario for the evolution of deep trophoblast invasion and an explanation for human pregnancy disorders: Reproduction, v. 141, no. 4, p. 391-396.

Carter, A. M., and Pijnenborg, R., 2011, Evolution of invasive placentation with special reference to non-human primates: Best Pract Res Clin Obstet Gynaecol, v. 25, no. 3, p. 249-257.

Cartwright, J. E., Fraser, R., Leslie, K., Wallace, A. E., and James, J. L., 2010, Remodelling at the maternal-fetal interface: relevance to human pregnancy disorders: Reproduction, v. 140, no. 6, p. 803-813.

Challis, J. R., Lockwood, C. J., Myatt, L., Norman, J. E., Strauss, J. F., 3rd, and Petraglia, F., 2009, Inflammation and pregnancy: Reprod Sci, v. 16, no. 2, p. 206-215.

Chaouat, G., 2008, Current knowledge on natural killer cells, pregnancy and pre-eclampsia. Introduction: Reprod Biomed Online, v. 16, no. 2, p. 170-172.

Chaouat, G., Rodde, N., Petitbarat, M., Bulla, R., Rahmati, M., Dubanchet, S., Zourbas, S., Bataillon, I., Coque, N., Hennuy, B., Martal, J., Munaut, C., Aubert, J., Serazin, V., Steffen, T., Jensenius, J. C., Foidart, J. M., Sandra, O., Tedesco, F., and Ledee, N., 2011, An insight into normal and pathological pregnancies using large-scale microarrays: lessons from microarrays: J Reprod Immunol, v. 89, no. 2, p. 163-172.

Check, J. H., Graziano, V., Cohen, R., Krotec, J., and Check, M. L., 2005, Effect of an abnormal sperm chromatin structural assay (SCSA) on pregnancy outcome following (IVF) with ICSI in previous IVF failures: Arch Androl, v. 51, no. 2, p. 121-124.

Chen, S. S., Huang, W. J., Chang, L. S., and Wei, Y. H., 2004, 8-hydroxy-2'-deoxyguanosine in leukocyte DNA of spermatic vein as a biomarker of oxidative stress in patients with varicocele: J Urol, v. 172, no. 4 Pt 1, p. 1418-1421.

Chen, J. Z., Wong, M. H., Brennecke, S. P., and Keogh, R. J., 2011, The effects of human chorionic gonadotrophin, progesterone and oestradiol on trophoblast function: Mol Cell Endocrinol, v. 342, no. 1-2, p. 73-80.

Cobo, A., and Diaz, C., 2011, Clinical application of oocyte vitrification: a systematic review and meta-analysis of randomized controlled trials: Fertil Steril, v. 96, no. 2, p. 277-285.

Cohen-Bacrie, P., Belloc, S., Menezo, Y. J., Clement, P., Hamidi, J., and Benkhalifa, M., 2009, Correlation between DNA damage and sperm parameters: a prospective study of 1,633 patients: Fertil Steril, v. 91, no. 5, p. 1801-1805.

Cole, L. A., 2010, Hyperglycosylated hCG, a review: Placenta, v. 31, no. 8, p. 653-664.

Cooper, T. G., Noonan, E., von Eckardstein, S., Auger, J., Baker, H. W., Behre, H. M., Haugen, T. B., Kruger, T., Wang, C., Mbizvo, M. T., and Vogelsong, K. M., 2010, World Health Organization reference values for human semen characteristics: Hum Reprod Update, v. 16, no. 3, p. 231-245.

Corry, G. N., Tanasijevic, B., Barry, E. R., Krueger, W., and Rasmussen, T. P., 2009, Epigenetic regulatory mechanisms during preimplantation development: Birth Defects Res C Embryo Today, v. 87, no. 4, p. 297-313.

Delbes, G., Hales, B. F., and Robaire, B., 2010, Toxicants and human sperm chromatin integrity: Mol Hum Reprod, v. 16, no. 1, p. 14-22.

Denney, J. M., Nelson, E. L., Wadhwa, P. D., Waters, T. P., Mathew, L., Chung, E. K., Goldenberg, R. L., and Culhane, J. F., 2011, Longitudinal modulation of immune system cytokine profile during pregnancy: Cytokine, v. 53, no. 2, p. 170-177.

Depa-Martynow, M., Jedrzejczak, P., Taszarek-Hauke, G., Josiak, M., and Pawelczyk, L., 2006, [The impact of cigarette smoking on oocytes and embryos quality during in vitro fertilization program]: Przegl Lek, v. 63, no. 10, p. 838-840.

Desai, N. R., Kesari, K. K., and Agarwal, A., 2009, Pathophysiology of cell phone radiation: oxidative stress and carcinogenesis with focus on male reproductive system: Reprod Biol Endocrinol, v. 7, p. 114.

Diaz-Gimeno, P., Horcajadas, J. A., Martinez-Conejero, J. A., Esteban, F. J., Alama, P., Pellicer, A., and Simon, C., 2011, A genomic diagnostic tool for human endometrial receptivity based on the transcriptomic signature: Fertil Steril, v. 95, no. 1, p. 50-60, 60 e51-15.

Dimova, T., Nagaeva, O., Stenqvist, A. C., Hedlund, M., Kjellberg, L., Strand, M., Dehlin, E., and Mincheva-Nilsson, L., 2011, Maternal Foxp3 expressing CD4+ CD25+ and CD4+ CD25- regulatory T-cell populations are enriched in human early normal pregnancy decidua: a phenotypic study of paired decidual and peripheral blood samples: Am J Reprod Immunol, v. 66 Suppl 1, p. 44-56.

Dirican, E. K., Ozgun, O. D., Akarsu, S., Akin, K. O., Ercan, O., Ugurlu, M., Camsari, C., Kanyilmaz, O., Kaya, A., and Unsal, A., 2008, Clinical outcome of magnetic activated cell sorting of non-apoptotic spermatozoa before density gradient centrifugation for assisted reproduction: J Assist Reprod Genet, v. 25, no. 8, p. 375-381.

Dong, M., Ding, G., Zhou, J., Wang, H., Zhao, Y., and Huang, H., 2008, The effect of trophoblasts on T lymphocytes: possible regulatory effector molecules--a proteomic analysis: Cell Physiol Biochem, v. 21, no. 5-6, p. 463-472.

Dosiou, C., and Giudice, L. C., 2005, Natural killer cells in pregnancy and recurrent pregnancy loss: endocrine and immunologic perspectives: Endocr Rev, v. 26, no. 1, p. 44-62.

Du Plessis, S. S., Cabler, S., McAlister, D. A., Sabanegh, E., and Agarwal, A., 2010, The effect of obesity on sperm disorders and male infertility: Nat Rev Urol, v. 7, no. 3, p. 153-161.

Duan, J. S., 1990, Production of granulocyte colony stimulating factor in decidual tissue and its significance in pregnancy: Osaka City Med J, v. 36, no. 2, p. 81-97.

Duran, E. H., Morshedi, M., Taylor, S., and Oehninger, S., 2002, Sperm DNA quality predicts intrauterine insemination outcome: a prospective cohort study: Hum Reprod, v. 17, no. 12, p. 3122-3128.

Ernerudh, J., Berg, G., and Mjosberg, J., 2011, Regulatory T helper cells in pregnancy and their roles in systemic versus local immune tolerance: Am J Reprod Immunol, v. 66 Suppl 1, p. 31-43.

Esfandiari, N., Javed, M. H., Gotlieb, L., and Casper, R. F., 2005, Complete failed fertilization after intracytoplasmic sperm injection--analysis of 10 years' data: Int J Fertil Womens Med, v. 50, no. 4, p. 187-192.

Evenson, D. P., Jost, L. K., Marshall, D., Zinaman, M. J., Clegg, E., Purvis, K., de Angelis, P., and Claussen, O. P., 1999, Utility of the sperm chromatin structure assay as a diagnostic and prognostic tool in the human fertility clinic: Hum Reprod, v. 14, no. 4, p. 1039-1049.

Evenson, D. P., and Wixon, R., 2008, Data analysis of two in vivo fertility studies using Sperm Chromatin Structure Assay-derived DNA fragmentation index vs. pregnancy outcome: Fertil Steril, v. 90, no. 4, p. 1229-1231.

Fahiminiya, S., Labas, V., Roche, S., Dacheux, J. L., and Gerard, N., 2011, Proteomic analysis of mare follicular fluid during late follicle development: Proteome Sci, v. 9, p. 54.

Fainboim, L., and Arruvito, L., 2011, Mechanisms involved in the expansion of Tregs during pregnancy: role of IL-2/STAT5 signalling: J Reprod Immunol, v. 88, no. 2, p. 93-98.

Fancsovits, P., Tothne, Z. G., Murber, A., Takacs, F. Z., Papp, Z., and Urbancsek, J., 2006, Correlation between first polar body morphology and further embryo development: Acta Biol Hung, v. 57, no. 3, p. 331-338.

Fang, C., Tang, M., Li, T., Peng, W. L., Zhou, C. Q., Zhuang, G. L., and Leong, M., 2007, Visualization of meiotic spindle and subsequent embryonic development in in vitro and in vivo matured human oocytes: J Assist Reprod Genet, v. 24, no. 11, p. 547-551.

Feuerstein, P., Cadoret, V., Dalbies-Tran, R., Guerif, F., Bidault, R., and Royere, D., 2007, Gene expression in human cumulus cells: one approach to oocyte competence: Hum Reprod, v. 22, no. 12, p. 3069-3077.

Fluhr, H., Bischof-Islami, D., Krenzer, S., Licht, P., Bischof, P., and Zygmunt, M., 2008a, Human chorionic gonadotropin stimulates matrix metalloproteinases-2 and -9 in cytotrophoblastic cells and decreases tissue inhibitor of metalloproteinases-1, -2, and -3 in decidualized endometrial stromal cells: Fertil Steril, v. 90, no. 4 Suppl, p. 1390-1395.

Fluhr, H., Carli, S., Deperschmidt, M., Wallwiener, D., Zygmunt, M., and Licht, P., 2008b, Differential effects of human chorionic gonadotropin and decidualization on insulin-like growth factors-I and -II in human endometrial stromal cells: Fertil Steril, v. 90, no. 4 Suppl, p. 1384-1389.

Fournier, T., Guibourdenche, J., Handschuh, K., Tsatsaris, V., Rauwel, B., Davrinche, C., and Evain-Brion, D., 2011, PPARgamma and human trophoblast differentiation: J Reprod Immunol, v. 90, no. 1, p. 41-49.

Fragouli, E., Bianchi, V., Patrizio, P., Obradors, A., Huang, Z., Borini, A., Delhanty, J. D., and Wells, D., 2010, Transcriptomic profiling of human oocytes: association of meiotic aneuploidy and altered oocyte gene expression: Mol Hum Reprod, v. 16, no. 8, p. 570-582.

Franco, J. G., Jr., Baruffi, R. L., Mauri, A. L., Petersen, C. G., Oliveira, J. B., and Vagnini, L., 2008, Significance of large nuclear vacuoles in human spermatozoa: implications for ICSI: Reprod Biomed Online, v. 17, no. 1, p. 42-45.

Franco Jr, J. G., Mauri, A. L., Petersen, C. G., Massaro, F. C., Silva, L. F., Felipe, V., Cavagna, M., Pontes, A., Baruffi, R. L., Oliveira, J. B., and Vagnini, L. D., 2011, Large nuclear vacuoles are indicative of abnormal chromatin packaging in human spermatozoa: Int J Androl.

Frydman, N., Prisant, N., Hesters, L., Frydman, R., Tachdjian, G., Cohen-Bacrie, P., and Fanchin, R., 2008, Adequate ovarian follicular status does not prevent the decrease in pregnancy rates associated with high sperm DNA fragmentation: Fertil Steril, v. 89, no. 1, p. 92-97.

Gandini, L., Lombardo, F., Paoli, D., Caruso, F., Eleuteri, P., Leter, G., Ciriminna, R., Culasso, F., Dondero, F., Lenzi, A., and Spano, M., 2004, Full-term pregnancies achieved with ICSI despite high levels of sperm chromatin damage: Hum Reprod, v. 19, no. 6, p. 1409-1417.

Garolla, A., Fortini, D., Menegazzo, M., De Toni, L., Nicoletti, V., Moretti, A., Selice, R., Engl, B., and Foresta, C., 2008, High-power microscopy for selecting spermatozoa for ICSI by physiological status: Reprod Biomed Online, v. 17, no. 5, p. 610-616.

Gianaroli, L., Magli, M. C., Collodel, G., Moretti, E., Ferraretti, A. P., and Baccetti, B., 2008, Sperm head's birefringence: a new criterion for sperm selection: Fertil Steril, v. 90, no. 1, p. 104-112.

Gianaroli, L., Magli, M. C., Ferraretti, A. P., Crippa, A., Lappi, M., Capitani, S., and Baccetti, B., 2010, Birefringence characteristics in sperm heads allow for the selection of reacted spermatozoa for intracytoplasmic sperm injection: Fertil Steril, v. 93, no. 3, p. 807-813.

Gleicher, N., Weghofer, A., and Barad, D. H., 2010a, Dehydroepiandrosterone (DHEA) reduces embryo aneuploidy: direct evidence from preimplantation genetic screening (PGS): Reprod Biol Endocrinol, v. 8, p. 140.

Gleicher N, Weghofer A, Barad DH., 2010b, Improvement in diminished ovarian reserve after dehydroepiandrosterone supplementation: Reprod Biomed Online, v. 21, no. 3, p. 360-365.

Gleicher, N., and Barad, D. H., 2011, Dehydroepiandrosterone (DHEA) supplementation in diminished ovarian reserve (DOR): Reprod Biol Endocrinol, v. 9, p. 67.

Govin, J., Gaucher, J., Ferro, M., Debernardi, A., Garin, J., Khochbin, S., and Rousseaux, S., 2011, Proteomic strategy for the identification of critical actors in reorganisation of the post-meiotic male genome: Mol Hum Reprod.

Griveau, J. F., and Le Lannou, D., 1997, Reactive oxygen species and human spermatozoa: physiology and pathology: Int J Androl, v. 20, no. 2, p. 61-69.

Gruber, I., Just, A., Birner, M., and Losch, A., 2008, Effect of a woman's smoking status on oocyte, zygote, and day 3 pre-embryo quality in in vitro fertilization and embryo transfer program: Fertil Steril, v. 90, no. 4, p. 1249-1252.

Guerif, F., Le Gouge, A., Giraudeau, B., Poindron, J., Bidault, R., Gasnier, O., and Royere, D., 2007, Limited value of morphological assessment at days 1 and 2 to predict blastocyst development potential: a prospective study based on 4042 embryos: Hum Reprod, v. 22, no. 7, p. 1973-1981.

Guerif, F., Lemseffer, M., Bidault, R., Gasnier, O., Saussereau, M. H., Cadoret, V., Jamet, C., and Royere, D., 2009, Single Day 2 embryo versus blastocyst-stage transfer: a prospective study integrating fresh and frozen embryo transfers: Hum Reprod, v. 24, no. 5, p. 1051-1058.

Guibourdenche, J., Handschuh, K., Tsatsaris, V., Gerbaud, P., Leguy, M. C., Muller, F., Brion, D. E., and Fournier, T., 2010, Hyperglycosylated hCG is a marker of early human trophoblast invasion: J Clin Endocrinol Metab, v. 95, no. 10, p. E240-244.

Hales, B. F., Grenier, L., Lalancette, C., and Robaire, B., 2011, Epigenetic programming: from gametes to blastocyst: Birth Defects Res A Clin Mol Teratol, v. 91, no. 8, p. 652-665.

Hamel, M., Dufort, I., Robert, C., Gravel, C., Leveille, M. C., Leader, A., and Sirard, M. A., 2008, Identification of differentially expressed markers in human follicular cells associated with competent oocytes: Hum Reprod, v. 23, no. 5, p. 1118-1127.

Hammoud, S. S., Nix, D. A., Hammoud, A. O., Gibson, M., Cairns, B. R., and Carrell, D. T., 2011, Genome-wide analysis identifies changes in histone retention and epigenetic modifications at developmental and imprinted gene loci in the sperm of infertile men: Hum Reprod, v. 26, no. 9, p. 2558-2569.

Harris, L. K., 2011, IFPA Gabor Than Award lecture: Transformation of the spiral arteries in human pregnancy: key events in the remodelling timeline: Placenta, v. 32 Suppl 2, p. S154-158.

Hassan-Ali, H., Hisham-Saleh, A., El-Gezeiry, D., Baghdady, I., Ismaeil, I., and Mandelbaum, J., 1998, Perivitelline space granularity: a sign of human menopausal gonadotrophin overdose in intracytoplasmic sperm injection: Hum Reprod, v. 13, no. 12, p. 3425-3430.

Hassold, T., and Hunt, P., 2001, To err (meiotically) is human: the genesis of human aneuploidy: Nat Rev Genet, v. 2, no. 4, p. 280-291.

Haugen, T. B., Egeland, T., and Magnus, O., 2006, Semen parameters in Norwegian fertile men: J Androl, v. 27, no. 1, p. 66-71.

Hazout, A., Menezo, Y., Madelenat, P., Yazbeck, C., Selva, J., and Cohen-Bacrie, P., 2008, [Causes and clinical implications of sperm DNA damages]: Gynecol Obstet Fertil, v. 36, no. 11, p. 1109-1117.

Henkel, R., Kierspel, E., Hajimohammad, M., Stalf, T., Hoogendijk, C., Mehnert, C., Menkveld, R., Schill, W. B., and Kruger, T. F., 2003, DNA fragmentation of spermatozoa and assisted reproduction technology: Reprod Biomed Online, v. 7, no. 4, p. 477-484.

Herrero, L., Martinez, M., and Garcia-Velasco, J. A., 2011, Current status of human oocyte and embryo cryopreservation: Curr Opin Obstet Gynecol, v. 23, no. 4, p. 245-250.

Horcajadas, J. A., Pellicer, A., and Simon, C., 2007, Wide genomic analysis of human endometrial receptivity: new times, new opportunities: Hum Reprod Update, v. 13, no. 1, p. 77-86.

Host, E., Lindenberg, S., and Smidt-Jensen, S., 2000, The role of DNA strand breaks in human spermatozoa used for IVF and ICSI: Acta Obstet Gynecol Scand, v. 79, no. 7, p. 559-563.

Houshdaran, S., Cortessis, V. K., Siegmund, K., Yang, A., Laird, P. W., and Sokol, R. Z., 2007, Widespread epigenetic abnormalities suggest a broad DNA methylation erasure defect in abnormal human sperm: PLoS One, v. 2, no. 12, p. e1289.

Huang, C. C., Lin, D. P., Tsao, H. M., Cheng, T. C., Liu, C. H., and Lee, M. S., 2005, Sperm DNA fragmentation negatively correlates with velocity and fertilization rates but might not affect pregnancy rates: Fertil Steril, v. 84, no. 1, p. 130-140.

Jackson, R. E., Bormann, C. L., Hassun, P. A., Rocha, A. M., Motta, E. L., Serafini, P. C., and Smith, G. D., 2010, Effects of semen storage and separation techniques on sperm DNA fragmentation: Fertil Steril, v. 94, no. 7, p. 2626-2630.

Jarkovska, K., Kupcova Skalnikova, H., Halada, P., Hrabakova, R., Moos, J., Rezabek, K., Gadher, S. J., and Kovarova, H., 2011, Development of ovarian hyperstimulation syndrome: interrogation of key proteins and biological processes in human follicular fluid of women undergoing in vitro fertilisation: Mol Hum Reprod.

Javed, M., Esfandiari, N., and Casper, R. F., 2010, Failed fertilization after clinical intracytoplasmic sperm injection: Reprod Biomed Online, v. 20, no. 1, p. 56-67.

Jonge, C. J. D., and Barratt, C. L. R., 2006, The sperm cell : production, maturation, fertilization, regeneration, Cambridge, UK ; New York, Cambridge University Press, xi, 359 p. p.:

Kajihara, T., Tochigi, H., Uchino, S., Itakura, A., Brosens, J. J., and Ishihara, O., 2011, Differential effects of urinary and recombinant chorionic gonadotropin on oxidative stress responses in decidualizing human endometrial stromal cells: Placenta, v. 32, no. 8, p. 592-597.

Kalinderis, M., Papanikolaou, A., Kalinderi, K., Ioannidou, E., Giannoulis, C., Karagiannis, V., and Tarlatzis, B. C., 2011, Elevated Serum Levels of Interleukin-6, Interleukin-1beta and Human Chorionic Gonadotropin in Pre-eclampsia: Am J Reprod Immunol.

Kane, N., Kelly, R., Saunders, P. T., and Critchley, H. O., 2009, Proliferation of uterine natural killer cells is induced by human chorionic gonadotropin and mediated via the mannose receptor: Endocrinology, v. 150, no. 6, p. 2882-2888.

Kashiwagi, A., DiGirolamo, C. M., Kanda, Y., Niikura, Y., Esmon, C. T., Hansen, T. R., Shioda, T., and Pru, J. K., 2007, The postimplantation embryo differentially regulates endometrial gene expression and decidualization: Endocrinology, v. 148, no. 9, p. 4173-4184.

Kayisli, U. A., Selam, B., Guzeloglu-Kayisli, O., Demir, R., and Arici, A., 2003, Human chorionic gonadotropin contributes to maternal immunotolerance and endometrial apoptosis by regulating Fas-Fas ligand system: J Immunol, v. 171, no. 5, p. 2305-2313.

Khan, N. A., Khan, A., Savelkoul, H. F., and Benner, R., 2001, Inhibition of diabetes in NOD mice by human pregnancy factor: Hum Immunol, v. 62, no. 12, p. 1315-1323.

Khazamipour, N., Noruzinia, M., Fatehmanesh, P., Keyhanee, M., and Pujol, P., 2009, MTHFR promoter hypermethylation in testicular biopsies of patients with non-obstructive azoospermia: the role of epigenetics in male infertility: Hum Reprod, v. 24, no. 9, p. 2361-2364.

Koldehoff, M., Katzorke, T., Wisbrun, N. C., Propping, D., Wohlers, S., Bielfeld, P., Steckel, N. K., Beelen, D. W., and Elmaagacli, A. H., 2011, Modulating impact of human chorionic gonadotropin hormone on the maturation and function of hematopoietic cells: J Leukoc Biol.

Koppers, A. J., De Iuliis, G. N., Finnie, J. M., McLaughlin, E. A., and Aitken, R. J., 2008, Significance of mitochondrial reactive oxygen species in the generation of oxidative stress in spermatozoa: J Clin Endocrinol Metab, v. 93, no. 8, p. 3199-3207.

Kort, H. I., Massey, J. B., Elsner, C. W., Mitchell-Leef, D., Shapiro, D. B., Witt, M. A., and Roudebush, W. E., 2006, Impact of body mass index values on sperm quantity and quality: J Androl, v. 27, no. 3, p. 450-452.

Kurosaka, K., Takahashi, M., Watanabe, N., and Kobayashi, Y., 2003, Silent cleanup of very early apoptotic cells by macrophages: J Immunol, v. 171, no. 9, p. 4672-4679.

Kwak-Kim, J. Y., Gilman-Sachs, A., and Kim, C. E., 2005, T helper 1 and 2 immune responses in relationship to pregnancy, nonpregnancy, recurrent spontaneous abortions and infertility of repeated implantation failures: Chem Immunol Allergy, v. 88, p. 64-79.

Lasiene, K., Vitkus, A., Valanciute, A., and Lasys, V., 2009, Morphological criteria of oocyte quality: Medicina (Kaunas), v. 45, no. 7, p. 509-515.

Leber, A., Teles, A., and Zenclussen, A. C., 2010, Regulatory T cells and their role in pregnancy: Am J Reprod Immunol, v. 63, no. 6, p. 445-459.

Ledee, N., 2005, Uterine receptivity and the two and three dimensions of ultrasound: Ultrasound Obstet Gynecol, v. 26, no. 7, p. 695-698.

Ledee, N., Chaouat, G., Serazin, V., Lombroso, R., Dubanchet, S., Oger, P., Louafi, N., and Ville, Y., 2008a, Endometrial vascularity by three-dimensional power Doppler ultrasound and cytokines: a complementary approach to assess uterine receptivity: J Reprod Immunol, v. 77, no. 1, p. 57-62.

Ledee, N., Lombroso, R., Lombardelli, L., Selva, J., Dubanchet, S., Chaouat, G., Frankenne, F., Foidart, J. M., Maggi, E., Romagnani, S., Ville, Y., and Piccinni, M. P., 2008b, Cytokines and chemokines in follicular fluids and potential of the corresponding embryo: the role of granulocyte colony-stimulating factor: Hum Reprod, v. 23, no. 9, p. 2001-2009.

Ledee, N., Munaut, C., Serazin, V., Perrier d'Hauterive, S., Lombardelli, L., Logiodice, F., Wainer, R., Gridelet, V., Chaouat, G., Frankenne, F., Foidart, J. M., and Piccinni, M. P., 2010, Performance evaluation of microbead and ELISA assays for follicular G-CSF: a non-invasive biomarker of oocyte developmental competence for embryo implantation: J Reprod Immunol, v. 86, no. 2, p. 126-132.

Ledee, N., Frydman, R., Osipova, A., Taieb, J., Gallot, V., Lombardelli, L., Logiodice, F., Petitbarat, M., Fanchin, R., Chaouat, G., Achour-Frydman, N., and Piccinni, M. P., 2011a, Levels of follicular G-CSF and interleukin-15 appear as noninvasive biomarkers of subsequent successful birth in modified natural in vitro fertilization/intracytoplasmic sperm injection cycles: Fertil Steril.

Ledee, N., Petitbarat, M., Rahmati, M., Dubanchet, S., Chaouat, G., Sandra, O., Perrier-d'Hauterive, S., Munaut, C., and Foidart, J. M., 2011b, New pre-conception immune biomarkers for clinical practice: interleukin-18, interleukin-15 and TWEAK on the endometrial side, G-CSF on the follicular side: J Reprod Immunol, v. 88, no. 2, p. 118-123.

Lédée, N., Munaut, C., Aubert, J., Sérazin, V., Rahmati, M., Chaouat, G., Sandra, O., and Foidart, J. M., 2011c, Specific and extensive endometrial deregulation is present before conception in IVF/ICSI repeated implantation failures (IF) or recurrent miscarriages.: The Journal of Pathology

Lee, T. H., Liu, C. H., Shih, Y. T., Tsao, H. M., Huang, C. C., Chen, H. H., and Lee, M. S., 2010, Magnetic-activated cell sorting for sperm preparation reduces spermatozoa with apoptotic markers and improves the acrosome reaction in couples with unexplained infertility: Hum Reprod, v. 25, no. 4, p. 839-846.

Lessey, B. A., 2011, Assessment of endometrial receptivity: Fertil Steril, v. 96, no. 3, p. 522-529.

Levi, A. J., Drews, M. R., Bergh, P. A., Miller, B. T., and Scott, R. T., Jr., 2001, Controlled ovarian hyperstimulation does not adversely affect endometrial receptivity in in vitro fertilization cycles: Fertil Steril, v. 76, no. 4, p. 670-674.

Li, L., Ferin, M., Sauer, M. V., and Lobo, R. A., 2011, Dehydroepiandrosterone in follicular fluid is produced locally, and levels correlate negatively with in vitro fertilization outcomes: Fertil Steril, v. 95, no. 5, p. 1830-1832.

Licht, P., Russu, V., Lehmeyer, S., Moll, J., Siebzehnrubl, E., and Wildt, L., 2002, Intrauterine microdialysis reveals cycle-dependent regulation of endometrial insulin-like growth factor binding protein-1 secretion by human chorionic gonadotropin: Fertil Steril, v. 78, no. 2, p. 252-258.

Lin, M. H., Kuo-Kuang Lee, R., Li, S. H., Lu, C. H., Sun, F. J., and Hwu, Y. M., 2008, Sperm chromatin structure assay parameters are not related to fertilization rates, embryo quality, and pregnancy rates in in vitro fertilization and intracytoplasmic sperm injection, but might be related to spontaneous abortion rates: Fertil Steril, v. 90, no. 2, p. 352-359.

Liu, J., Nagy, Z., Joris, H., Tournaye, H., Smitz, J., Camus, M., Devroey, P., and Van Steirteghem, A., 1995, Analysis of 76 total fertilization failure cycles out of 2732 intracytoplasmic sperm injection cycles: Hum Reprod, v. 10, no. 10, p. 2630-2636.

Loft, S., Kold-Jensen, T., Hjollund, N. H., Giwercman, A., Gyllemborg, J., Ernst, E., Olsen, J., Scheike, T., Poulsen, H. E., and Bonde, J. P., 2003, Oxidative DNA damage in human sperm influences time to pregnancy: Hum Reprod, v. 18, no. 6, p. 1265-1272.

Lorquet, S., Pequeux, C., Munaut, C., and Foidart, J. M., 2010, Aetiology and physiopathology of preeclampsia and related forms: Acta Clin Belg, v. 65, no. 4, p. 237-241.

Lucifero, D., Chaillet, J. R., and Trasler, J. M., 2004, Potential significance of genomic imprinting defects for reproduction and assisted reproductive technology: Hum Reprod Update, v. 10, no. 1, p. 3-18.

Lunghi, L., Ferretti, M. E., Medici, S., Biondi, C., and Vesce, F., 2007, Control of human trophoblast function: Reprod Biol Endocrinol, v. 5, p. 6.

Macklon, N. S., Geraedts, J. P., and Fauser, B. C., 2002, Conception to ongoing pregnancy: the 'black box' of early pregnancy loss: Hum Reprod Update, v. 8, no. 4, p. 333-343.

Maheshwari, A., Stofberg, L., and Bhattacharya, S., 2007, Effect of overweight and obesity on assisted reproductive technology--a systematic review: Hum Reprod Update, v. 13, no. 5, p. 433-444.

Mahfouz, R., Sharma, R., Lackner, J., Aziz, N., and Agarwal, A., 2009, Evaluation of chemiluminescence and flow cytometry as tools in assessing production of hydrogen peroxide and superoxide anion in human spermatozoa: Fertil Steril, v. 92, no. 2, p. 819-827.

Marchesi, D. E., Biederman, H., Ferrara, S., Hershlag, A., and Feng, H. L., 2010, The effect of semen processing on sperm DNA integrity: comparison of two techniques using the novel Toluidine Blue Assay: Eur J Obstet Gynecol Reprod Biol, v. 151, no. 2, p. 176-180.

Mayes, M. A., and Sirard, M. A., 2001, The influence of cumulus-oocyte complex morphology and meiotic inhibitors on the kinetics of nuclear maturation in cattle: Theriogenology, v. 55, no. 4, p. 911-922.

Meseguer, M., Santiso, R., Garrido, N., Garcia-Herrero, S., Remohi, J., and Fernandez, J. L., 2011, Effect of sperm DNA fragmentation on pregnancy outcome depends on oocyte quality: Fertil Steril, v. 95, no. 1, p. 124-128.

Miao, Y. L., Kikuchi, K., Sun, Q. Y., and Schatten, H., 2009, Oocyte aging: cellular and molecular changes, developmental potential and reversal possibility: Hum Reprod Update, v. 15, no. 5, p. 573-585.

Micinski, P., Pawlicki, K., Wielgus, E., Bochenek, M., and Tworkowska, I., 2009, The sperm chromatin structure assay (SCSA) as prognostic factor in IVF/ICSI program: Reprod Biol, v. 9, no. 1, p. 65-70.

Mieusset, R., Bujan, L., Mondinat, C., Mansat, A., Pontonnier, F., and Grandjean, H., 1987, Association of scrotal hyperthermia with impaired spermatogenesis in infertile men: Fertil Steril, v. 48, no. 6, p. 1006-1011.

Miller, D., Brinkworth, M., and Iles, D., 2010, Paternal DNA packaging in spermatozoa: more than the sum of its parts? DNA, histones, protamines and epigenetics: Reproduction, v. 139, no. 2, p. 287-301.

Monqaut, A. L., Zavaleta, C., Lopez, G., Lafuente, R., and Brassesco, M., 2011, Use of high-magnification microscopy for the assessment of sperm recovered after two different sperm processing methods: Fertil Steril, v. 95, no. 1, p. 277-280.

Moon, J. H., Hyun, C. S., Lee, S. W., Son, W. Y., Yoon, S. H., and Lim, J. H., 2003, Visualization of the metaphase II meiotic spindle in living human oocytes using the Polscope enables the prediction of embryonic developmental competence after ICSI: Hum Reprod, v. 18, no. 4, p. 817-820.

Munro, S. K., Farquhar, C. M., Mitchell, M. D., and Ponnampalam, A. P., 2010, Epigenetic regulation of endometrium during the menstrual cycle: Mol Hum Reprod, v. 16, no. 5, p. 297-310.

Muriel, L., Meseguer, M., Fernandez, J. L., Alvarez, J., Remohi, J., Pellicer, A., and Garrido, N., 2006, Value of the sperm chromatin dispersion test in predicting pregnancy outcome in intrauterine insemination: a blind prospective study: Hum Reprod, v. 21, no. 3, p. 738-744.

Nagano, M., Katagiri, S., and Takahashi, Y., 2006, Relationship between bovine oocyte morphology and in vitro developmental potential: Zygote, v. 14, no. 1, p. 53-61.

Nagy, Z. P., Joris, H., Liu, J., Staessen, C., Devroey, P., and Van Steirteghem, A. C., 1993, Intracytoplasmic single sperm injection of 1-day-old unfertilized human oocytes: Hum Reprod, v. 8, no. 12, p. 2180-2184.

Navarro, P. A., de Araujo, M. M., de Araujo, C. M., Rocha, M., dos Reis, R., and Martins, W., 2009, Relationship between first polar body morphology before intracytoplasmic sperm injection and fertilization rate, cleavage rate, and embryo quality: Int J Gynaecol Obstet, v. 104, no. 3, p. 226-229.

Nevers, T., Kalkunte, S., and Sharma, S., 2011, Uterine Regulatory T cells, IL-10 and hypertension: Am J Reprod Immunol, v. 66 Suppl 1, p. 88-92.

Norris, W., Nevers, T., Sharma, S., and Kalkunte, S., 2011, Review: hCG, preeclampsia and regulatory T cells: Placenta, v. 32 Suppl 2, p. S182-185.

Noyes, N., Liu, H. C., Sultan, K., Schattman, G., and Rosenwaks, Z., 1995, Endometrial thickness appears to be a significant factor in embryo implantation in in-vitro fertilization: Hum Reprod, v. 10, no. 4, p. 919-922.

Oliveira, J. B., Massaro, F. C., Baruffi, R. L., Mauri, A. L., Petersen, C. G., Silva, L. F., Vagnini, L. D., and Franco, J. G., Jr., 2010, Correlation between semen analysis by motile sperm organelle morphology examination and sperm DNA damage: Fertil Steril, v. 94, no. 5, p. 1937-1940.

Paiva, P., Hannan, N. J., Hincks, C., Meehan, K. L., Pruysers, E., Dimitriadis, E., and Salamonsen, L. A., 2011, Human chorionic gonadotrophin regulates FGF2 and other cytokines produced by human endometrial epithelial cells, providing a mechanism for enhancing endometrial receptivity: Hum Reprod, v. 26, no. 5, p. 1153-1162.

Palermo, G., Joris, H., Devroey, P., and Van Steirteghem, A. C., 1992, Pregnancies after intracytoplasmic injection of single spermatozoon into an oocyte: Lancet, v. 340, no. 8810, p. 17-18.

Parmegiani, L., Cognigni, G. E., Bernardi, S., Troilo, E., Ciampaglia, W., and Filicori, M., 2010a, "Physiologic ICSI": hyaluronic acid (HA) favors selection of spermatozoa without DNA fragmentation and with normal nucleus, resulting in improvement of embryo quality: Fertil Steril, v. 93, no. 2, p. 598-604.

Parmegiani, L., Cognigni, G. E., Ciampaglia, W., Pocognoli, P., Marchi, F., and Filicori, M., 2010b, Efficiency of hyaluronic acid (HA) sperm selection: J Assist Reprod Genet, v. 27, no. 1, p. 13-16.

Patrizio, P., Fragouli, E., Bianchi, V., Borini, A., and Wells, D., 2007, Molecular methods for selection of the ideal oocyte: Reprod Biomed Online, v. 15, no. 3, p. 346-353.

Paul, C., Melton, D. W., and Saunders, P. T., 2008, Do heat stress and deficits in DNA repair pathways have a negative impact on male fertility?: Mol Hum Reprod, v. 14, no. 1, p. 1-8.

Paulson, R. J., 2011, Hormonal induction of endometrial receptivity: Fertil Steril, v. 96, no. 3, p. 530-535.

Pellestor, F., Anahory, T., and Hamamah, S., 2005, The chromosomal analysis of human oocytes. An overview of established procedures: Hum Reprod Update, v. 11, no. 1, p. 15-32.

Perdrix, A., Travers, A., Chelli, M. H., Escalier, D., Do Rego, J. L., Milazzo, J. P., Mousset-Simeon, N., Mace, B., and Rives, N., 2011, Assessment of acrosome and nuclear abnormalities in human spermatozoa with large vacuoles: Hum Reprod, v. 26, no. 1, p. 47-58.

Perrier D'hauterive, S., Charlet-Renard, C., Goffin, F., Foidart, M., and Geenen, V., 2002, [The implantation window]: J Gynecol Obstet Biol Reprod (Paris), v. 31, no. 5, p. 440-455.

Perrier d'Hauterive, S., Charlet-Renard, C., Berndt, S., Dubois, M., Munaut, C., Goffin, F., Hagelstein, M. T., Noel, A., Hazout, A., Foidart, J. M., and Geenen, V., 2004, Human chorionic gonadotropin and growth factors at the embryonic-endometrial interface control leukemia inhibitory factor (LIF) and interleukin 6 (IL-6) secretion by human endometrial epithelium: Hum Reprod, v. 19, no. 11, p. 2633-2643.

Perry, M. J., 2008, Effects of environmental and occupational pesticide exposure on human sperm: a systematic review: Hum Reprod Update, v. 14, no. 3, p. 233-242.

Petitbarat, M., Serazin, V., Dubanchet, S., Wayner, R., de Mazancourt, P., Chaouat, G., and Ledee, N., 2010, Tumor necrosis factor-like weak inducer of apoptosis (TWEAK)/fibroblast growth factor inducible-14 might regulate the effects of interleukin 18 and 15 in the human endometrium: Fertil Steril, v. 94, no. 3, p. 1141-1143.

Plaisier, M., 2011, Decidualisation and angiogenesis: Best Pract Res Clin Obstet Gynaecol, v. 25, no. 3, p. 259-271.

Polak de Fried, E., and Denaday, F., 2010, Single and twin ongoing pregnancies in two cases of previous ART failure after ICSI performed with sperm sorted using annexin V microbeads: Fertil Steril, v. 94, no. 1, p. 351 e315-358.

Poplinski, A., Tuttelmann, F., Kanber, D., Horsthemke, B., and Gromoll, J., 2010, Idiopathic male infertility is strongly associated with aberrant methylation of MEST and IGF2/H19 ICR1: Int J Androl, v. 33, no. 4, p. 642-649.

Rajender, S., Avery, K., and Agarwal, A., 2011, Epigenetics, spermatogenesis and male infertility: Mutat Res, v. 727, no. 3, p. 62-71.

Ramu, S., Acacio, B., Adamowicz, M., Parrett, S., and Jeyendran, R. S., 2011, Human chorionic gonadotropin from day 2 spent embryo culture media and its relationship to embryo development: Fertil Steril, v. 96, no. 3, p. 615-617.

Rao, C. V., 2006, Physiological and pathological relevance of human uterine LH/hCG receptors: J Soc Gynecol Investig, v. 13, no. 2, p. 77-78.

Rashid, N. A., Lalitkumar, S., Lalitkumar, P. G., and Gemzell-Danielsson, K., 2011, Endometrial receptivity and human embryo implantation: Am J Reprod Immunol, v. 66 Suppl 1, p. 23-30.

Rawe, V. Y., Boudri, H. U., Alvarez Sedo, C., Carro, M., Papier, S., and Nodar, F., 2010, Healthy baby born after reduction of sperm DNA fragmentation using cell sorting before ICSI: Reprod Biomed Online, v. 20, no. 3, p. 320-323.

Reich, A., Klatsky, P., Carson, S., and Wessel, G., 2011, The transcriptome of a human polar body accurately reflects its sibling oocyte: J Biol Chem.

Rienzi, L., Martinez, F., Ubaldi, F., Minasi, M. G., Iacobelli, M., Tesarik, J., and Greco, E., 2004, Polscope analysis of meiotic spindle changes in living metaphase II human oocytes during the freezing and thawing procedures: Hum Reprod, v. 19, no. 3, p. 655-659.

Rienzi, L., Ubaldi, F. M., Iacobelli, M., Minasi, M. G., Romano, S., Ferrero, S., Sapienza, F., Baroni, E., Litwicka, K., and Greco, E., 2008, Significance of metaphase II human oocyte morphology on ICSI outcome: Fertil Steril, v. 90, no. 5, p. 1692-1700.

Rienzi, L., Vajta, G., and Ubaldi, F., 2011, Predictive value of oocyte morphology in human IVF: a systematic review of the literature: Hum Reprod Update, v. 17, no. 1, p. 34-45.

Robertson, S. A., Chin, P. Y., Glynn, D. J., and Thompson, J. G., 2011, Peri-conceptual cytokines--setting the trajectory for embryo implantation, pregnancy and beyond: Am J Reprod Immunol, v. 66 Suppl 1, p. 2-10.

Robker, R. L., Akison, L. K., Bennett, B. D., Thrupp, P. N., Chura, L. R., Russell, D. L., Lane, M., and Norman, R. J., 2009, Obese women exhibit differences in ovarian metabolites, hormones, and gene expression compared with moderate-weight women: J Clin Endocrinol Metab, v. 94, no. 5, p. 1533-1540.

Rockett, J. C., Mapp, F. L., Garges, J. B., Luft, J. C., Mori, C., and Dix, D. J., 2001, Effects of hyperthermia on spermatogenesis, apoptosis, gene expression, and fertility in adult male mice: Biol Reprod, v. 65, no. 1, p. 229-239.

Rousseaux, S., Reynoird, N., Escoffier, E., Thevenon, J., Caron, C., and Khochbin, S., 2008, Epigenetic reprogramming of the male genome during gametogenesis and in the zygote: Reprod Biomed Online, v. 16, no. 4, p. 492-503.

Rousseaux, S., Boussouar, F., Gaucher, J., Reynoird, N., Montellier, E., Curtet, S., Vitte, A. L., and Khochbin, S., 2011, Molecular models for post-meiotic male genome reprogramming: Syst Biol Reprod Med, v. 57, no. 1-2, p. 50-53.

Royere, D., Feuerstein, P., Cadoret, V., Puard, V., Uzbekova, S., Dalbies-Tran, R., Teusan, R., Houlgatte, R., Labas, V., and Guerif, F., 2009, [Non invasive assessment of embryo quality: proteomics, metabolomics and oocyte-cumulus dialogue]: Gynecol Obstet Fertil, v. 37, no. 11-12, p. 917-920.

Rudak, E., Jacobs, P. A., and Yanagimachi, R., 1978, Direct analysis of the chromosome constitution of human spermatozoa: Nature, v. 274, no. 5674, p. 911-913.

Ruder, E. H., Hartman, T. J., Blumberg, J., and Goldman, M. B., 2008, Oxidative stress and antioxidants: exposure and impact on female fertility: Hum Reprod Update, v. 14, no. 4, p. 345-357.

Ruder, E. H., Hartman, T. J., and Goldman, M. B., 2009, Impact of oxidative stress on female fertility: Curr Opin Obstet Gynecol, v. 21, no. 3, p. 219-222.

Rutella, S., Zavala, F., Danese, S., Kared, H., and Leone, G., 2005, Granulocyte colony-stimulating factor: a novel mediator of T cell tolerance: J Immunol, v. 175, no. 11, p. 7085-7091.

Sakkas, D., Seli, E., Manicardi, G. C., Nijs, M., Ombelet, W., and Bizzaro, D., 2004, The presence of abnormal spermatozoa in the ejaculate: did apoptosis fail?: Hum Fertil (Camb), v. 7, no. 2, p. 99-103.

Sakkas, D., and Alvarez, J. G., 2010, Sperm DNA fragmentation: mechanisms of origin, impact on reproductive outcome, and analysis: Fertil Steril, v. 93, no. 4, p. 1027-1036.

Saleh, R. A., Agarwal, A., Sharma, R. K., Nelson, D. R., and Thomas, A. J., Jr., 2002, Effect of cigarette smoking on levels of seminal oxidative stress in infertile men: a prospective study: Fertil Steril, v. 78, no. 3, p. 491-499.

Salmassi, A., Schmutzler, A. G., Huang, L., Hedderich, J., Jonat, W., and Mettler, L., 2004, Detection of granulocyte colony-stimulating factor and its receptor in human follicular luteinized granulosa cells: Fertil Steril, v. 81 Suppl 1, p. 786-791.

Santos, M. A., Kuijk, E. W., and Macklon, N. S., 2010, The impact of ovarian stimulation for IVF on the developing embryo: Reproduction, v. 139, no. 1, p. 23-34.

Saragusty, J., and Arav, A., 2011, Current progress in oocyte and embryo cryopreservation by slow freezing and vitrification: Reproduction, v. 141, no. 1, p. 1-19.

Sartorius, G. A., and Nieschlag, E., 2010, Paternal age and reproduction: Hum Reprod Update, v. 16, no. 1, p. 65-79.

Saunders, C. M., Larman, M. G., Parrington, J., Cox, L. J., Royse, J., Blayney, L. M., Swann, K., and Lai, F. A., 2002, PLC zeta: a sperm-specific trigger of Ca(2+) oscillations in eggs and embryo development: Development, v. 129, no. 15, p. 3533-3544.

Scarpellini, F., and Sbracia, M., 2009, Use of granulocyte colony-stimulating factor for the treatment of unexplained recurrent miscarriage: a randomised controlled trial: Hum Reprod, v. 24, no. 11, p. 2703-2708.

Schatten, H., and Sun, Q. Y., 2011, Centrosome dynamics during mammalian oocyte maturation with a focus on meiotic spindle formation: Mol Reprod Dev.

Schultz, R. M., 2002, The molecular foundations of the maternal to zygotic transition in the preimplantation embryo: Hum Reprod Update, v. 8, no. 4, p. 323-331.

Schultz RM., 2005, From egg to embryo: a peripatetic journey: Reproduction, v. 130, no. 6, p. 825-828.

Schumacher, A., Brachwitz, N., Sohr, S., Engeland, K., Langwisch, S., Dolaptchieva, M., Alexander, T., Taran, A., Malfertheiner, S. F., Costa, S. D., Zimmermann, G., Nitschke, C., Volk, H. D., Alexander, H., Gunzer, M., and Zenclussen, A. C., 2009, Human chorionic gonadotropin attracts regulatory T cells into the fetal-maternal interface during early human pregnancy: J Immunol, v. 182, no. 9, p. 5488-5497.

Sepaniak, S., Forges, T., and Monnier-Barbarino, P., 2006, [Cigarette smoking and fertility in women and men]: Gynecol Obstet Fertil, v. 34, no. 10, p. 945-949.

Setti, A. S., Figueira, R. C., Braga, D. P., Colturato, S. S., Iaconelli, A., Jr., and Borges, E., Jr., 2011, Relationship between oocyte abnormal morphology and intracytoplasmic sperm injection outcomes: a meta-analysis: Eur J Obstet Gynecol Reprod Biol.

Shapiro, B. S., Daneshmand, S. T., Garner, F. C., Aguirre, M., Hudson, C., and Thomas, S., 2011a, Evidence of impaired endometrial receptivity after ovarian stimulation for in vitro fertilization: a prospective randomized trial comparing fresh and frozen-thawed embryo transfer in normal responders: Fertil Steril, v. 96, no. 2, p. 344-348.

Shapiro BS, Daneshmand ST, Garner FC, Aguirre M, Hudson C, Thomas S., 2011b, Evidence of impaired endometrial receptivity after ovarian stimulation for in vitro fertilization: a prospective randomized trial comparing fresh and frozen-thawed embryo transfers in high responders: Fertil Steril, v. 96, no. 2, p. 516-518.

Shaw, J. L., and Horne, A. W., 2011, The paracrinology of tubal ectopic pregnancy: Mol Cell Endocrinol.

Shen, H. M., Chia, S. E., and Ong, C. N., 1999, Evaluation of oxidative DNA damage in human sperm and its association with male infertility: J Androl, v. 20, no. 6, p. 718-723.

Shen, Y., Stalf, T., Mehnert, C., De Santis, L., Cino, I., Tinneberg, H. R., and Eichenlaub-Ritter, U., 2006, Light retardance by human oocyte spindle is positively related to pronuclear score after ICSI: Reprod Biomed Online, v. 12, no. 6, p. 737-751.

Sherwin, J. R., Sharkey, A. M., Cameo, P., Mavrogianis, P. M., Catalano, R. D., Edassery, S., and Fazleabas, A. T., 2007, Identification of novel genes regulated by chorionic gonadotropin in baboon endometrium during the window of implantation: Endocrinology, v. 148, no. 2, p. 618-626.

Shi, Q., Spriggs, E., Field, L. L., Ko, E., Barclay, L., and Martin, R. H., 2001, Single sperm typing demonstrates that reduced recombination is associated with the production of aneuploid 24,XY human sperm: Am J Med Genet, v. 99, no. 1, p. 34-38.

Simon, L., Brunborg, G., Stevenson, M., Lutton, D., McManus, J., and Lewis, S. E., 2010, Clinical significance of sperm DNA damage in assisted reproduction outcome: Hum Reprod, v. 25, no. 7, p. 1594-1608.

Singh, N. P., Muller, C. H., and Berger, R. E., 2003, Effects of age on DNA double-strand breaks and apoptosis in human sperm: Fertil Steril, v. 80, no. 6, p. 1420-1430.

Singh, M., Chaudhry, P., and Asselin, E., 2011, Bridging endometrial receptivity and implantation: network of hormones, cytokines, and growth factors: J Endocrinol, v. 210, no. 1, p. 5-14.

Smit, M., Romijn, J. C., Wildhagen, M. F., Veldhoven, J. L., Weber, R. F., and Dohle, G. R., 2010, Decreased sperm DNA fragmentation after surgical varicocelectomy is associated with increased pregnancy rate: J Urol, v. 183, no. 1, p. 270-274.

Smith, R., Kaune, H., Parodi, D., Madariaga, M., Rios, R., Morales, I., and Castro, A., 2006, Increased sperm DNA damage in patients with varicocele: relationship with seminal oxidative stress: Hum Reprod, v. 21, no. 4, p. 986-993.

Smith, R., Kaune, H., Parodi, D., Madariaga, M., Morales, I., Rios, R., and Castro, A., 2007, [Extent of sperm DNA damage in spermatozoa from men examined for infertility. Relationship with oxidative stress]: Rev Med Chil, v. 135, no. 3, p. 279-286.

Soares, S. R., and Melo, M. A., 2008, Cigarette smoking and reproductive function: Curr Opin Obstet Gynecol, v. 20, no. 3, p. 281-291.

Somers, C. M., and Cooper, D. N., 2009, Air pollution and mutations in the germline: are humans at risk?: Hum Genet, v. 125, no. 2, p. 119-130.

Sonmezer, M., Ozmen, B., Cil, A. P., Ozkavukcu, S., Tasci, T., Olmus, H., and Atabekoglu, C. S., 2009, Dehydroepiandrosterone supplementation improves ovarian response and cycle outcome in poor responders: Reprod Biomed Online, v. 19, no. 4, p. 508-513.

Spano, M., Bonde, J. P., Hjollund, H. I., Kolstad, H. A., Cordelli, E., and Leter, G., 2000, Sperm chromatin damage impairs human fertility. The Danish First Pregnancy Planner Study Team: Fertil Steril, v. 73, no. 1, p. 43-50.

Speyer, B. E., Pizzey, A. R., Ranieri, M., Joshi, R., Delhanty, J. D., and Serhal, P., 2010, Fall in implantation rates following ICSI with sperm with high DNA fragmentation: Hum Reprod, v. 25, no. 7, p. 1609-1618.

Stewart, C. L., 1994, The role of leukemia inhibitory factor (LIF) and other cytokines in regulating implantation in mammals: Ann N Y Acad Sci, v. 734, p. 157-165.

Styne-Gross, A., Elkind-Hirsch, K., and Scott, R. T., Jr., 2005, Obesity does not impact implantation rates or pregnancy outcome in women attempting conception through oocyte donation: Fertil Steril, v. 83, no. 6, p. 1629-1634.

Sugihara, K., Kabir-Salmani, M., Byrne, J., Wolf, D. P., Lessey, B., Iwashita, M., Aoki, D., Nakayama, J., and Fukuda, M. N., 2008, Induction of trophinin in human endometrial surface epithelia by CGbeta and IL-1beta: FEBS Lett, v. 582, no. 2, p. 197-202.

Talebi, A. R., Vahidi, S., Aflatoonian, A., Ghasemi, N., Ghasemzadeh, J., Firoozabadi, R. D., and Moein, M. R., 2011, Cytochemical evaluation of sperm chromatin and DNA integrity in couples with unexplained recurrent spontaneous abortions: Andrologia.

Tarozzi, N., Nadalini, M., Stronati, A., Bizzaro, D., Dal Prato, L., Coticchio, G., and Borini, A., 2009, Anomalies in sperm chromatin packaging: implications for assisted reproduction techniques: Reprod Biomed Online, v. 18, no. 4, p. 486-495.

Tavalaee, M., Razavi, S., and Nasr-Esfahani, M. H., 2009, Influence of sperm chromatin anomalies on assisted reproductive technology outcome: Fertil Steril, v. 91, no. 4, p. 1119-1126.

Teklenburg, G., Salker, M., Heijnen, C., Macklon, N. S., and Brosens, J. J., 2010a, The molecular basis of recurrent pregnancy loss: impaired natural embryo selection: Mol Hum Reprod, v. 16, no. 12, p. 886-895.

Teklenburg, G., Salker, M., Molokhia, M., Lavery, S., Trew, G., Aojanepong, T., Mardon, H. J., Lokugamage, A. U., Rai, R., Landles, C., Roelen, B. A., Quenby, S., Kuijk, E. W., Kavelaars, A., Heijnen, C. J., Regan, L., Brosens, J. J., and Macklon, N. S., 2010b, Natural selection of human embryos: decidualizing endometrial stromal cells serve as sensors of embryo quality upon implantation: PLoS One, v. 5, no. 4, p. e10258.

Tesarik, J., Greco, E., and Mendoza, C., 2004, Late, but not early, paternal effect on human embryo development is related to sperm DNA fragmentation: Hum Reprod, v. 19, no. 3, p. 611-615.

Thomson, L. K., Fleming, S. D., Aitken, R. J., De Iuliis, G. N., Zieschang, J. A., and Clark, A. M., 2009, Cryopreservation-induced human sperm DNA damage is predominantly mediated by oxidative stress rather than apoptosis: Hum Reprod, v. 24, no. 9, p. 2061-2070.

Tremellen, K., 2008, Oxidative stress and male infertility--a clinical perspective: Hum Reprod Update, v. 14, no. 3, p. 243-258.

Tsampalas, M., Gridelet, V., Berndt, S., Foidart, J. M., Geenen, V., and Perrier d'Hauterive, S., 2010, Human chorionic gonadotropin: a hormone with immunological and angiogenic properties: J Reprod Immunol, v. 85, no. 1, p. 93-98.

Tuckerman, E., Mariee, N., Prakash, A., Li, T. C., and Laird, S., 2010, Uterine natural killer cells in peri-implantation endometrium from women with repeated implantation failure after IVF: J Reprod Immunol, v. 87, no. 1-2, p. 60-66.

Umapathy, E., Simbini, T., Chipata, T., and Mbizvo, M., 2001, Sperm characteristics and accessory sex gland functions in HIV-infected men: Arch Androl, v. 46, no. 2, p. 153-158.

van Montfoort, A. P., Geraedts, J. P., Dumoulin, J. C., Stassen, A. P., Evers, J. L., and Ayoubi, T. A., 2008, Differential gene expression in cumulus cells as a prognostic indicator of embryo viability: a microarray analysis: Mol Hum Reprod, v. 14, no. 3, p. 157-168.

Van Steirteghem, A. C., Nagy, Z., Joris, H., Liu, J., Staessen, C., Smitz, J., Wisanto, A., and Devroey, P., 1993, High fertilization and implantation rates after intracytoplasmic sperm injection: Hum Reprod, v. 8, no. 7, p. 1061-1066.

Venkatesh, S., Singh, A., Shamsi, M. B., Thilagavathi, J., Kumar, R., D, K. M., and Dada, R., 2011, Clinical significance of sperm DNA damage threshold value in the assessment of male infertility: Reprod Sci, v. 18, no. 10, p. 1005-1013.

Visioli, F., and Hagen, T. M., 2011, Antioxidants to enhance fertility: Role of eNOS and potential benefits: Pharmacol Res, v. 64, no. 5, p. 431-437.

Wang, W. H., Meng, L., Hackett, R. J., Odenbourg, R., and Keefe, D. L., 2001, The spindle observation and its relationship with fertilization after intracytoplasmic sperm injection in living human oocytes: Fertil Steril, v. 75, no. 2, p. 348-353.

Ward, W. S., 2010, Function of sperm chromatin structural elements in fertilization and development: Mol Hum Reprod, v. 16, no. 1, p. 30-36.

Warriach, H. M., and Chohan, K. R., 2004, Thickness of cumulus cell layer is a significant factor in meiotic competence of buffalo oocytes: J Vet Sci, v. 5, no. 3, p. 247-251.

Wattanakumtornkul, S., Damario, M. A., Stevens Hall, S. A., Thornhill, A. R., and Tummon, I. S., 2003, Body mass index and uterine receptivity in the oocyte donation model: Fertil Steril, v. 80, no. 2, p. 336-340.

Weng, S. L., Taylor, S. L., Morshedi, M., Schuffner, A., Duran, E. H., Beebe, S., and Oehninger, S., 2002, Caspase activity and apoptotic markers in ejaculated human sperm: Mol Hum Reprod, v. 8, no. 11, p. 984-991.

Wilcox, A. J., Weinberg, C. R., and Baird, D. D., 1998, Post-ovulatory ageing of the human oocyte and embryo failure: Hum Reprod, v. 13, no. 2, p. 394-397.

World Health Organization., 2010, WHO laboratory manual for the examination and processing of human semen, Geneva, World Health Organization, xiv, 271 p. p.:

Xia, P., 1997, Intracytoplasmic sperm injection: correlation of oocyte grade based on polar body, perivitelline space and cytoplasmic inclusions with fertilization rate and embryo quality: Hum Reprod, v. 12, no. 8, p. 1750-1755.

Xiong, H., Zhou, C., and Qi, G., 2010, Proportional changes of CD4+CD25+Foxp3+ regulatory T cells in maternal peripheral blood during pregnancy and labor at term and preterm: Clin Invest Med, v. 33, no. 6, p. E422.

Yannaki, E., Athanasiou, E., Xagorari, A., Constantinou, V., Batsis, I., Kaloyannidis, P., Proya, E., Anagnostopoulos, A., and Fassas, A., 2005, G-CSF-primed hematopoietic stem cells or G-CSF per se accelerate recovery and improve survival after liver injury, predominantly by promoting endogenous repair programs: Exp Hematol, v. 33, no. 1, p. 108-119.

Yerushalmi, G. M., Maman, E., Yung, Y., Kedem, A., and Hourvitz, A., 2011, Molecular characterization of the human ovulatory cascade-lesson from the IVF/IVM model: J Assist Reprod Genet, v. 28, no. 6, p. 509-515.

Zhang, J., Chen, Z., Smith, G. N., and Croy, B. A., 2011, Natural killer cell-triggered vascular transformation: maternal care before birth?: Cell Mol Immunol, v. 8, no. 1, p. 1-11.

Zini, A., Meriano, J., Kader, K., Jarvi, K., Laskin, C. A., and Cadesky, K., 2005, Potential adverse effect of sperm DNA damage on embryo quality after ICSI: Hum Reprod, v. 20, no. 12, p. 3476-3480.

Zini, A., and Sigman, M., 2009, Are tests of sperm DNA damage clinically useful? Pros and cons: J Androl, v. 30, no. 3, p. 219-229.

Zini, A., 2011a, Are sperm chromatin and DNA defects relevant in the clinic?: Syst Biol Reprod Med, v. 57, no. 1-2, p. 78-85.

Zini, A., 2011b, Sperm chromatin : biological and clinical applications in male infertility and assisted reproduction, New York, Springer.

6

The Role of Macrophages in the Placenta

Grace Pinhal-Enfield, Nagaswami S. Vasan and Samuel Joseph Leibovich
UMDNJ – New Jersey Medical School,
Department of Cell Biology and Molecular Medicine,
Newark, New Jersey,
USA

1. Introduction

Placenta formation occurs through a complex and coordinated effort between the fetus's extraembryonic tissues and the gravid endometrial tissues. Many macrophages are present in the placenta throughout pregnancy and have been detected as early as day 10 of pregnancy (Chang et al., 1993). Placental macrophages include Hofbauer cells of the fetal chorionic villi and decidual macrophages of the maternal decidua basalis (Figure 1) (Bulmer & Johnson, 1984). Functions of placental macrophages include the production of substances that regulate local immune reactions (such as factors that regulate maternal immunological tolerance and fetal protection) and that promote angiogenesis in the placenta during its development (Mues et al., 1989; Sevala et al., 2007). Although they represent a significant presence on both the maternal and fetal sides of the placenta, placental macrophage functions have not been completely elucidated and still remain a significantly studied area of investigation.

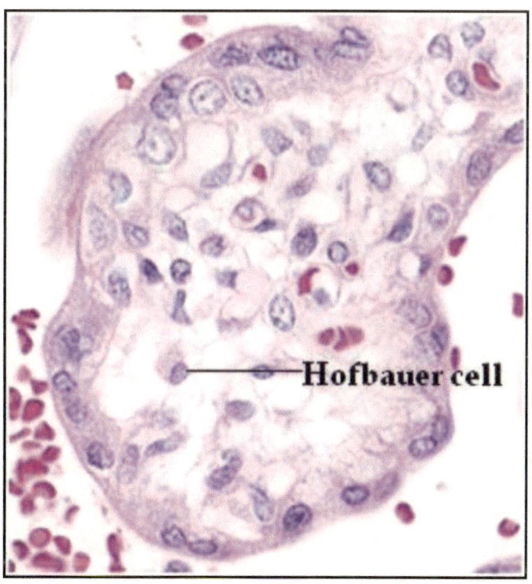

Fig. 1. The mesenchymal core of this fetal tertiary chorionic villus contains a Hofbauer cell, seen as a large, pale-staining placental macrophage with an eccentric nucleus.

2. General placenta function

The placenta provides an interface for communication between the maternal blood and the developing fetus. Substances are selectively transported between the separate but closely approximated maternal and fetal vascular systems. Substances exchanged between the fetal blood (contained within fetal vessels of chorionic villi) and maternal blood (located within sinusoids surrounding villi) include nutrients, waste products, oxygen, carbon dioxide, hormones, cytokines, immunoglobulins, drugs, and microbes.

Signaling molecules found in the placenta, such as hormones and cytokines, are important for the maintenance of pregnancy, maternal-fetal tolerance, parturition, lactation, and barrier to infection. Placental cells that produce these mediators include fetal trophoblasts and placental macrophages. This article is not intended to describe the placenta and all of its components in detail, but to discuss one facet of this complex, transient, vital organ of pregnancy, the placental macrophage.

3. Macrophage phenotypes overview

Human monocytes and macrophages are players in the innate immune system. Macrophages are bone marrow/hematopoietic-derived cells that migrate through the blood circulation to home to and take up residence within various tissues, where they play a pivotal role in coordinating processes such as development and host defense by secretion of cytokines, chemokines and growth factors and by phagocytosis. Furthermore, macrophages link the innate and adaptive immune systems through phagocytosis, digestion, and antigen presentation (macrophage expression of Human Leukocyte Antigen (HLA) class II molecules and co-stimulatory receptors such as CD80/CD86) to T lymphocytes. The role of macrophages in the orchestration of inflammation, immune responses, and wound healing/tissue remodeling has been a field of intensive study.

Macrophages exhibit great plasticity in their ability to transform between phenotypes. While unstimulated, resting macrophages are relatively quiescent and express low levels of cytokines and growth factors, activation by agents in the surrounding micro-environmental milieu polarizes macrophages to yield a broad spectrum of macrophage phenotypes based on location and micro-environmental influences. Environmental stimuli that influence the expression of inducible genes and polarization of macrophages may include activating agents, pathogens, cytokines, hypoxia, and ischemia (Pinhal-Enfield et al., 2003; Ramanathan et al., 2007).

Based on exposure to surrounding micro-environmental cues, macrophages are induced to alter their expression profiles. Two major classifications have been used in the past to describe activated macrophages as polarized towards pro-inflammatory or anti-inflammatory phenotypes. Although categorization of macrophage phenotypes can be muddled by assorted combinations of activating agents, macrophages respond to micro-environmental stimuli and are typically described as "classically"-activated (M1) pro-inflammatory macrophages or "alternatively"-activated (M2) wound healing macrophages. Based on exposure time and the types of activating stimuli, macrophages may switch activation state to other phenotypes (Martinez et al., 2009; Daley et al., 2010; Porcheray et al., 2005; Classen et al., 2009). M1 macrophages induce processes during injury and infection that promote inflammatory processes and promote Th1- type immune responses leading to enhanced killing of microbes and tumor cells. M2 macrophages on the other hand, display immunosuppressive effects, resolve inflammation through production of anti-inflammatory cytokines, enhance phagocytosis and elimination of debris, secrete growth and angiogenic factors that stimulate tissue remodeling and repair, and stimulate Th2-type immune responses. M2 macrophages are characterized by a distinctive molecular repertoire and have been implicated in the down-regulation of inflammation, tissue remodeling, repair, parasite killing, allergic reactions, tumor promotion, and placenta formation (Pinhal-Enfield & Leibovich, 2011; Leibovich et al., 2010). These descriptions of polarized M1 macrophages or M2 macrophages only represent macrophages at the far ends of the spectrum. Macrophage phenotypes might be better described as a series of gradations and nuances within a broad spectrum between the pro-inflammatory macrophage at one far end and the anti-inflammatory macrophage at the other end.

3.1 Induction of macrophage phenotypes

Heterogenous macrophage populations are found within tissues. These differentiated macrophages are plastic and retain the capacity to transform their phenotype in response to micro-environmental cues and immune-specific influences. Activated macrophages switch phenotypes to develop phagocytic and secretory profiles directed for specific functional activities that are in alignment with the original stimuli (Daley et al., 2010).

Unstimulated macrophages express low levels of inflammatory cytokines and growth factors such as tumor necrosis factor-α (TNF-α), interleukin (IL)-12, and vascular endothelial growth factor (VEGF). Stimulation of macrophages with products of pathogenic agents such as Toll-like Receptor (TLR) agonists (e.g., lipopolysaccharide (LPS), flagellins, or lipoproteins), either with or without interferon-γ (IFN-γ) induces "classical" M1 activation, characterized by production of TNF-α and IL-12.

M2 "alternatively activated" macrophages are commonly characterized as being induced in response to IL-4 and IL-13 through the IL-4 receptor-α (IL-4Rα), and up-regulating expression of IL-10, TGF-β, and VEGF while down-regulating expression of TNF-α and IL-12 (Martinez et al., 2009; Classen et al., 2009). M2 macrophages also display a series of cell surface and intracellular markers (Table 1). Further investigation, however, has revealed that IL-4 and IL-13 are not necessary for induction of the M2-like phenotype and mouse wounds lacking IL-4/IL-13 still contain M1 and M2-like macrophages (Daley et al., 2010). The IL-4/IL-13-independent pathways of alternative macrophage activation that induce phenotypes resembling M2 macrophages represent an intense field of study as the functional traits of these M2 macrophages and the different activating agents that induce them are characterized.

3.1.1 M1 macrophages

M1 macrophages promote and coordinate inflammatory processes during injury and infection that are essential for intracellular pathogen removal. These macrophages are induced by pathogen associated molecular patterns (PAMPs), such as the TLR4 agonist, LPS, damage associated molecular patterns (DAMPs), such as those derived from mitochondria of apoptotic neutrophils, and IFN-γ (produced by activated Th1, Tc1, NK cells). Pro-inflammatory and cytotoxic responses against intracellular pathogens and transformed cells (such as cancer cells) are directedthrough induction of cytokines (e.g., IL-1, IL-6, IL-12, TNF-α) and inflammatory mediators (e.g., nitric oxide (NO) through inducible NO synthase (iNOS)), and increases in leukocyte recruitment and phagocytic and antigen presenting activity. These initial, pro-inflammatory M1 macrophages predominate in the early wound, which becomes a dynamic cauldron of mediators that are in constant flux as interactions, inductions, and adaptations take place (Martinez et al., 2009; Classen et al., 2009; VanGinderarchter et al., 2006). In contrast, M2 macrophages are critical for the resolution of inflammation and the induction of angiogenesis, tissue remodeling, and repair through production of anti-inflammatory cytokines, growth and angiogenic factors, and by phagocytosis and elimination of debris (Martinez et al., 2009; Daley et al., 2010; Classen et al., 2009; Taylor et al., 2005; Rubartelli et al., 2007; VanGinderarchter et al., 2006). M2 macrophages are discussed in further detail in the next section.

	M1	M2
INDUCING AGENTS	IFN-γ, LPS	IL-4, IL-13 IC + IL-1β or TLR agonists GC, IL-10, TGF-β TLR and AR agonists
UP-REGULATED EXPRESSION	TNF-α, IL-12, IL-1, IL-6, IL-23, IP-10, MIP-1α, MHC II, CD80, collagenase?, $A_{2A}R$, MMP9, NO/iNOS/respiratory burst	IL 10, IL-1R antagonist, TGF-β, CD206, CD23, AMAC-1. CCL22, CCL17, CCL2/MCP-1, IL-1Rα, IL-1 decoy receptor, TGF-β, Arg-1, Fizz-1, Ym1/Ym2, dectin-1, VEGF
DOWN-REGULATED EXPRESSION		IL-12, TNF-α
GENERAL PROFILE	pro-inflammatory	anti-inflammatory and wound healing

Table 1. Inducing agents involved in the regulation of gene expression contribute to the development of M1 and M2 macrophage phenotypes. (Martines et al., 2009; Leibovich et al., 2011).

3.1.2 M2 macrophages

Macrophage phenotypes can vary and switch based on temporal and environmental cues, making their investigation and classification complex. Wound macrophages in the early stages of healing have a more M1-like profile with elevated expression of TNF-α and IL-6 and less TGF-β, while those in later stages of healing have a more M2-like profile, with less pro-inflammatory cytokine expression, no induction of iNOS, and elevated markers of alternative activation, including CD206, dectin-1, arginase-1 (Arg-1), and chitinase 3-like 3 (Ym1) (Daley et al., 2010; Black et al., 2008) It is worth noting that the complexity of studying alternatively-activated M2 macrophage phenotypes is also attributed to divergent profiles in different animal models. While Arg-1 has become a reliable marker of murine M2 macrophages, it is not a good marker for human M2 macrophages.

The M2 macrophage phenotype describes a broad category of anti-inflammatory, wound healing macrophages that down-regulate effects of pro-inflammatory M1 macrophages and have often been described as IL-4/IL-13-dependent. However, both M1 and M2-like macrophage phenotypes were also observed in wounds in mice where IL-4 and IL-13 signaling is absent. Stimulation of macrophages with IL-4 and IL-13 (cytokines produced by induced CD4+ Th2 and CD8+ Tc2 cells, NK cells, basophils, mast cells, eosinophils) results in "alternative activation" of macrophages . These "alternatively activated" macrophages were first termed M2 macrophages, have anti-inflammatory and wound healing properties, and are associated with allergic and anti-parasitic responses (Martinez et al., 2009). M2 macrophages express low levels of pro-inflammatory mediators (such as TNF-α, IL-12, IL-1β, IL-8, and oxygen radicals) and express elevated levels of anti-inflammatory cytokines, growth factors, and phagocytic receptors such as IL-10, VEGF, TGF-β, CD206 mannose receptor, and CD204 scavenger receptor (Martinez et al., 2009; Classen et al., 2009; Stein et al., 1992; Abramson and Gallin, 1990). M2 macrophages tailor immune responses by chemokine secretion patterns that attract specific sets of leukocytes (monocytes, basophils, memory T cells, Th2 cells, and eosinophils). See Table 1 for a list of some M2 macrophage markers (Martinez et al., 2009; Classen et al., 2009; Mantovani et al., 2006). In addition to their key role in resolving inflammation and mediating repair, M2 macrophages also play a role in some pathological processes. Up-regulation of alternative macrophage gene expression leads to stifling of Th1 responses and consequent vulnerability to infection, allergic responses in asthma, tumor progression assisted by tumor-associated M2-like macrophages (TAMs), and fibroproliferative complications of infection and inflammation (VanGinderarchter et al., 2006; Mantovani et al., 2006; Luo et al., 2006; Yang et al., 2009; Porta et al., 2009; Macedo et al., 2007).

Further characterization of macrophages has lead to the exploration of other types of alternatively activated M2-like macrophages involved in wound healing that are IL-4/IL-13-independent. The broad category of M2 macrophages has been subdivided based on inducing agents and subsequent expression patterns. M2 macrophage subtypes include the more commonly investigated M2a macrophage (activated by IL-4 or IL-13), M2b macrophage (induced by immune complexes (ICs) and IL-1β or TLR agonists), M2c macrophage (stimulated by IL-10, transforming growth factor-β (TGF-β), or glucocorticoids), and M2d macrophage (a new subtype induced by switching M1 macrophages into M2-like macrophages through initial TLR activation followed by adenosine A_{2A} receptor ($A_{2A}R$) activation) (Figure 2) (Martinez et al., 2009; Leibovich et al., 2011).

In contrast to the relatively well-characterized M2a macrophage subtype, many M2 macrophages (M2b, M2c, M2d) are not associated with a Th2 immune response or IL-4/IL-13 activation. Evidence of this is seen with M2 macrophage activation in IL-4Rα KO mice and in the presence of the IL-13Rα2 decoy receptor in spite of inhibition of IL-13-dependent phosphorylation of downstream STAT6 and the absence of IL-4 or IL-13 in the wound environment. These IL-4/IL-13-deficient mice exhibit macrophages with both M1 and M2 overlapping and non-overlapping expression patterns (Daley et al., 2010; Leibovich et al., 2011).

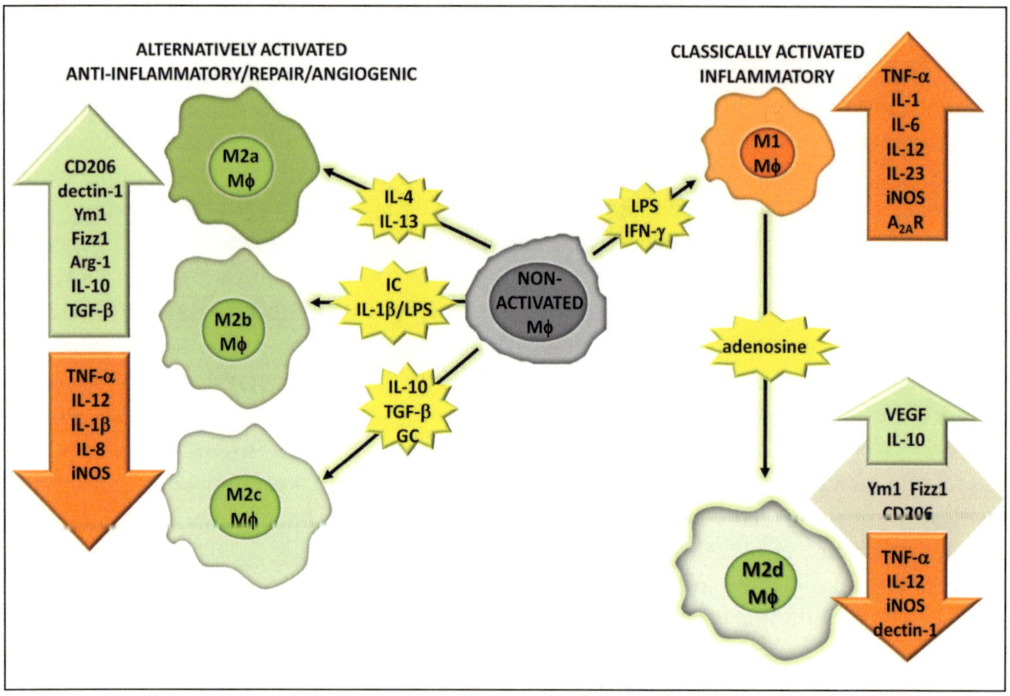

Fig. 2. The broad spectrum of macrophage phenotypes induced by various agents show overlapping and non-overlapping patterns of gene expression.

Investigation of the influence of multiple co-stimulators on macrophages allows for some insight on downstream signaling pathways and the resulting phenotypic macrophage profiles. For example, in the inflammatory environment, the retaliatory metabolite adenosine rapidly accumulates extracellularly as an ATP breakdown product and TLRs and A_{2A}Rs are both expressed on macrophages as important regulators of inflammation and repair. Understanding the downstream effects from simultaneous activation of these receptors may provide insight on how to enhance or block M2 macrophage effects by regulating signaling mediators or using receptor agonists or antagonists. Characterization of M2 macrophages by simultaneous induction with multiple agents may more accurately mimic true micro-environmental conditions. Co-stimulation of TLRs and A_{2A}Rs may be a more relevant reflection of macrophage surroundings, where adenosine accumulation occurs in inflammatory, ischemic, and hypoxic settings, leading to phenotype switching from TLR agonist-induced pro-inflammatory M1 macrophages (secrete TNF-α and IL-12) to adenosine-induced wound healing and anti-inflammatory M2d macrophages (down-regulation of TNF-α and IL-12 and secretion of IL-10 and VEGF) in a temporally defined manner. This induction of $A_{2A}Rs$ plays a key role in switching pro-inflammatory M1 to an angiogenic M2-like phenotype. Solitary activation of these receptors may not show expression patterns seen in co-stimulation models. This example of receptor co-stimulation on macrophages shows that macrophage activation and its role in wound healing is a complex event involving multiple micro-environmental stimuli (Leibovich et al., 2011; Macedo et al., 2007). Knowledge of the processes leading to the phenotypic switches occurring in macrophages may yield potential therapeutic targets for promoting wound healing or for dampeningabnormal wound healing. Although macrophages have varying and dynamic roles that adapt to activating agents in the surrounding milieu, a clear understanding of how to manipulate and switch these phenotypic profiles may provide potential benefits.

4. Placental macrophages

In adult tissues, macrophages coordinate tissue homeostasis and orchestrate host defense and wound healing. During fetal development, macrophages are abundantly present in the pregnant uterus and placenta. These placental macrophages play a key role in synthesis of vital mediators involved in the establishment and maintenance of pregnancy, parturition, lactation, local immune reactions, and maternal-fetal tolerance.

Although they exhibit their own, unique phenotypic profile, placental macrophages display key features of many other tissue macrophages. These characteristics include growth with stimulation by colony-stimulating factor-1, expression of Fc receptors and phagocytosis, and expression of macrophage markers (CD14, CD11b/Cd18, and F4/80) (Chang et al., 1993).

4.1 Origins of placental macrophages

In the human placenta, macrophages have been detected as early as day 10 of pregnancy and many of these macrophages are present throughout pregnancy (Chang et al., 1993) Placental macrophages reside within fetal chorionic villi and uterine decidua during pregnancy, where they assist placental development, manage host defense, and maintain pregnancy.

Fetal placental macrophages and maternal placental macrophages reach the placenta by different routes. During prenatal development, monocyte progenitors originate in the hypoblast-derived fetal yolk sac and migrate to the mesenchymal core of fetal chorionic villi in the placenta. These placental macrophages on the fetal side are called Hofbauer cells. Placental macrophages on the maternal side of the placenta are derived from hematopoietic pluripotent stem cells that differentiate into monocyte progenitors, which will mature and leave the bone marrow to enter the blood. After an approximate 8 hour migration through the circulatory system, the maturing monocyte will enter various organs and tissues, such as the uterine endometrial stroma, where tissue-specific factors will further induce differentiation to form a tissue macrophage (Figure 3) (McIntire et al., 2006).

4.2 General functions of placental macrophages

Macrophages are present in many tissues, including the placenta, and make up a dynamic and heterogenous population of cells. Macrophage phenotypes and functions, which are induced at least in part by the surrounding micro-environmental milieu, are crucial in inflammation, immunity, and wound healing. Macrophages are induced by environmental cues to regulate expression of growth factors, chemokines, cytokines, proteolytic enzymes, adhesion molecules, and phagocytic receptors (Martinez et al., 2009).

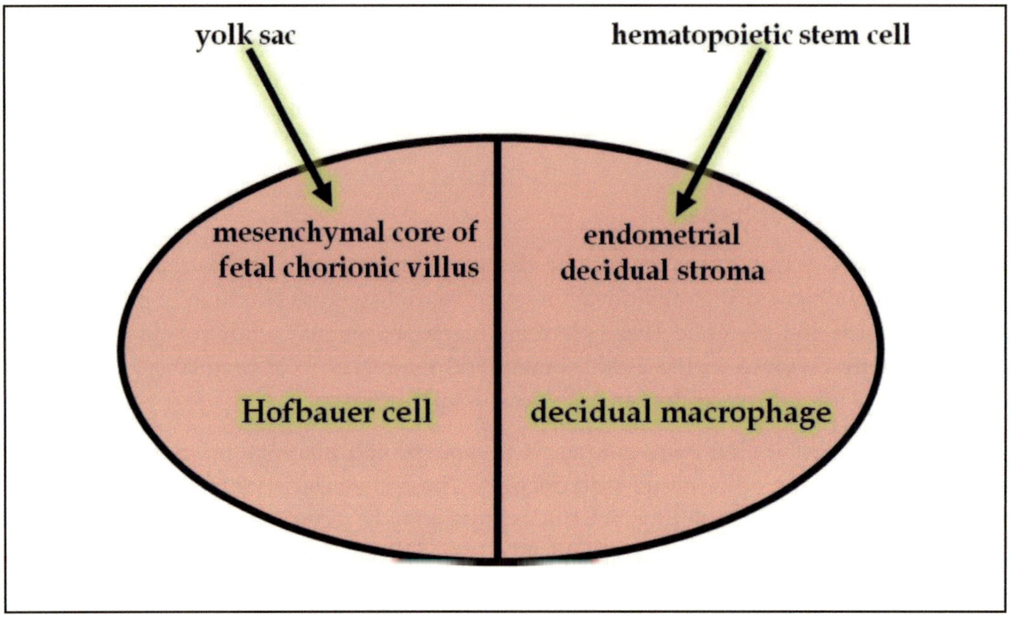

Fig. 3. Placental macrophages originate from two different sources. Hofbauer cells originate from yolk sac cells that migrate to ultimately take up residence in the fetal chorionic villi of the placenta. Decidual macrophages are bone marrow-derived and take up residence in the uterine endometrial stroma of the placenta's decidua basalis.

Macrophages are abundant in the placenta and play a large role through regulation of immune cells and tissue growth. Placental macrophages are influential in many processes, including implantation, trophoblast function, placenta construction (via induction of angiogenesis), maternal-fetal tolerance, fetal defense/protection from infection, and parturition (Bulmer & Johnson, 1984; Mues et al., 1989; Sevala et al., 2007; Kzhyshkowska et al., 2008; Nagamatsu & Schust, 2010; Garcia et al., 1993). Placental macrophages are pivotal to many of these processes because of their plasticity and ability to switch phenotypes based on surrounding micro-environmental cues. For example, decidual placental macrophages respond to placental stimuli that control maternal immune reponses to the fetus (maternal-fetal tolerance) as well as to pathogenic stimuli during infections (host defense). During pregnancy, placental macrophages are exposed to local stimuli that induce an array of macrophage phenotypes with M1 and M2 characteristics. Placental macrophages express surface markers and secrete mediators such as cytokines, prostaglandins, proteolytic enzymes, and chemokines (McIntire et al., 2006). It should be noted that while microenvironmental modulation of placental macrophages plays an important role in regulating macrophage phenotype, it is also possible that selective recruitment of distinct monocyte sub-populations might also play role. The nature of the monocyte precursor subpopulations that give rise to resident placental macrophages has not been studied in detail, and the role of local proliferation of macrophages in the placenta versus recruitment of new macrophages from the circulation in the maintenance of macrophage homeostasis is not yet clear.

4.2.1 Placental macrophages in placenta construction

Macrophages are abundant in fetal tissue and these macrophages play a role in placental development. Angiogenesis is critical for vascular development in the fetal chorionic villi of the placenta and is induced by secretion of VEGF from nearby Hofbauer cells located in the chorionic villus mesenchyme. VEGF binds the fms-like tyrosine kinase 1 receptor (Flt-1), which is expressed on trophoblast and endothelial cells during early pregnancy (Clark et al., 1986). Interestingly, the placenta produces elevated levels of soluble Flt-1 receptor (sFlt-1) in women with pre-eclampsia. This sFlt-1 sequesters VEGF and results in dysfunctional angiogenesis in the placenta (Luttun and Carmeliet, 2003) Placental development progresses (apoptosis, phagocytosis, matrix degradation, remodeling) via continued interactions between maternal decidual placental macrophages and fetal-derived trophoblasts (Nagamatsu & Schust, 2010).

In addition to secretion of angiogenic factors by fetal Hofbauer cells that promote angiogenesis during placental development, maternal decidual placental macrophages also contribute to angiogenesis in the placenta by the abundant expression of the stabilin-1 scavenger receptor. Stabilin-1 is a receptor for placental lactogen (PL), a prolactin growth hormone secreted by trophoblasts. PL is important for placental angiogenesis, as seen by the correlation between low PL concentrations and aberrant placental angiogenesis leading to intrauterine growth restriction. PL is also involved in post-natal growth and mammogenesis/lactogenesis (Kzhyshkowska et al., 2008). Stabilin-1 binds and endocytoses PL. Internalized PL within the decidual macrophage can then proceed through a degradation pathway in which PL within lysosomes is enzymatically digested or through a storage pathway in which PL migrates to the *trans*-Golgi network for processing to be stored within secretory vesicles until called upon for release by low extracellular PL concentrations. Therefore, stabilin-1 expression on decidual M2 macrophages may serve as a sensor to regulate extracellular PL (through endocytosis and intracellular sorting of PL) and result in further regulation of factors involved in angiogenesis during placental growth (Kzhyshkowska et al., 2008).

4.2.2 Placental macrophages in maternal-fetal tolerance

There are multiple factors involved in maternal-fetal tolerance. Decidual placental macrophages are an integral component for suppression of the maternal immunologic response to the allogenic placenta and fetus. Macrophages are antigen presenting cells (APCs) that often activate T cells to initiate an adaptive immune response. However, although decidual placental macrophages constitutively express major histocompatibility complex class II molecules, they have a decreased ability to present antigens to T cells. As a result, T lymphocyte activation does not occur and the allogenic fetus is protected from rejection by the maternal immune system. Another result of placental macrophage inability to present antigen to T cells is the inability of placental macrophages to induce an efficient immune response against *Listeria monocytogene* despite expressing MHC class II molecules (Chang et al., 1993; Nagamatsu & Schust, 2010) Furthermore, placental macrophages express higher levels of indole 2,3 dioxygenase (IDO), which catabolizes L-tryptophan. L-tryptophan is required for T-cell activation and inhibition of IDO correlated to rejection of the fetus by the mother (Gordon et al., 2003).

4.2.3 Placental macrophages as a barrier to infection

The main defenders against infection of the placenta are placental macrophages. Among many other functions, macrophages play a vital role in removal of nonself as well as unwanted self-ligands through expression of scavenger receptors and other receptors that are targeted for pathogen binding, such as Toll-like receptors (e.g., TLR4 bind LPS) and chemokine-CC motif-receptor 5 (CCR5) (HIV-1 infection requires surface expression of CCR5 and CD4). While sometimes displaying a weak barrier to some pathogens like *Listeria monocytogenes*, placental macrophages often show a strong barrier to numerous bacterial and viral infections (Garcia-Crespo et al., 2009; Kzhyshkowska et al., 2008).

Although macrophages are primary targets for viral infection, placental macrophages are less susceptible to HIV-1 infection (and other viral infections) than other macrophages. The placenta is often a barrier to viral replication, and viral transmission from mother to infant often occurs during the birth. Placental macrophages show lower levels of susceptibility to HIV-1 infection than monocyte-derived macrophages and this may be attributed to differing phenotypes. Although there are lower levels of CD4 and CCR5 expression in placental macrophages, it has been suggested that the mode of HIV-1 restriction in the placenta occurs at the transcriptional level. HIV-1-infected placental macrophages and monocyte-derived macrophages show different protein secretion profiles. Peroxiredoxins are antioxidant molecules that are more abundantly associated with placental macrophages compared to monocyte-derived macrophages. Also, some peroxiredoxins (1,2, and 4) have been shown to have anti-viral activity against HIV-1. Compared to monocyte-derived macrophages, placental macrophages show abundant secretion of peroxiredoxin 5. The up-regulated secretion of peroxiredoxin and its protective role against HIV-1 in the placenta may be through suppression of TNF-α-induced transcriptional activity of nuclear factor-kappa light chain enhancer of activated B cells (NF-κB). In fact, although there appears to be no difference in viral integration for placental macrophages versus other monocyte-derived macrophages, there is reduced viral protein expression in placental macrophages. Therefore, HIV-1 restriction appears to be occuring in placental macrophages at the transcriptional level. NF-κB is required for viral transcription of HIV-1 and suppression may lead to inefficient HIV-1 replication in placental macrophages. TNF-α-induced transcriptional activity of NF-κB may also be thwarted in placental macrophages by down-regulation of cystatin B. Cystatin B induces NO production in macrophages, which up-regulates TNF-α. Reduced HIV-1 replication has been correlated with cystatin B down-regulation and increased HIV-1 replication is enhanced by NO synthesis in macrophages. These factors (among others) distinguish placental macrophages from other macrophages and are influential in the restriction of HIV-1 (and perhaps other viruses) in the placenta (Garcia et al, 2009; Garcia-Crespo et al., 2010).

4.2.4 Placental macrophages as a mediator in parturition and other signaling/hormone

Placental macrophages exhibit varying phenotypes throughout pregnancy. Placental macrophages are stimulated by changes in the micro-environment to switch phenotypes at varying stages of pregnancy as well as during aberrant activity. During late pregnancy, placental macrophage activity is switched to play a role in parturition through production of pro-inflammatory cytokines and prostaglandin E_2. Altered placental macrophage phenotypes are also linked to the pathologic disease such as preeclampsia (Nagamatsu & Schust, 2010; Wetzka et al., 1997).

During most of pregnancy placental macrophages have an M2-like immunosuppressive gene expression profile (such as IL-10 secretion). However, decidual placental macrophages may expess pro-inflammatory characteristics. For example, $CD11c^{HI}$ decidual placental macrophages may express pro-inflammatory and antigen-presenting genes while $CD11c^{LO}$ decidual placental macrophages express anti-inflammatory and tissue repair genes (Houser et al., 2011). It is important to recognize that classification of placental macrophages is complex and dependent on multiple factors, such as location within the placenta, stage of pregnancy, and factors (e.g., hormones and hypoxia) that change the surrounding micro-environment. Differential recruitment of distinct monocyte sub-populations may also play a role, as mentioned earlier.

4.3 M2-type placental macrophages

Placental macrophages display polarization towards an alternatively-activated, M2 phenotype. Hofbauer cells (fetal placental macrophages within chorionic villi) and decidual macrophages (maternal placental macrophages within the endometrial decidual stroma) are both classified as M2/alternatively-activated macrophages (Joerink et al., 2011; Gustafsson et al., 2011). The role of IL-4 in polarization of placental macrophages is blurred by variable placental macrophage phenotypic profiles. Although IL-4 is involved in endocytosis of stabilin-1-bound PL by placental macrophages, decidual macrophages induced via IL-4-independent means display different expression profiles than those seen with IL-4 stimulation, indicating that a Th2 environment is not required for decidual macrophage development (Svensson et al., 2011; Kzhyshkowska et al., 2008).

As M2 macrophages, the placental macrophage profile typically shows down-regulation of pro-inflammatory mediators and up-regulation of mediators that are immunosuppressive and anti-inflammatory. However, as with many other tissue macrophages, a placental macrophage does not exhibit a phenotype extreme that solely expresses M2 markers. Placental macrophages may express M1 markers, but the representative expression pattern emphasizes many M2 markers. The placental macrophage M2 marker expression is regulated by activating agents (such as cytokines, hormones, and adenosine) in the surrounding micro-environment in a manner that yields an M2 phenotype profile that is distinct from other M2 macrophages. The exact nature and combinations of activating agents to achieve this unique profile has yet to be completely elucidated. Analyses of M2 placental macrophages from normal pregnancies and pregnancies with complications have shed some light on dysregulation. In fact, a correlation between dysregulation of some placental macrophage markers and preeclampsia (MR, CCL2, IGF-1, MMP9), pre-term labor (CCL18) and intrauterine growth restriction (IGF-1) has been shown (Gustafsson et al., 2011).

Placental macrophages typically lack or exhibit low expression of pro-inflammatory M1 markers (such as CX3CR1, IL-7R and CCR7), as described in Hofbauer cells by Joerink et al. (2011). Many placental macrophage M2 markers have been investigated. A number of M2 markers have been identified in placental macrophages such as DC-SIGN (attachment factor for HIV-1 entry), CD163 and mannose receptor/CD206, stabilin-1, CCL18 (involved in tolerance of T lymphocytes with reduced expression after the onset of labor), CD209, fibronectin-1 and insulin-like growth factor-1 (IGF-1), TREM-2, alpha 2 macroglobulin (A2M), prostaglandin D_2 synthase (PGDS), and IL-10 (Kzhyshkowska et al., 2008; Gustafsson et al., 2011; Svensson et al., 2011; Joerink et al., 2011). Recall that placental macrophages are not at the extreme end of the M2 spectrum of macrophages and that M1-type markers can also be seen in placental macrophage populations. For example, placental macrophages express IL-6 and CCL2 (aka monocyte-chemoattractant protein-1, MCP-1), which are typically pro-inflammatory M1 markers. It has been suggested that CCL2 expression by placental macrophages may be important for monocyte recruitment to decidua and that macrophage phenotypes may then be altered by other environmental factors upon arrival (Svensson et al., 2011; Gustafsson et al., 2011). Furthermore, Houser et al. (2011) have characterized decidual placental macrophages as having profiles associated with M1 and M2 macrophages. As mentioned previously, $CD11c^{HI}$ decidual macrophages are pro-inflammatory while $CD11c^{LO}$ decidual macrophages are anti-inflammatory. Thus, decidual macrophages have varying phenotypes that help to coordinate maternal-fetal interactions at the placenta (Houser et al., 2011).

5. Research and clinical implications

5.1 Advantages and limitations of experimental models/materials

It should be emphasized that some limitations to consider in the characterization of alternatively-activated macrophages include species-species variation in gene expression patterns and variable effects of the micro-environment surrounding macrophages. Investigation of M1 and M2 marker expression has been focused more predominantly using murine macrophages and human homologs for some murine markers have not been found (Martinez et al., 2009).

Furthermore, agents in the environment surrounding macrophages influence macrophage activation, and while *in vitro* models are useful, analyses obtained from them are limited. Macrophages in culture enable investigators to highlight activation factors and gene expression patterns; however, absence of *in vivo* models may overlook interactions with other cell types and the mediators that they may elaborate. It is crucial to investigate interactions of expressed proteins that may be different in an *in vivo* context in which there may be interaction with other cells or mediators. Macrophage activation is a complex and dynamic enterprise with characteristics that vary with time to result in early and late markers. For example, expression of CCL18 in decidual macrophages declines with onset of labor (Gustaffson et al., 2011). Macrophage phenotype switching is driven in a temporally-orchestrated manner that depends upon alterations in the micro-environment, such as that seen in the changing placenta.

5.2 Clinical implications and future research

M1 and M2 macrophages are pivotal players in host defense, wound repair, and pregnancy and further investigation into these polarized activated macrophage phenotypes may elucidate normal and variant processes. Knowledge of processes associated with placental macrophages may shed light on pathological mechanisms and may present attractive potential pharmaceutical targets to regulate disease. For example, decidual macrophages with overexpression of M1 markers have been implicated in parturition and early embryo loss (Gustaffson et al., 2011). Thus, manipulation of placental macrophage phenotypes by modulating signaling pathways in macrophages or by altering the surrounding environment may provide therapeutic benefits. For example, because of their angiogenic effects, induction of M2 macrophages phenotypes may enhance placental growth.

6. Conclusion

Macrophages are found throughout the body and exhibit a broad spectrum of phenotypes based in large part on the surrounding microenvironment. Placental macrophages exhibit a phenotypic profile more characterisitic of M2 macrophages that are immunomodulatory and promote angiogenesis for placental development. Investigation of placental macrophages may provide insight into normal development as well as possible causes of embryo loss.

7. Acknowledgment

Dr. Leibovich's research on macrophage phenotypes has been supported by grants from the US Public Health Service National Institute of General Medical sciences (NIGMS)(5-RO1-GM068636).

8. References

Abramson SL, Gallin JI. (1990) IL-4 inhibits superoxide production by human mononuclear phagocytes. *Journal of Immunology*, Vol. 145, pp. 435-1439.

Black SG, Wilson JM, Ernst PB, Smith MF. (2008) A_{2A} adenosine receptor stimulation enhances arginase I expression in macrophages resulting in a phenotypically unique macrophage. *FASEB*, Vol. 22, 1065.25. (Abstract)

Bulmer JN, Johnson PM. (1984) Macrophage populations in the human placenta and amniochorion. *Clinical Experimental Immunology*, Vol. 57, pp. 393-403.

Chang MDY, Pollard JW, Khalili H, Goyert SM, Diamond B. (1993) Mouse placental macrophages have a decreased ability to present antigen. *Proceedings of the National Academy of Sciences*, Vol. 90, pp. 462-466.

Classen A, Lloberas J, Celada A. (2009) Macrophage activation: classical versus alternative. *Methods in Molecular Biology*, Vol. 531, pp. 29-43.

Clark D, Chaput A, Tutton D. (1986) Active suppression of host-vs-graft reaction in pregnant mice. VII. Spontaneous abortion of allogeneic CBA/J x DBA/2 fetuses in the uterus of CBA/J mice correlates with deficient non-T suppressor cell activity. *Journal of Immunology,* Vol. 136, pp.1668–1675.

Daley JM, Brancato SK, Thomay AA, Reichner JS, Albina JE. (2010) The phenotype of murine wound macrophages. *Journal of Leukocyte Biology,* Vol. 87, pp. 59-67.

Garcia-Crespo K, Cadilla C, Skolasky R., Melendez LM (2009) Restricted HIV-1 replication in placental macrophages is caused by inefficient viral transcription. *Journal of Leukocyte Biology,* Vol. 87, 633-636.

Garcia K, Garcia V, Laspiur JP, Duan F, Melendez LM. (2009) Characterization of the Placental Macrophage Secretome: Implications for Antiviral Activity. *Placenta,* Vol. 30, pp. 149-155.

Gustafsson C, Mjösberg J, Matussek A, Geffers R, Matthiesen L, Berg G, Sharma S, Buer J, and Ernerudh J. (2011) Gene Expression Profiling of Human Decidual Macrophages: Evidence for Immunosuppressive Phenotype. *PLoS ONE,* Vol. 3, pp. e2078.

Houser BL, Tilburgs T, Hill J, Nicotra ML, Strominger JL. (2011) Two Unique Human Decidual Macrophage Populations. *Journal Of Immunology,* Vol. 186, pp. 2633-2642.

Joerink M, Rindsjo E, van Riel B, Alm J, Papadogiannakis N. (2011) Placental macrophage (Hofbauer cell) polarization is independent of maternal allergen-sensitization and presence of chorioamnionitis. *Placenta,* Vol. 32, pp. 380-385.

Kzhyshkowska J, Gratchev A, Schmuttermaier C, Brundiers H, Krusell L, Mamidi S, Zhang J, Workman G, Sage EH, Anderle C, Sedlmayr P, Goerdt S. (2008) Alternatively activated macrophages regulate extracellular levels of the hormone placental lactogen via receptor-mediated uptake and transcytosis. *Journal of Immunology,* Vol. 180, pp. 3028-3037.

Leibovich SJ and Pinhal-Enfield G . (2011) Macrophage heterogeneity and wound healing. *Advances in Wound Care,* Vol. 2, pp. 89-95.

Luo Y, Zhou H, Krueger C, Kaplan S, Lee SH, Dolman C, Markowitz D, Wu W, Liu C, Reisfeld RA, Xiang R. (2006) Targeting tumor-associated macrophages as a novel strategy against breast cancer. *Journal of Clinical Investigation,* Vol. 116, pp. 2132-2141.

Luttun A and Carmeliet P. (2003) Soluble VEGF receptor Flt1: the elsuive preeclampsia factor discovered? *The Journal of Clinical Investigation,* Vol. 111, pp. 600-602.

Macedo L, Pinhal-Enfield G, Alshits V, Elson G, Cronstein BN, Leibovich SJ. (2007) Wound healing is impaired in MyD88-deficient mice: A role for MyD88 in the regulation of wound healing by adenosine A_{2A} receptors. *American Journal of Pathology,* Vol. 171, pp. 1774-1788.

Mantovani A. (2006) Macrophage diversity and polarization: in vivo veritas. *Blood,* Vol. 108, pp. 408-409.

Mantovani A, Sozzani S, Locati M, Allavena P, Sica A. (2004) Macrophage polarization: tumor-associated macrophages as a paradigm for polarized M2 mononuclear phagocytes. *Trends in Immunology,* Vol. 25, pp. 677-686.

Martinez FO, Helming L, Gordon S. (2009) Alternative activation of macrophages: an immunologic functional perspective. *Annual Reviews in Immunology,* Vol. 27, pp. 451-483.

McIntire RH, Petroff MG, Phillips TA, Junt JS. (2006) In vitro models for studying human uterine and placental macrophages. *Methods in Molecular Medicine* , Vol. 122, pp. 123-145. *Placenta and Trophoblast: Methods and Protocols*, Vol. 2, edited by MJ Soares and JS Hunt, Humana Press Inc

Mues B, Langer D, Zwaldo D, Sorg C. (1989) Phenotypic Characterization of Macrophages in Human Term Placenta. *Immunology*, Vol. 67, pp. 303-307.

Nagamatsu T, Schust DJ. (2010) Review: The Immunomodulatory Roles of Macrophages at the Maternal—Fetal Interface. *Reproductive Sciences*, Vol. 17, pp. 209-218.

Pinhal-Enfield G, Leibovich SJ. (2011) Macrophage Heterogeneity and Wound Healing. *Advances in Wound Care*, Vol. 2, pp. 89-95.

Pinhal-Enfield G, Ramanathan M, Hasko G, Vogel SN, Salzman AL, Boons GJ, Leibovich SJ. (2003) An angiogenic switch in macrophages involving synergy between Toll-like receptors 2, 4, 7, and 9 and adenosine A(2A) receptors. *American Journal of Pathology*, Vol. 163, pp. 711-721.

Porcheray F, Viaud S, Rimaniol AC, Leone C, Samah B, Dereuddre-Bosquet N, Dormont D, Gras G. (2005) Macrophage activation switching: an asset for the resolution of inflammation. *Clinical Experimental Immunology*, Vol. 142, pp. 481-489.

Porta C, Rimoldi M, Raes G, Brys L, Ghezzi P, Di Liberto D, Dieli F, Ghisletti S, Natoli G, De Baetselier P, Mantovani A, Sica A. (2009) Tolerance and M2 (alternative) macrophage polarization are related processes orchestrated by p50 nuclear factor B. *Proceedings of the National Academies of Science*, Vol. 106, pp. 14978-14983.

Ramanathan M, Luo W, Csoka B, Hasko G, Lukashev D, Sitkovsky M, Leibovich SJ. (2009) Differential regulation of HIF-1α isoforms in murine macrophages by TLR4 and adenosine A_{2A} receptor agonists. *Journal of Leukocyte Biology*, Vol. 86, pp. 681-689.

Ramanathan M, Pinhal- Enfield G, Hao I, Leibovich SJ. (2007) Synergistic up-regulation of vascular endothelial growth factor (VEGF) expression in macrophages by adenosine A2A receptor agonists and endotoxin involves transcriptional regulation via the hypoxia response element in the VEGF promoter. *Molecular Biology of the Cell*, Vol. 18, pp. 14-23.

Rubartelli A, Lotze MT. (2007) Inside, outside, upside down: damage-associated molecular-pattern molecules (DAMPs) and redox. *Trends in Immunology*, Vol. 28, pp. 429-436.

Sevala Y, Korguna ET, Demir R. (2007) Hofbauer Cells in Early Human Placenta: Possible Implications in Vasculogenesis and Angiogenesis. *Placenta*, Vol. 28, pp. 841-845.

Stein M, Keshav S, Harris N, Gordon S. (1992) Interleukin 4 potently enhances murine macrophage mannose receptor activity: a marker of alternative immunologic macrophage activation. *Journal of Experimental Medicine*, Vol. 176, pp. 287-292.

Svensson J, Jenmalm MC, Matussek A, Geffers R, Berg G, Ernerudh J. (2011) Macrophages at the fetal–maternal interface express markers of alternative activation and are induced by M-CSF and IL-10. *Journal of Immunology*, Vol. 87, pp. 3671-3682.

Taylor PR, Martinez-Pomares L, Stacey M, Lin HH, Brown GD, Gordon S. (2005) Macrophage receptors and immune recognition. *Annual Reviews in Immunology*; Vol. 23, pp. 901-944.

Van Ginderachter JA, Movahedi K, Ghassabeh GH, Meerschaut S, Beschin A, Raes G, Baetselier P. (2006) Classical and alternative activation of mononuclear phagocytes: Picking the best of both worlds for tumor promotion. *Immunobiology,* Vol. 211, pp. 487-501.

Wetzka B, Clark DE, Charnock-Jone DS, Zahradnik HP, Smith SK. (1997) Isolation of macrophages (Hofbauer cells) from human term placenta and their prostaglandin E_2 and thromboxane production. *Human Reproduction,* Vol. 12, pp. 847-852.

Yang HZ, Cui B, Liu HZ, Chen ZR, Yan HM, Hua F, Hu ZW. (2009) Targeting TLR2 attenuates pulmonary inflammation and fibrosis by reversion of suppressive immune microenvironment. *Journal of Immunology, Vol.* 182, pp. 692-702.

7

DNA Methylation in Development

Xin Pan, Roger Smith and Tamas Zakar
Mothers and Babies Research Centre, Hunter Medical Research Institute, University of Newcastle, Newcastle, NSW, Australia

1. Introduction

Early embryonic development is a very precise and complicated process. When a sperm meets an egg, a series of well-orchestrated changes take place, which end up with distinct types of cells that make up an organism. Cells start from a pluripotent state and differentiate without changes in DNA sequence. A differentiated cell shares the same DNA sequence with the zygote from which it is descended (mammalian B and T cells being an exception). The diverse functions of different cells are due to tissue-specific patterns of gene expression, which are established during development; once the fates of the cells are decided, they will be maintained faithfully through cell divisions. Hence it is reasonable to assert that development is, by definition, an epigenetic process (Reik, 2007). The specific gene expression programs in differentiated cells are regulated by a more flexible system, which dynamically switches on and off the genes for maintaining homeostasis or responding to environmental changes.

Epigenetics is defined as "the study of heritable changes in genome function that occur without alterations to the DNA sequence" (Probst, et al., 2009). Epigenetics has been suggested as the key regulatory system in early development. Mechanistically, epigenetic regulation involves the covalent modification of chromatin components such as DNA methylation and histone modifications (acetylation, methylation and phosphorylation are the best characterized). Short and long non-coding RNAs are also part of the epigenetic regulatory system because of their role in targeting the chromatin modifications within the genome (Hawkins & Morris, 2008; Morris, 2009a). DNA methylation at the cytosine residue of CpG dinucleotides is the most studied epigenetic modification in mammals. Its effects on genome function underlie a number of physiological phenomena such as genomic imprinting and X chromosome inactivation, and it also contributes to the genesis of human cancers and to aging. CpG methylation was the only known chemical modification of mammalian genomic DNA with an epigenetic role before the discovery of 5-hydroxymethylcytosine that will be discussed later (Haluskova, 2010; Ohgane, et al., 2008). CpG methylation is stable, heritable and reversible, which fulfils the requirement for a dynamic regulation system for development.

DNA methylation is most vulnerable to the environment during early development, because the genome methylation pattern is established during this stage and the DNA synthetic rate is very high in the early embryo. In mammals, proper DNA methylation is essential for normal development. Aberrant methylation patterns are involved in various developmental pathological phenomena and even diseases in adult life that are known under the rubric: the Developmental Origin of Health and Disease (DOHaD) (Waterland & Michels, 2007).

In this chapter, we will discuss the biochemistry of DNA CpG methylation including the enzymes catalyzing the process and the controversial pathways of DNA demethylation. The dynamics of DNA methylation in early development will be covered as well as the role of methylation in cell-lineage determination, imprinting and the genesis of germ cells. We will also review the evidence supporting the importance of DNA methylation in DOHaD.

2. The biochemical mechanism of DNA methylation

DNA methylation occurs via covalent modification of cytosine by adding a methyl group to the carbon-5 of the pyrimidine ring mainly in a 5′-CG-3′ dinucleotide pattern (CpG: C phosphodiester G). It is performed by the DNA methyltranferase (Dnmt) enzyme family. All identified Dnmts use S-Adenosylmethionine (SAM) as the methyl group (CH_3) donor. SAM, a biological sulfonium compound, is a major methyl donor involved in a number of essential reactions, including DNA, RNA and protein methylation (Lin, 2011).

The mechanism of DNA cytosine 5-methylation was analyzed for the prokaryotic Dnmt M.HhaI, which recognizes the specific sequence of 5′-GCGC-3′ and methylates the first cytosine (J. C. Wu & Santi, 1985, 1987). Prokaryotic and eukaryotic Dnmts share a number of conserved primary amino acid motifs that are believed to be important both structurally and functionally.

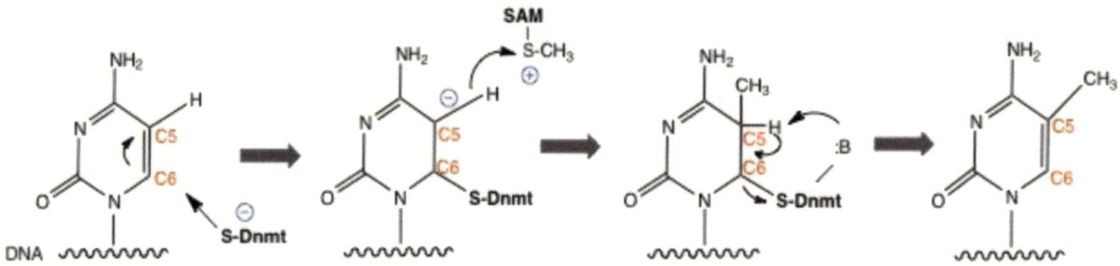

Fig. 1. Catalytic mechanism of cytosine methylation by Dnmt.

The Dnmt performs a nucleophilic attack on the C6, which leads to the formation of a covalent intermediate and activation of the C5. The activated C5 performs a nucleophilic attack on the methyl donor S-Adenosylmethionine (SAM) to acquire a methyl group from SAM. Following the methyl transfer, the C5 proton is eliminated and the intermediate is released, yielding the ultimate product: cytosine methylated on C5. Adapted from Kumar et al (1994).

The key features of the catalytic mechanism involve the nucleophilic attack on the carbon-6 of the target cytosine by a conserved cysteine residue of the enzyme and formation of a covalent intermediate (Kumar, et al., 1994) (Fig 1). The process is induced by binding of the enzyme to DNA, which turns the target cytosine 180° away from the DNA double helix and positions it in the active site with little disturbance to the rest of the DNA duplex. The details of this process, termed base flipping, are poorly understood (Gerasimaite, et al., 2011; Klimasauskas, et al., 1994; Matje, et al., 2011). The base flipping pulls the target base into closer contact with the enzyme, allowing for the accurate recognition of the extrahelical base and the subsequent chemical reactions. The catalytic loop of the enzyme, which contains six highly conserved residues, including the catalytic nucleophile cysteine residue, moves towards the DNA and stabilizes the extrahelical base for methylation (Hermann, Gowher, et al., 2004; Matje, et al., 2011). The nucleophilic attack on carbon-6 is performed by the thiol group of the cysteine residue, whereas the carbon-5 is relatively unreactive. The positively charged sulfonium ion in SAM makes the methyl group that is bonded to the sulfur atom chemically reactive to nucleophilic attack (Hermann, Gowher, et al., 2004). The attack leads to the formation of a covalent intermediate between the enzyme and the carbon-6 atom of the flipped out target cytosine, which results in activation of the carbon-5 atom. The activated carbon-5 performs a nucleophilic attack on SAM to acquire a methyl group from SAM. Following the methyl transfer, the carbon-5 proton is eliminated and the intermediate is released, yielding the ultimate product: cytosine methylated on carbon-5 (Hermann, Gowher, et al., 2004; Klimasauskas, et al., 1994; Kumar, et al., 1994; Yoo & Medina-Franco, 2011).

3. DNA methyltranferases (Dnmts)

In mammals, the DNA methylation pattern is set up by three active members of the Dnmt family: Dnmt1, Dnmt3a and Dnmt3b. Dnmt3a and Dnmt3b are responsible for the *de novo* methylation that establishes the initial CpG methylation pattern during embryonic development, while Dnmt1 is required for maintenance of this pattern, copying the information to newly synthesized DNA during replication (Hermann, Goyal, et al., 2004). Although structurally similar to other Dnmts, the fourth member, Dnmt2, acts more as an RNA methyltransferase (Goll, et al., 2006). Dnmt3-Like protein (Dnmt3L) is a Dnmt-related protein that is catalytically inactive, but physically associates with Dnmt3a and Dnmt3b and modulates their catalytic activity (Hata, et al., 2002).

3.1 Dnmt1

Dnmt1 was the first discovered mammalian DNA methyltransferase (Gruenbaum, et al., 1982). It is highly conserved among eukaryotes. Dnmt1 is a large single polypeptide that comprises 1620 amino acid residues. It contains a large N-terminal domain and a smaller C-terminal part, which are connected by a linker of glycine-lysine repeats (Frauer, et al., 2011). The N-terminal domain comprises several motifs and is responsible for intracellular targeting and regulation of catalytic activity. The C-terminal part harbours all ten catalytic motifs that have been identified in bacterial Dnmts (Goll & Bestor, 2005). However, studies have shown Dnmt1 exhibited no enzyme activity without the presence of its N-terminal domain (Fatemi, et al., 2001; Margot, et al., 2000; Zimmermann, et al., 1997). In agreement with this, direct interactions between the C-terminal and N-terminal parts have been observed and found necessary for Dnmt1 enzyme activity (Fatemi, et al., 2001; Margot, et al., 2003).

Dnmt1 is responsible predominantly for the maintenance of DNA methylation information through cell divisions. Targeted mutation of the Dnmt1 gene in mice was lethal with significantly reduced levels of DNA methylation (E. Li, et al., 1992). The preference of Dnmt1 for hemimethylated CpG sites that appear after DNA replication is the key property for this function. After DNA replication, methylated and unmethylated CpG sites are transformed into hemimethylated and unmethylated sites, respectively. Dnmt1 shows a 15- to 50- fold preference for methylation at hemimethylated sites compared with unmethylated sites (Fatemi, et al., 2001; Pradhan, et al., 1999; Zucker, et al., 1985). Dnmt1 acts on hemimethylated sites to ensure maintenance of DNA methylation patterns from parental to daughter genomes, which also makes it a major target of current pharmaceutical interest. 5-aza-cytidine, which is an inhibitor of Dnmt1, has been approved by US Food and Drug Administration (FDA) for the treatment of all subtypes of myelodysplatic syndromes (MDS) (Kaminskas, et al., 2005).

Dnmt1 was also found to have *de novo* methylation activity in mouse embryo lysates (Yoder, et al., 1997) and CpG island *de novo* methylation is related to Dnmt1 (Feltus, et al., 2003; Jair, et al., 2006). Additionally, Dnmt1 is required for maintenance of methylation in non-CpG DNA methylation (Grandjean, et al., 2007). Dnmt1 also interacts with a number of proteins, such as methyl-CpG binding proteins, histone deacetylases and histone methyltransferases, forming a complicated network regulating chromatin organization and gene expression (Kimura & Shiota, 2003).

3.2 Dnmt2

Dnmt2 was identified through the analysis of expressed sequence tags (EST) databases during a search for *de novo* DNA methyltransferases in 1998 (Van den Wyngaert, et al., 1998; Yoder & Bestor, 1998). Although the existence of all the consensus catalytic motifs strongly suggests DNA methyltranferase activity, no catalytic enzyme activity could be detected in initial assays. Dnmt2-deficient mouse embryonic stem cells were viable and no obvious DNA methylation changes were observed (Okano, et al., 1998b). The function of Dnmt2 remained enigmatic until a comparatively weak DNA methyltransferase activity was demonstrated both *in vitro* and *in vivo* (Hermann, et al., 2003; Kunert, et al., 2003; Liu, et al., 2003; Tang, et al., 2003).

The weak enzyme activity, however, is not sufficient explanation for the surprisingly extensive conservation of Dnmt2 among different species ranging from *Schizosaccharomyces pombe* to Homo sapiens, which indicates an important biological role. Moreover, the presence of Dnmt2 is not always accompanied by the presence of DNA methylation, which suggests Dnmt2 has additional functions (Schaefer & Lyko, 2009). Studies have uncovered a robust transfer RNA (tRNA) methyltransferase activity of Dnmt2 through a DNA methyltransferase-like catalytic mechanism (Goll, et al., 2006; Jurkowski, et al., 2008). The functional significance of the dual DNA and RNA methyltransferase activity is still undiscovered. More work needs to be done to force the closed door of Dnmt2 open.

3.3 Dnmt3 family

The Dnmt3 family involves two active enzymes (Dnmt3a and Dnmt3b) and another Dnmt-like protein (Dnmt3L). Dnmt3a and Dnmt3b were identified in the human and the mouse through a search of EST databases using full-length bacterial type II cytosine-5 methyltransferase sequences. They share similar domain arrangements, including a variable region at the N-terminal, which is involved in enzyme targeting, and a C-terminal catalytic domain (Chen, et al., 2004).

Dnmt3a and Dnmt3b are highly expressed in undifferentiated embryonic stem (ES) cells. Though Dnmt3a expression can be readily detected in most adult tissues, Dnmt3b is expressed at very low levels in most tissues except testis, thyroid and bone marrow (Okano, et al., 1998a; Xie, et al., 1999). The methylation activities of Dnmt3a and Dnmt3b for unmethylated DNA and hemimethylated DNA are comparable (Okano, et al., 1998a). Inactivation of Dnmt3a and Dnmt3b in mouse ES cells impairs *de novo* methylation and causes postnatal and embryonic lethality respectively. Consequently, they were assigned to be responsible for the establishment of DNA methylation pattern during embryogenesis (Okano, et al., 1999).

Encoded by different genes, Dnmt3a and Dnmt3b are functionally overlapping. The embryonic defects in mouse ES cells deficient in both Dnmt3a and Dnmt3b are more severe than in single mutant mice (Okano, et al., 1999). The two enzymes, however, possesses different functions during embryogenesis. Dnmt3b is the major *de novo* DNA methyltransferase detected at embryonic day 4.5-7.0 (E4.5–7.0) in mouse embryonic cells, whereas Dnmt3a is significantly and ubiquitously detectable after E10.5 when Dnmt3b is below the detection level (Watanabe, et al., 2002). Mutation of Dnmt3b is identified to be responsible for a rare autosomal recessive disorder, which is termed as ICF syndrome, characterized by immunodeficiency, facial anomalies and centromere instability (Hansen, et al., 1999). Dnmt3b also participates in maintaining aberrant hypermethylation in colorectal cancer cells, acting cooperatively with Dnmt1 (Rhee, et al., 2002). Lately, high recurrence of Dnmt3a mutations has been reported in acute myeloid leukemia (Ley, et al., 2010; Shah & Licht, 2011). In addition, certain CpG sites within the Fgf-1 gene locus have been proved to be selectively methylated by Dnmt3a *in vivo* but not by Dnmt3b (Oka, et al., 2006). The mechanism and significance of this activity are still undiscovered.

Although the third member of Dnmt3 family, Dnmt3L, shows a high degree of structural similarities of its N-terminal PHD-like zinc finger domain to Dnmt3a and Dnmt3b, it has been shown to have no catalytic activity. Key catalytic residues in C-terminal that are necessary for DNA methyltransferase activity are missing in Dnmt3L (Aapola, et al., 2000; Cheng & Blumenthal, 2008). However, Dnmt3L was reported to play an important role in establishment of DNA methylation in maternally imprinted genes in mice (Hata, et al., 2002). Deletion of the Dnmt3L gene results in aberrant *de novo* methylation of dispersed repeated DNA sequences in male germ cells (Bourc'his & Bestor, 2004). Besides, Dnmt3L has been shown to bind and colocalize with Dnmt3a and Dnmt3b in the nuclei of mammalian cells. It also stimulates *de novo* methylation by Dnmt3a in human cell lines (Chedin, et al., 2002; Hata, et al., 2002).

4. CpG content and methylation pattern in the DNA of mammalian cells and relationship to gene activity

In mammalian genomes, the abundance of CpG dinucleotides is less than expected on the basis of GC content. In human DNA, CpG incidence is only about 25% of what is expected considering the base composition (Saxonov, et al., 2006). Moreover, the CpG dinucleotides are unevenly distributed in the genome, forming clusters called CpG islands. According to the complete genomic sequence of human chromosomes 21 and 22, the CpG islands are characterized as DNA regions >500 bp with a GC content >55% and observed

CpG/expected CpG of 0.65 (Takai & Jones, 2002). CpG islands occur at or near about 40% of mammalian promoters, and play crucial roles in the regulation of gene expression (Fatemi, et al., 2005). In accordance with this, bioinformatic evaluation showed a bimodal distribution of promoters segregating them into high CpG content and low CpG content classes (HCP and LCP, respectively) (Nagae, et al., 2011; Saxonov, et al., 2006). The two distributions overlap, and promoters in the overlapping region form a class of intermediate CpG content promoters (ICP, or "weak" CpG-promoters) (Weber, et al., 2007).

The majority of CpG sites are cytosine-methylated in genomic DNA with percentages between 60%~90% depending on the source reports (Gruenbaum, et al., 1981; Razin, et al., 1984). The exceptions are CpG islands, where CpG sites are usually unmethylated. It follows that HCPs are usually undermethylated while LCPs are highly methylated at a level not different from the rest of genomic DNA. According to a genome-wide study (Weber, et al., 2007), 65% of established human promoters belong to the HCP class, but only 25% of the hypermethylated promoters are from this category. On the other hand, only 25% of promoters classify as LCP, but 42% of hypermethylated promoters belong to the LCP group. Notably, matching the promoter CpG density and methylation data with gene expression information in public databases has revealed that the undermethylated HCP-controlled genes often perform general "housekeeping" functions and are expressed in many tissues, while the highly methylated LCPs are frequently associated with tissue–specific functions and exhibit more restricted expression (Nagae, et al., 2011; Saxonov, et al., 2006; Weber, et al., 2007). Genome-scale analyses have also demonstrated that a small percentage of HCPs (3-4%) is hypermethylated in somatic cells (Meissner, et al., 2008; Shen, et al., 2007; Weber, et al., 2007). More detailed examination of this subgroup has shown that it is enriched in germline-specific genes and in germline-derived cells, where the genes function, these HCPs are undermethylated. Collectively the above relationships between CpG density, methylation level and expression suggest that hypermethylation of HCPs is incompatible with gene activity and controls the tissue specific expression of a small group of genes in differentiated cells. The experimentally determined genome-wide relationship between CpG rich (HCP and ICP) promoter activity and methylation level supports this possibility (Weber, et al., 2007). The role of CpG methylation in controlling LCP-associated gene activity may be similar, but has been demonstrated only recently as discussed below.

Illingworth et al. (Illingworth, et al., 2008) devised a method to detect CpG islands experimentally and determine their methylation levels. They have found that about half of the islands overlap with annotated transcription start sites, the rest being intragenic and intergenic. Interestingly, these non-promoter CpG islands were methylated at a much higher frequency than annotated HCPs, and tissue-specific methylation was also more frequent. In another genome-wide study, methylation of intragenic CpG islands was found to correlate positively with the level of expression of the host gene (Straussman, et al., 2009). Therefore the possibility exists that many "non-promoter" CpG islands actually span unannotated transcription start sites and function as HCPs controlling the production of regulatory (suppressive) non-coding RNAs (Morris, 2009a, 2009b) in a tissue-specific and methylation-sensitive fashion. Thus, some or most of the unmethylated HCPs may be regulated indirectly by methylation-sensitive CpG island promoters in a manner analogous to imprinted genes (Latos & Barlow, 2009).

Immunoprecipitation-based genome-wide analysis of promoters showed relatively high methylation levels of LCPs without significant relationship to promoter activity (Weber, et al., 2007). Since LCPs control many genes with tissue specific expression and function, this issue required further study. Nagae et al. used a sensitive beadarray technique (HumanMethylation27 BeadChip, Illumina) to determine the methylation level of 27 587 individual CpG sites in 14 475 promoters in 21 normal human tissue samples (Nagae, et al., 2011). As expected, HCPs were generally hypomethylated in all tissues. The CpG poor LCPs were methylated more extensively; however, the methylation level of individual LCPs varied widely between tissues. Gene ontology (GO) analysis showed that the hypomethylated, but not hypermethylated, LCPs in each tissue belonged to genes closely related to tissue specific functions. Furthermore, the expression of representative sets of tissue specific hypomethylated genes was increased in the corresponding tissue, while hypermethylation (relative to the average) had no relationship to expression. These methylation-sensitive LCPs were particularly enriched in transcription factor recognition motifs. Thus, LCPs are also regulated negatively by methylation, possibly by blocking transcription factor binding/activity at critical CpG sites. These results, together with the additional observation that many tissue-specific hypomethylated LCPs were fully methylated in embryonic stem cells and induced pluripotent cells (Nagae, et al., 2011), raise the possibility that terminal differentiation is associated with the demethylation of tissue-specific LCP-driven genes.

Genomic imprinting, which suppresses either the paternal or the maternal alleles of defined gene clusters, and X chromosome inactivation in females are the best characterised examples of gene silencing by CpG methylation (Heard, et al., 1997; Illingworth, et al., 2008; Latos & Barlow, 2009). Both of these processes are integrated with development and will be discussed further below.

Finally, small, but detectable, non-CpG (notably CpA and CpT) methylation has been reported in mammalian genomes (Kouidou, et al., 2006; Ramsahoye, et al., 2000; White, et al., 2002). Bernard H. et al (Ramsahoye, et al., 2000) reported that 15~20% of total cytosine methylation in ESCs (embryonic stem cells) is at non-CpG sites, mostly at CpA and, to a less extent, at CpT. However, the establishment and maintenance of non-CpG methylation is unclear. It is reported that similar to CpG site methylation, non-CpG site methylation is also mediated by Dnmt3 and maintained by Dnmt1 (Grandjean, et al., 2007; Ramsahoye, et al., 2000; White, et al., 2002).

5. Mechanisms of DNA methylation-mediated transcriptional repression

As described in the previous section, DNA methylation is strongly associated with the suppression of gene expression. Two main mechanisms are involved in the repression and they are biologically relevant. First, DNA methylation has been shown to directly interfere with the binding of some transcription factors to the CpG sites. It is known that many factors bind CpG-containing regions to stimulate gene expression, and a number of these consensus binding motifs (CRE, ETS, NRF-1, E-Box, AP2, etc, see ref. (Rozenberg, et al., 2008) and references therein) fail to bind or function if the CpG sites are methylated (Campanero, et al., 2000; Pierard, et al.; Sunahori, et al., 2009).

The second mechanism involves recruitment of methyl-CpG-binding domain (MBD) proteins, which selectively recognize methylated CpG sites and silence associated genes by recruiting transcriptional corepressor complexes involving histone deacetylases (HDACs) and histone methyltransferases (P. A. Jones & Laird, 1999; Munro, et al.; Prendergast & Ziff, 1991). A family of MBD proteins has been identified that include MeCP2, MBD1, MBD2, MBD3 and MBD4 (Hendrich & Bird, 1998). The MBD domains show homology among these proteins, whereas the transcription repression domains (TRD) identified in MeCP2, MBD1 and MBD2 are non-conserved.

MeCP2 is the founder of the MBD protein family, which represses gene expression by recruiting the corepressor mSin3A that interacts with HDAC1 (P. L. Jones, et al., 1998; Nan, et al., 1998). Mutation in the MeCP2 gene in the X-chromosome accounts for the vast majority of RTT (Rett) syndrome cases, which is a postnatal neurodevelopment disorder characterized by mental retardation, ataxia, hand stereotypes, seizures, and breathing irregularities (Amir, et al., 1999). MeCP2 represses gene expression by recruiting mSin3A, which interacts with HDAC1 (P. L. Jones, et al., 1998; Nan, et al., 1998). Mammalian MBD3 does not directly bind to CpG sites because of a mutation of the MBD (Saito & Ishikawa, 2002). However, it is essential for embryogenesis, since targeted deletion of MBD3 results in mouse embryo lethality (Hendrich, et al., 2001). There is little evidence of MBD4 acting as a transcription repressor, instead, it is an important thymine DNA glycosylase involved in DNA repair (Hendrich, et al., 1999).

6. 5-hydroxymethylcytosine (5hmC), the sixth base of mammalian DNA

Besides 5-methylcytosine (5mC), mammalian genomic DNA also contains 5-hydroxymethylcytosine (5hmC), which is the oxidative modification of 5mC and is recognized as the sixth base of DNA. The first report of 5hmC in mammals was in 1972 (Penn, et al., 1972). Little attention, however, was attracted because of the lack of consistently reproducible data (Kothari & Shankar, 1976). Two recent reports, however, brought 5hmC back into the limelight (Kriaucionis & Heintz, 2009; Tahiliani, et al., 2009). Kriaucionis and Heintz detected 5hmC in mouse Purkinje cells and granule cells as 0.6% and 0.2% of total nucleotides, respectively. Tahiliani et al detected an even lower percentage of 5hmC in mouse ES cells, only 0.03% of all bases. After these discoveries, a number of methods were developed to distinguish 5hmC from 5mC, since traditional bisulfite sequencing or methylation-sensitive restriction digestion do not differentiate between the two modified bases (Huang, et al., 2010; Jin, et al., 2010).

Munzel et al (Munzel, et al., 2010) have found that 5hmC is widely distributed in mouse brain. Relatively high levels of 5hmC were found in areas responsible for higher cognitive functions, such as in hippocampus and cortex. The level of 5hmC in mouse hippocampus increased by approximately 75% in 90-day-old mice compared to in one-day-old mice, which was unrelated to oxidative DNA damage due to aging. Coincidentally, Song et al (Song, et al., 2010) reported an increase of 5hmc level from 0.1% of total nucleotides in post-natal day 7 to 0.4% in adult age in mouse cerebellum tissue. 5hmC is speculated to play an important role in central nervous system development.

The generation of 5hmC is based on the pre-existence of 5mC (Ficz, et al., 2011; Szwagierczak, et al., 2010; Williams, et al., 2011). The conversion of 5mC to 5hmC is achieved by the ten-eleven translocation (TET) enzyme family that has been identified in a homology search using the sequences of JBP1 and JPB2, which are responsible for hydroxylation and glucosylation of the 5-methyl group of thymine (Tahiliani, et al., 2009; Yu, et al., 2007). Three members of TET family have been uncovered, TET1, TET2 and TET3. Lately, the TET protein has been reported to catalyse the generation of 5-carboxylcytosine (5caC) and 5-formylcytosine (5fC), which might be involved in the active demethylation of DNA (He, et al., 2011; Ito, et al., 2011).

The functional consequences of the 5hmC modification are still unclear. Like other modifications, it may alter chromatin organization by recruiting or excluding factors that influence transcription (S. C. Wu & Zhang, 2010). Importantly, it is proposed to be involved in the enigmatic process of DNA demethylation. First, 5hmC cannot be recognised by Dnmt1. Therefore, it interferes with the Dnmt1-mediated maintenance of DNA methylation, which results in passive demethylation during replication (Valinluck & Sowers, 2007). Second, it may be a key intermediate in the controversial process of active demethylation. We discuss this in detail in the section on DNA demethylation below. Additionally, it was reported that the MBD family, including MeCP2, MBD1, MBD2 and MBD4, have significantly reduced affinity for CpG sites that are hydroxymethylated (Jin, et al., 2010; Valinluck, et al., 2004). Given the repressive transcriptional role of MBD proteins and their interactions with histone modifying enzymes, the conversion from 5mC to 5hmC could reduce MDB activity at the affected sites and change epigenetic patterns to promote transcriptional activity.

7. DNA demethylation in the zygote

DNA methylation is a stable but not irreversible epigenetic mark. The reverse process, DNA demethylation, has been observed in specific contexts. Genome-wide DNA demethylation occurs after fertilization, when the existing DNA methylation pattern of the paternal and maternal genomes is erased and the epigenetic state of the early embryo is reset (Hajkova, et al., 2002; Mayer, et al., 2000).

The loss of DNA methylation is achieved in two distinct ways, passive demethylation and active demethylation. Passive demethylation occurs due to replication-dependent inhibition or absence of Dnmt1, which results in lack of DNA methylation in the newly synthesized DNA. Active demethylation refers to actively removing the methyl group, the methylated cytosine or the whole nucleotide; the mechanisms in mammals are still poorly understood and controversial.

Various mechanisms of active demethylation in mammals have been proposed, including direct enzymatic removal of the methyl group from 5mC, nucleotide-replacement reaction and further modifications of 5mC followed by a base excision repair (BER) pathway.

Direct removal of the methyl group from 5mC seems to be the most direct way to achieve demethylation, however, a robust enzyme activity of a "demethylase"is needed to break the strong carbon-carbon bond. One of the MBD proteins, MBD2, is the first to be identified to serve as a DNA demethylase. MBD2 was shown *in vitro* to remove the methyl group and release in the form of methanol (Bhattacharya, et al., 1999). However, this mechanism was strongly contested because several groups failed to reproduce the enzyme activity in different genes (Boeke, et al., 2000; H. H. Ng, et al., 1999). Moreover, it has been revealed that MBD2-null mice still exhibit normal pattern of DNA methylation (Hendrich, et al., 2001), and the active demethylation of the paternal genome still occurs normally in zygote deficient in MBD2 (Santos, et al., 2002), which suggests that the major demethylation mechanisms do not involve MBD2.

The nucleotide-replacement reaction is commonly involved in substantial DNA repair following damage due to exposure to radiation or chemicals. The basic steps include dual incisions flanking the lesion, release of a 24~32 nucleotide oligomer and replacement with new synthesized nucleotides followed by ligation (Sancar, et al., 2004). A DNA repair protein, growth arrest and DNA-damage-inducible 45a (Gadd45a), was reported to participate in active demethylation. Knockdown of Gadd45a leads to hypermethylation and gene silencing, whereas overexpression results in activation of loci-specific gene and global demethylation (Barreto, et al., 2007). Again, other studies failed to confirm the role of Gadd45a in active DNA demethylation, which raises doubts on this pathway (Engel, et al., 2009; Jin, et al., 2008).

The DNA glycosylase and base excision repair (BER) system mediated active demethylation is the main mechanism utilized by plants. The key steps involve a group of 5mC glycosylases (including ROS1, DME, DML2 and DML3) removing the methylated cytosine, followed by a BER pathway to fill in the gap with an unmethylated cytosine (J. K. Zhu, 2009). However, evidence supporting this mechanism in mammals seems less convincing due to the lack of 5mC glycosylase in mammalian cells. Although MBD4 and thymine DNA glycosylase (TDG), two confirmed T•G mismatch glycosylases, have been reported to have weak activity against 5mC: G *in vitro,* this activity is 30~40 fold lower compared to the T: G mismatch (B. Zhu, Zheng, Angliker, et al., 2000; B. Zhu, Zheng, Hess, et al., 2000). In addition, deletion of MBD4 does not affect the active demethylation of the paternal genome in zygotes, which challenges the possibility that MBD4 is the principal DNA demethylase operating after fertilization (Santos & Dean, 2004).

It has also been proposed that active demethylation can be achieved by the deamination of 5mC to generate T followed by the BER pathway to replace the T with an unmethylated cytosine. The activation-induced deaminase (AID) and apolipoprotein B mRNA-editing catalytic polypeptides (Apobec) family are able to deaminate 5mC (Morgan, et al., 2004). Recent work has also shown that AID is necessary for promoter demethylation of the OCT4 and NANOG genes in the reprogramming of human somatic cells to pluripotent stem cells (Bhutani, et al., 2010). However, AID is active on single-stranded, but not on double-stranded DNA and its affinity to 5mC is 10-fold lower than to unmethylated cytosine (Bransteitter, et al., 2003). Surprisingly, Dnmt3a and Dnmt3b have also been shown to possess 5mC deaminase activity *in vitro,* which may be involved in the CpG methylation/demethylation cycling observed at the pS2/TFF1 gene promoter after estrogen stimulation (Metivier, et al., 2008). More work needs to be done to uncover the mechanism and significance of these dual functions.

Recently, an oxidative-deamination-BER pathway (Fig.2) of active demethylation has been described with 5hmC as the key intermediate. The TET proteins have been shown to oxidize 5mC to generate 5hmC that is acted upon by the AID and Apobec family deaminases to generate 5hmU for subsequent processing by BER (Guo, et al., 2011). TDG exhibits robust glycosylase activity on the 5hmU•G mismatch to remove 5hmU and the gap is filled by an unmethylated cytosine. Both knockout and inactive point mutations of TDG result in mice embryo lethality and developmental defect (Cortellino, et al., 2011). In addition, it has been demonstrated that 5mC and 5hmC can be further modified to 5-formylcytosine (5fC) and 5-carboxylcytosine (5caC) by TET proteins. The existence of 5fC and 5caC has been identified in mice ES cells (Ito, et al., 2011). TDG is able to recognize and cleave 5caC and 5fC to initiate the BER pathway (He, et al., 2011; Maiti & Drohat, 2011). It is also proposed that an unknown decarboxylase enzyme is involved in the conversion from 5caC to C, without the participation of BER (Ito, et al., 2011).

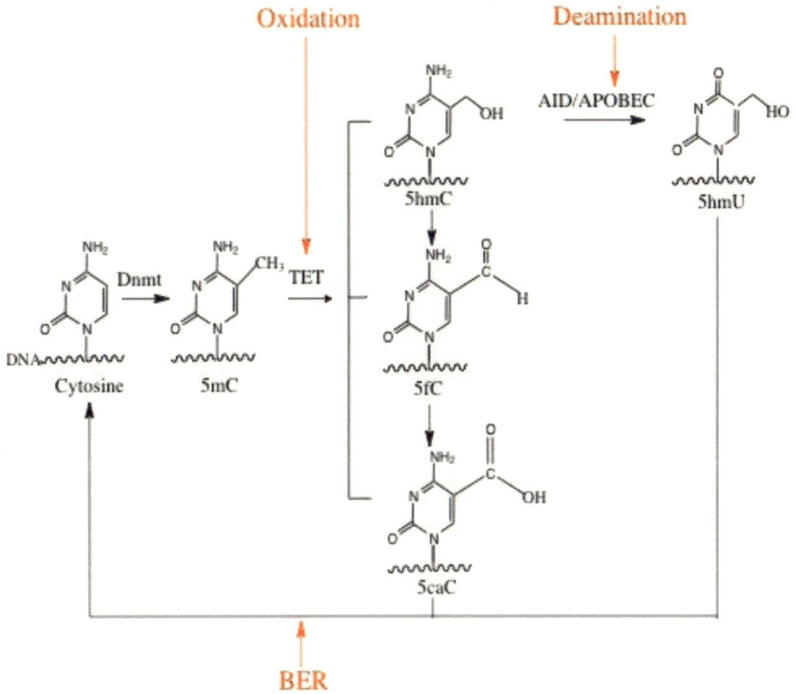

Fig. 2. Pathways of active DNA demethylation.

5mC can be hydroxylated by TET to generate 5hmC, which can be further oxidized to 5fC and 5caC. Alternatively, 5hmC can be deaminased by AID and Apobec family members to generate 5hmU. Base excision repair (BER) will be initiated by modified bases, eventually replacing them with cytosine.

8. Dynamic of DNA methylation in development

Extensive reprogramming of epigenomes occurs during early development, allowing cells to attain a totipotent developmental potential. Remodeling of DNA methylation patterns, including a genome-wide demethylation with imprinted genes as exceptions, is observed in mammalian zygotes, primordial germ cells and early embryos. This process is essential for erasure of the previous germ line DNA methylation pattern and establishment of new marks.

8.1 Asymmetric genome-wide demethylation

The genomes of sperm and oocytes are organised differently. The genome of sperm is packaged with protamines, while oocytes are packaged with histones. At fertilization, a rapid exchange from protamine to histone occurs in the paternal genome. As early as a few hours after fertilization, a remarkable asymmetric DNA demethylation occurs in the two parental genomes, although they are exposed to the same oocyte cytoplasm. The paternal genome undergoes a genome-wide demethylation before the first cell division cycle, which is therefore considered active demethylation, while the maternal genome is protected from this reprogramming (Oswald, et al., 2000). Notably, some regions of the paternal genome are still resistant to demethylation, including imprinted genes.

Recent work from two laboratories has linked this loss of 5mC to its conversion to 5hmC (Iqbal, et al., 2011; Wossidlo, et al., 2011). Analysis of mouse, rabbit and bovine embryos showed a strong presence of 5hmC in the paternal pronucleus, while the amount of 5mC decreased. On the other hand, 5hmC in the maternal pronucleus was nearly undetectable, compared to the abundance of 5mC. This suggests that 5mC in the paternal, but not in the maternal, pronuclei is converted to 5hmC possibly as part of the active demethylation pathway outlined above. The distribution differences of 5hmC in the two parental pronucleus were maintained towards the two-cell stage. It was also shown that TET3 contributed significantly to this conversion. Tet3 exhibits extremely high expression levels in oocytes and zygotes compared to the very low levels of TET1 and TET2. Levels of TET3 decrease dramatically at the two-cell stage, which stops the further increase of 5hmC levels. Knockdown of TET3 results in substantial reduction of 5hmC and a strong increase of 5mC in the paternal pronucleus. Consistent with the lack of active demethylation, the maternal pronucleus is inaccessible to this TET3-mediated conversion from 5mC to 5hmC. Knockdown of TET3 has little effect on the 5mC signal in the maternal pronucleus.

An intriguing question is how the maternal pronucleus is protected from active demethylation. It has been proposed that the PGC7/Stella gene is important for this protection, since the maternal genome of the PGC7/Stella-null zygote undergoes significant demethylation (Coward, et al., 2009). It has also been reported that the absence of PGC7/Stella leads to the increase of 5hmC in the maternal pronucleus, which suggests PGC7/Stella as a protector of 5mC against oxidation (Wossidlo, et al., 2011). The mechanism of the protection, however, remains unclear. A further question is why the presence of PGC7/Stella does not protect the paternal pronucleus from demethylation. Structural differences may be the answer.

The maternal genome undergoes gradual passive demethylation during the several cell divisions leading towards blastocyst formation due to a lack of maintenance of methylation in the zygote. Imprinted genes are resistant to this process, as well as to the active demethylation in the paternal genome.

8.2 Genomic imprinting

Although extensive DNA methylation reprogramming occurs in early development, a number of genes are resistant to this change including imprinted genes. Gene imprinting is an enigmatic phenomenon that causes genes to be expressed in a parental-origin-specific way, wherein either the paternally or the maternally inherited allele is silenced.

Approximately 60 imprinted genes have been identified in the human (R. K. Yuen, et al., 2011). Genomic imprinting plays a significant role in normal development through regulating embryo growth, placental function and neurobehavioral processes (McGrath & Solter, 1984; Surani, et al., 1984). Abnormal expression of imprinted genes has been demonstrated to be associated with a few human diseases, including the Prader-Willi syndrome (PWS), the Angelman syndrome (AS), the Silver-Russell syndrome (SRS) and the Beckwith-Wiedemann syndrome (BWS) (Paulsen & Ferguson-Smith, 2001).

DNA methylation is the key feature of imprinted gene clusters. The regions at imprinted loci marked by DNA methylation (DMRs) are methylated only on one allele. The product of one gene in an imprinted cluster is often a non-coding RNA, which regulates the expression of the other members of the cluster *in cis*. The DMRs established through the germ line often coincide with crucial imprinting control regions (ICRs), which if methylated, suppress the expression of the regulatory non-coding RNA. During development of primordial germ cells, DNA methylation is erased and then re-established in the process of oogenesis and spermatogenesis to generate gender specific germline imprinting patterns. Once established, the imprinted status will be stably maintained in all somatic lineages. Other DMRs acquire this methylation status after fertilization and are often tissue-specific (J. R. Mann, 2001). The mechanisms of the resistance of imprinted genes to epigenetic reprogramming post-fertilization are complex and still not fully characterised (see ref. (Weaver, et al., 2009)for review). A few factors have been suggested to be involved, including the DNMT1 isoforms DNMT1o and DNMT1s, MDB3 and PGC7/Stella, which also participates in the protection of the maternal genome from post-fertilization active demethylation (Nakamura, et al., 2007). ZFP57, which is a KRAB zinc finger protein, has been demonstrated to be required for maintenance of maternal methylation imprints at the *Snrpn* imprinted region. It is also important for maintenance of genomic imprinting at multiple DMRs (X. Li, et al., 2008).

Dysregulation of imprinted genes are associated with a number of human syndromes that will be discussed in the following sections.

8.3 DNA methylation in germ cell development

After implantation of the blastocyst, a subset of pluripotent cells in the epiblast generate primordial germ cells (PGCs), the origin of both oocytes and spermatozoa, in responding to signals from extraembryonic ectoderm and visceral endoderm. Two critical transcription factors, B-lymphocyte maturation-induced protein1 (BLIMP1, also known as PR domain zinc finger protein1)) and PRDM14 are essential for specification of PGCs and suppression of somatic gene expression (Ohinata, et al., 2005; Saitou, 2009).

When the germ cell fate is established, the level of global DNA methylation is similar to that observed in surrounding epiblast cells (Seki, et al., 2005). The cells then gradually migrate to the genital ridge accompanied by significant reprogramming of epigenetic marks. A substantial reduction of genome-wide DNA methylation has been observed after specification of germ cells. However, it is not until arriving at genital ridge that DNA methylation including parental imprinting marks is extensively erased (Hajkova, et al., 2002). Due to the presence of Dnmt1 for most of the period, this widespread demethylation is considered to be active, replication-independent. BER pathway has been reported to be involved in the process. Interestingly, significant expression of TET1 gene has been detected in mouse PGCs, indicating a role of 5hmC as an intermediate in this active demethylation (Hajkova, et al., 2010).

In female mammals, one of the two copies of X chromosomes is inactivated to balance the X-linked gene dosage between females and males. The process is termed X chromosome inactivation and is stably maintained in female somatic cells by the DNA methylation mark at the silenced genes. In female germ cell development, the inactivated X chromosome is reactivated. This reactivation occurs during migration and gonadal colonization, which coincides with genome-wide DNA demethylation.

In the global demethylation of PGCs, parental imprinting marks are erased and reset in a gender specific fashion. In male germ cells, the new imprints are progressively established during late fetal development and are not completed until the mature sperm stage (J. Y. Li, et al., 2004). In contrast, the initiation of methylation imprints occurs postnatally during the oocyte growth phase (Lucifero, et al., 2004). The Dnmt3 family is responsible for the *de novo* methylation of imprinted DMRs. Knockout of Dnmt3a and Dnmt3b in mouse germ cell line revealed that although they are both involved, Dnmt3a plays the major role in establishment of imprinting with Dnmt3L as a critical co-factor (Kaneda, et al., 2004; Kato, et al., 2007).

8.4 DNA methylation in cellular differentiation during development

The evidence summarized above demonstrates that DNA (CpG) methylation is a critically important component of all aspects of cellular differentiation from fertilization to senescence. At the molecular level, CpG methylation of DNA results in gene silencing. X-chromosome inactivation and genomic imprinting are well characterised, specific examples of this phenomenon. In the developmental time line, the paternally and maternally inherited DNA methylation patterns are quickly erased in the zygote after fertilization by active and passive demethylation, respectively. DNA methylation levels in the genome reach a nadir at the morula stage (Abdalla, et al., 2009; Weaver, et al., 2009). In the subsequent phase of blastocyst formation, a primary differentiation event occurs, giving rise to two distinct cell lineages, the inner cell mass (ICM) and the trophectoderm. The ICM will generate the embryo proper as well as extraembryonic endoderm and mesoderm, while the trophectoderm will form the chorion and the placenta (Fleming, 1987). During blastocyst differentiation, a wave of *de novo* methylation is carried out by Dnmt3 family members, which leads to different methylation patterns between the ICM and trophectoderm. This occurs around the time of implantation. The main *de novo* Dnmt at the blastocyst stage, Dnmt3b, has been specifically localized in the ICM of mouse blastocysts (Watanabe, et al., 2002). Thus the pluripotent stem cells of the ICM show much higher levels of DNA methylation compared to the trophectoderm; and this lineage-based methylation difference persists into the later stage of development, placenta and embryo proper (Santos, et al., 2002). The embryonic stem cells lose cell lineage restriction and differentiate into trophoblast cells after the deletion of Dnmt1, which implicates DNA methylation as the key factor to achieve cell lineage stability. At the gene expression level, the robust difference in methylation of a transcription factor encoding gene, *Elf5*, has been shown to be critical in cell lineage fate. It is hypermethylated and repressed in the embryonic cell lineage but hypomethylated and expressed in the trophoblast compartment (R. K. Ng, et al., 2008).

The general genomic pattern of CpG methylation is established during ICM differentiation in the blastocyst at the time of implantation. Thus, CpG-rich promoters (HCPs and ICPs) remain unmethylated, probably due to the protective effect of cis-acting sequence motifs (Straussman, et al., 2009). On the other hand, CpG-poor promoters (LCPs), often controlling genes with tissue-specific expression are highly methylated possibly to prevent untimely expression in the pluripotent embryonic stem cells (ESCs) (Nagae, et al., 2011). The exceptions include the pluripotency and ESC-self renewal-associated genes, which often belong to the LCP class but are unmethylated in ESCs (e.g. *Nanog, Pou5f1, Gdf3* (Straussman, et al., 2009)). Recent work suggests that the demethylation of *Nanog* is a TET1-dependent active process maintaining Nanog expression in mouse ESCs (Ito, et al., 2010). This example raises the possibility that other LCP class promoters also undergo targeted active demethylation in ESCs.

Differentiation of ESCs to various somatic cell types is also associated with changing DNA methylation of functionally relevant promoters. A small proportion of HCPs, controlling germline-specific genes, becomes methylated *de novo* in all cell types except for germline-derived cells (Shen, et al., 2007; Weber, et al., 2007), where they are expressed. Similarly, ESC-specific HCP-controlled genes are shut down in differentiated cells by *de novo* methylation of the promoter (Straussman, et al., 2009). CpG-island sequences at gene loci controlling development, such as *OSR1, PAX6* and *HOXC* are frequently methylated in normal tissues; however, these methylations do not appear to correlate closely with gene expression levels (Illingworth, et al., 2008). The mechanisms that select these HCPs for methylation are unclear, but may involve other epigenetic marks such as histone modifications (Meissner, 2010). CpG-poor promoters (LCPs) undergo tissue-specific demethylation during differentiation, which correlates with the increased activity of functionally relevant genes (Nagae, et al., 2011). The mechanism of this gene and tissue selective process is unclear; it may involve the newly discovered TET-dependent active demethylation, or it may be passive demethylation driven by transcription factor binding to the promoter, which may prevent maintenance methylation after cell division.

The pattern of DNA methylation is considered stable after differentiation as long as cellular identity is maintained. Recent studies, however, have shown that the CpG methylation pattern changes during life. Christensen et al., (Christensen, et al., 2009) analysed 1413 CpG sites at 773 genes in 217 healthy individuals, and observed that CpG island methylation levels increased, while non-CpG island methylation levels decreased with age. Environmental factors, such as tobacco smoking, also affected CpG methylation levels in individuals. Yuen et al. (R. K. C. Yuen, et al., 2011) compared the methylation status of 1315 CpG loci in 752 genes from five somatic tissues between normal second trimester fetuses and adults. They found tissue-specific differentially methylated regions in 195 loci in the fetuses, but only 17% of these were maintained in adults. Further, the methylation status of about 10% of the examined sites changed more than 40% between the fetus and the adult. These data indicate the plasticity of the tissue-specific DNA methylation patterns and their susceptibility to environmental conditions.

9. DNA methylation and human diseases

Age and the environment have key importance in many disease processes; therefore it is not surprising that aberrant DNA methylation contributes to a variety of pathologies. Epigenetic changes have been firmly established as components of malignant transformation; however, this large area of investigation has been recently reviewed (Brait & Sidransky, 2011) and is beyond the scope of this chapter (Esteller, 2011). In recent years, the relationship between adverse intrauterine conditions at early development and increased risk of post-natal diseases has been recognized.

Chronic diseases including cardiovascular disease, type 2 diabetes and obesity are associated with abnormal intrauterine growth and development (Hales & Barker, 1992; Osmond, et al., 1993; Rich-Edwards, et al., 2005). This relationship has been termed as the Developmental Origins of Health and Disease (DOHaD). The hypothesis of DOHaD indicates a high degree of phenotypic plasticity during development. Epigenetic regulation have been suggested to be the most attractive mediator between transient environmental exposures and sustained changes at gene, cell or tissues levels. Especially, aberrant DNA methylation of key disease-associated genes has been suggested to be involved in DOHaD.

9.1 Role of maternal diet

Maternal malnutrition is a comparatively well characterized factor influencing DNA methylation of genes in the context of DOHaD (See ref. (Burdge & Lillycrop, 2010) for review). Further, several studies have reported that altered maternal intake of folate is linked to change of DNA methylation in the offspring (Kim, et al., 2009; Waterland, et al., 2006; Waterland & Jirtle, 2003). Folate deficiency has been strongly associated with incomplete closure of the neural tube. Suppression of methylation cycles, which can be the consequence of insufficient maternal intake of folate, causes a high risk of neural tube defect in mice (Dunlevy, et al., 2006). Recently, it has been reported that reduced maternal folate status may contribute to the development of colorectal cancer in adult offspring (McKay, et al., 2011). Maternal folate depletion is associated with a locus-specific drop of DNA methylation in the *Slc39a4* gene in fetal gut. The *Slc39a4* gene has been shown to be hypomethylated in colorectal cancer. Thus the deficiency of folate in pregnancy might have consequence for colorectal cancer development if the altered methylation is sustained into adulthood.

Folate supplementation during pregnancy has been widely used to reduce the incidence of neural tube defects. It has been suspected, however, that folate may have adverse effects inducing allergic diseases such as asthma and eczema by altering the methylation status of DNA in the offspring (Dunstan, et al., 2011; Hollingsworth, et al., 2008).

Apart from nutrition, maternal behaviour (level of care, depression) has also been shown to alter the methylation of steroid receptor gene promoters in the offspring and influence steroid hormone sensitivity later in life (Champagne, et al., 2006; Oberlander, et al., 2008).

9.2 Infection and inflammation

Maternal infections such as urinary tract and periodontal infection have been correlated with pregnancy complications (reviewed in ref. (Conde-Agudelo, et al., 2008)). Treatment of periodontal disease during pregnancy was reported to reduce the rate of preterm birth and lower incidence of low birth weight (Polyzos, et al., 2009). The mechanisms underlying this relationship has been linked to abnormal DNA methylation patterns (Bobetsis, et al., 2007).

Specifically, it has been shown in murine placenta tissues that maternal oral infection caused by *C. rectus* can induce hypermethylation in the promoter of imprinted insulin-like growth factor 2 (*Igf2)* gene. Deficiency of Igf2 gene expression leads to reduction of placental growth and restricted fetal growth (Constancia, et al., 2002). This suggests that DNA altered methylation (and expression) of key genes can contribute to infection-associated adverse pregnancy outcomes such as intrauterine growth restriction (IUGR).

Loss of DNA methylation in T cells has been shown in systemic lupus erythematosus (SLE), an autoimmune disease affecting multiple organs. A significant decrease of genomic DNA methylation with reduced Dnmt1 levels has been reported (Richardson, et al., 1990). Specific genes relevant to the SLE phenotype are also hypomethylated in T cells of SLE patients compared to their normal counterparts (Lu, et al., 2002; Oelke, et al., 2004). The DNA methylation inhibitor 5-aza-2′-deoxycytidine causes autoreactivity of T cells *in vitro* as well as an SLE-like disease *in vivo*, suggesting that T cell DNA hypomethylation is involved in the autoantibody response in SLE (Quddus, et al., 1993).

9.3 Endometriosis

Dysregulation of DNA methylation in several genes has been reported in endometriosis, a common gynecological condition affecting women of reproductive age. The promoter of progesterone receptor B (PR-B) has been shown to be hypermethylated in endometriosis, which may be responsible for PR-B down-regulation and the notable progesterone resistance (Y. Wu, et al., 2006). The promoter of estrogen receptor 2 (ESR2) is hypomethylated in endometriosis, which may result in the significantly increased level of estrogen receptor expression compared to the stromal cells in endometrium (Y. Wu, et al., 2006).

9.4 Disruptions of imprinting

The human chromosome 11p15.5 harbours a cluster of imprinted genes including paternally expressed insulin-like growth factor 2 (IGF2) and maternally expressed H19 and KCNQ1 (KVLQT1) genes (Paulsen, et al., 1998). DNA methylation abnormalities at 11p15.5 can cause two distinct growth disorders, the Beckwith–Wiedemann (BWS) and the Silver–Russell (SRS) syndromes. BWS is characterized by fetal and postnatal overgrowth, macroglossia, neonatal hypoglycaemia and an increased incidence of childhood tumors. SRS is characterized by severe intrauterine and postnatal growth retardation, dysmorphic facial features, feeding difficulties, and body and limb asymmetry. DNA methylation defects account for approximately 60–70% of BWS and SRS patients (Demars, et al., 2011). Two imprinting control regions, ICR1 and ICR2, control the differential expression of imprinted genes at 11p15.5. Both the paternal and maternal alleles of ICR1 and ICR2 are methylated in normal cells. The DNA methylation defects at ICR1 is usually the hypermethylation of the maternal allele, leading to both BWS and SRS, whereas DNA methylation defects at ICR2 usually involve loss of maternal-allele-specific DNA methylation, which results in only BWS (Robertson, 2005).

Prader-Willi syndrome (PWS) and Angelman syndrome (AS) are two distinct neurogenetic disorders in which the same domain on chromosome 15 is affected. AS is caused by the loss of the maternally expressed gene, UBE3A, which is only imprinted in the brain. Loss of maternal DNA methylation or maternal ICR deletion is involved in the imprinting defects, which account for ~5% of AS.

9.5 ARTs and imprinting defects

In the setting of infertility, the use of assisted reproductive technologies (ARTs) has been growing. Although the majorities of the children conceived with ARTs develop normally, recent studies have suggested a possible link between ARTs and genomic imprinting disorders. Cox et al. (2002) have reported that two children who were conceived by intracytoplasmic sperm injection (ICSI) develop AS and a third case has been reported by Orstavik et al. (2003). A loss of methylation on the maternal allele at *SNRPN* locus have been shown in all three children, while the paternal allele has normal methylation pattern, suggesting a relationship between AS and an ICSI associated imprinting defect. A few studies have also pointed to the association between the occurrence of BWS and *in vitro* fertilization (IVF) and ICSI. Hypomethylation of KCNQ1OT1 due to a chromosome 11p15.5 ICR2 defect as well as an abnormal methylation pattern of H19 have been observed in children conceived via ARTs (DeBaun, et al., 2003).

It has been proposed that *in vitro* manipulations of several steps involved in conception contribute to alteration of the normal imprinting processes. In current ARTs protocols, *in vitro* culturing of embryos is extended until the blastocyst stage before embryo transfer to allow high pregnancy rate and reduce the risk of multiple pregnancy. It has been demonstrated that culture of preimplantation embryos influences genomic imprinting marks (Khosla, et al., 2001; M. R. Mann, et al., 2004). ARTs also involve induced ovulation via hormonal stimulation and *in vitro* maturation of oocytes, which might interrupt the natural development of oocytes and the genomic imprinting marks of the maturing oocytes (Iliadou, et al., 2011).

The number of pathological conditions where aberrant DNA methylation is a contributor will likely increase as more refined technologies become available for the sensitive, accurate and high-throughput determination of CpG methylation at affordable prices. It is reasonable to expect that the complex mechanisms underlying CpG methylation, demethylation, methylation targeting and methylation protection will offer new therapeutic targets to alleviate the consequences of aberrant methylation in disease.

10. References

Aapola, U., Kawasaki, K., Scott, H. S., et al. (2000). Isolation and initial characterization of a novel zinc finger gene, DNMT3L, on 21q22.3, related to the cytosine-5-methyltransferase 3 gene family. *Genomics, 65*(3), 293-298.

Abdalla, H., Yoshizawa, Y., & Hochi, S. (2009). Active demethylation of paternal genome in mammalian zygotes. *J Reprod Dev, 55*(4), 356-360.

Amir, R. E., Van den Veyver, I. B., Wan, M., et al. (1999). Rett syndrome is caused by mutations in X-linked MECP2, encoding methyl-CpG-binding protein 2. *Nat Genet, 23*(2), 185-188.

Barreto, G., Schafer, A., Marhold, J., et al. (2007). Gadd45a promotes epigenetic gene activation by repair-mediated DNA demethylation. *Nature, 445*(7128), 671-675.

Bhattacharya, S. K., Ramchandani, S., Cervoni, N., & Szyf, M. (1999). A mammalian protein with specific demethylase activity for mCpG DNA. *Nature, 397*(6720), 579-583.

Bhutani, N., Brady, J. J., Damian, M., et al. (2010). Reprogramming towards pluripotency requires AID-dependent DNA demethylation. *Nature, 463*(7284), 1042-1047.

Bobetsis, Y. A., Barros, S. P., Lin, D. M., et al. (2007). Bacterial infection promotes DNA hypermethylation. *J Dent Res, 86*(2), 169-174.

Boeke, J., Ammerpohl, O., Kegel, S., Moehren, U., & Renkawitz, R. (2000). The minimal repression domain of MBD2b overlaps with the methyl-CpG-binding domain and binds directly to Sin3A. *J Biol Chem, 275*(45), 34963-34967.

Bourc'his, D., & Bestor, T. H. (2004). Meiotic catastrophe and retrotransposon reactivation in male germ cells lacking Dnmt3L. *Nature, 431*(7004), 96-99.

Brait, M., & Sidransky, D. (2011). Cancer epigenetics: above and beyond. *Toxicol Mech Methods, 21*(4), 275-288.

Bransteitter, R., Pham, P., Scharff, M. D., & Goodman, M. F. (2003). Activation-induced cytidine deaminase deaminates deoxycytidine on single-stranded DNA but requires the action of RNase. *Proc Natl Acad Sci U S A, 100*(7), 4102-4107.

Burdge, G. C., & Lillycrop, K. A. (2010). Nutrition, epigenetics, and developmental plasticity: implications for understanding human disease. [Research Support, Non-U.S. Gov't Review]. *Annual review of nutrition, 30*, 315-339.

Campanero, M. R., Armstrong, M. I., & Flemington, E. K. (2000). CpG methylation as a mechanism for the regulation of E2F activity. *Proc Natl Acad Sci U S A, 97*(12), 6481-6486.

Champagne, F. A., Weaver, I. C., Diorio, J., et al. (2006). Maternal care associated with methylation of the estrogen receptor-alpha1b promoter and estrogen receptor-alpha expression in the medial preoptic area of female offspring. *Endocrinology, 147*(6), 2909-2915.

Chedin, F., Lieber, M. R., & Hsieh, C. L. (2002). The DNA methyltransferase-like protein DNMT3L stimulates de novo methylation by Dnmt3a. *Proc Natl Acad Sci U S A, 99*(26), 16916-16921.

Chen, T., Tsujimoto, N., & Li, E. (2004). The PWWP domain of Dnmt3a and Dnmt3b is required for directing DNA methylation to the major satellite repeats at pericentric heterochromatin. *Mol Cell Biol, 24*(20), 9048-9058.

Cheng, X., & Blumenthal, R. M. (2008). Mammalian DNA methyltransferases: a structural perspective. *Structure, 16*(3), 341-350.

Christensen, B. C., Houseman, E. A., Marsit, C. J., et al. (2009). Aging and environmental exposures alter tissue-specific DNA methylation dependent upon CpG island context. [Research Support, N.I.H., Extramural Research Support, Non-U.S. Gov't]. *PLoS genetics, 5*(8), e1000602.

Conde-Agudelo, A., Villar, J., & Lindheimer, M. (2008). Maternal infection and risk of preeclampsia: systematic review and metaanalysis. *Am J Obstet Gynecol, 198*(1), 7-22.

Constancia, M., Hemberger, M., Hughes, J., et al. (2002). Placental-specific IGF-II is a major modulator of placental and fetal growth. *Nature, 417*(6892), 945-948.

Cortellino, S., Xu, J., Sannai, M., et al. (2011). Thymine DNA glycosylase is essential for active DNA demethylation by linked deamination-base excision repair. *Cell, 146*(1), 67-79.

Coward, W. R., Watts, K., Feghali-Bostwick, C. A., Knox, A., & Pang, L. (2009). Defective histone acetylation is responsible for the diminished expression of cyclooxygenase 2 in idiopathic pulmonary fibrosis. *Mol Cell Biol, 29*(15), 4325-4339.

DeBaun, M. R., Niemitz, E. L., & Feinberg, A. P. (2003). Association of in vitro fertilization with Beckwith-Wiedemann syndrome and epigenetic alterations of LIT1 and H19. *Am J Hum Genet, 72*(1), 156-160.

Demars, J., Rossignol, S., Netchine, I., et al. (2011). New insights into the pathogenesis of beckwith-wiedemann and silver-russell syndromes: Contribution of small copy number variations to 11p15 imprinting defects. *Hum Mutat, 32*(10), 1171-1182.

Dunlevy, L. P., Burren, K. A., Mills, K., et al. (2006). Integrity of the methylation cycle is essential for mammalian neural tube closure. *Birth Defects Res A Clin Mol Teratol, 76*(7), 544-552.

Dunstan, J. A., West, C., McCarthy, S., et al. (2011). The relationship between maternal folate status in pregnancy, cord blood folate levels, and allergic outcomes in early childhood. *Allergy*.

Engel, N., Tront, J. S., Erinle, T., et al. (2009). Conserved DNA methylation in Gadd45a(-/-) mice. *Epigenetics, 4*(2), 98-99.

Esteller, M. (2011). Cancer Epigenetics for the 21st Century. *Genes & Cancer, 2*(6), 604-606.

Fatemi, M., Hermann, A., Pradhan, S., & Jeltsch, A. (2001). The activity of the murine DNA methyltransferase Dnmt1 is controlled by interaction of the catalytic domain with the N-terminal part of the enzyme leading to an allosteric activation of the enzyme after binding to methylated DNA. *J Mol Biol, 309*(5), 1189-1199.

Fatemi, M., Pao, M. M., Jeong, S., et al. (2005). Footprinting of mammalian promoters: use of a CpG DNA methyltransferase revealing nucleosome positions at a single molecule level. *Nucleic Acids Res, 33*(20), e176.

Feltus, F. A., Lee, E. K., Costello, J. F., Plass, C., & Vertino, P. M. (2003). Predicting aberrant CpG island methylation. *Proc Natl Acad Sci U S A, 100*(21), 12253-12258.

Ficz, G., Branco, M. R., Seisenberger, S., et al. (2011). Dynamic regulation of 5-hydroxymethylcytosine in mouse ES cells and during differentiation. *Nature, 473*(7347), 398-402.

Fleming, T. P. (1987). A quantitative analysis of cell allocation to trophectoderm and inner cell mass in the mouse blastocyst. *Dev Biol, 119*(2), 520-531.

Frauer, C., Rottach, A., Meilinger, D., et al. (2011). Different binding properties and function of CXXC zinc finger domains in Dnmt1 and Tet1. *PLoS One, 6*(2), e16627.

Gerasimaite, R., Merkiene, E., & Klimasauskas, S. (2011). Direct observation of cytosine flipping and covalent catalysis in a DNA methyltransferase. *Nucleic Acids Res, 39*(9), 3771-3780.

Goll, M. G., & Bestor, T. H. (2005). Eukaryotic cytosine methyltransferases. *Annu Rev Biochem, 74*, 481-514.

Goll, M. G., Kirpekar, F., Maggert, K. A., et al. (2006). Methylation of tRNAAsp by the DNA methyltransferase homolog Dnmt2. *Science, 311*(5759), 395-398.

Grandjean, V., Yaman, R., Cuzin, F., & Rassoulzadegan, M. (2007). Inheritance of an epigenetic mark: the CpG DNA methyltransferase 1 is required for de novo establishment of a complex pattern of non-CpG methylation. *PLoS One, 2*(11), e1136.

Gruenbaum, Y., Cedar, H., & Razin, A. (1982). Substrate and sequence specificity of a eukaryotic DNA methylase. *Nature, 295*(5850), 620-622.

Gruenbaum, Y., Stein, R., Cedar, H., & Razin, A. (1981). Methylation of CpG sequences in eukaryotic DNA. *FEBS Lett, 124*(1), 67-71.

Guo, J. U., Su, Y., Zhong, C., Ming, G. L., & Song, H. (2011). Hydroxylation of 5-methylcytosine by TET1 promotes active DNA demethylation in the adult brain. *Cell, 145*(3), 423-434.

Hajkova, P., Erhardt, S., Lane, N., et al. (2002). Epigenetic reprogramming in mouse primordial germ cells. *Mech Dev, 117*(1-2), 15-23.

Hajkova, P., Jeffries, S. J., Lee, C., et al. (2010). Genome-wide reprogramming in the mouse germ line entails the base excision repair pathway. *Science, 329*(5987), 78-82.

Hales, C. N., & Barker, D. J. (1992). Type 2 (non-insulin-dependent) diabetes mellitus: the thrifty phenotype hypothesis. *Diabetologia, 35*(7), 595-601.

Haluskova, J. (2010). Epigenetic studies in human diseases. *Folia Biol (Praha), 56*(3), 83-96.

Hansen, R. S., Wijmenga, C., Luo, P., et al. (1999). The DNMT3B DNA methyltransferase gene is mutated in the ICF immunodeficiency syndrome. *Proc Natl Acad Sci U S A, 96*(25), 14412-14417.

Hata, K., Okano, M., Lei, H., & Li, E. (2002). Dnmt3L cooperates with the Dnmt3 family of de novo DNA methyltransferases to establish maternal imprints in mice. *Development, 129*(8), 1983-1993.

Hawkins, P. G., & Morris, K. V. (2008). RNA and transcriptional modulation of gene expression. *Cell Cycle, 7*(5), 602-607.

He, Y. F., Li, B. Z., Li, Z., et al. (2011). Tet-Mediated Formation of 5-Carboxylcytosine and Its Excision by TDG in Mammalian DNA. *Science.*

Heard, E., Clerc, P., & Avner, P. (1997). X-chromosome inactivation in mammals. [Review]. *Annual Review of Genetics, 31*, 571-610.

Hendrich, B., & Bird, A. (1998). Identification and characterization of a family of mammalian methyl-CpG binding proteins. *Mol Cell Biol, 18*(11), 6538-6547.

Hendrich, B., Guy, J., Ramsahoye, B., Wilson, V. A., & Bird, A. (2001). Closely related proteins MBD2 and MBD3 play distinctive but interacting roles in mouse development. *Genes Dev, 15*(6), 710-723.

Hendrich, B., Hardeland, U., Ng, H. H., Jiricny, J., & Bird, A. (1999). The thymine glycosylase MBD4 can bind to the product of deamination at methylated CpG sites. *Nature, 401*(6750), 301-304.

Hermann, A., Gowher, H., & Jeltsch, A. (2004). Biochemistry and biology of mammalian DNA methyltransferases. *Cell Mol Life Sci, 61*(19-20), 2571-2587.

Hermann, A., Goyal, R., & Jeltsch, A. (2004). The Dnmt1 DNA-(cytosine-C5)-methyltransferase methylates DNA processively with high preference for hemimethylated target sites. *J Biol Chem, 279*(46), 48350-48359.

Hermann, A., Schmitt, S., & Jeltsch, A. (2003). The human Dnmt2 has residual DNA-(cytosine-C5) methyltransferase activity. *J Biol Chem, 278*(34), 31717-31721.

Hollingsworth, J. W., Maruoka, S., Boon, K., et al. (2008). In utero supplementation with methyl donors enhances allergic airway disease in mice. *J Clin Invest, 118*(10), 3462-3469.

Huang, Y., Pastor, W. A., Shen, Y., et al. (2010). The behaviour of 5-hydroxymethylcytosine in bisulfite sequencing. *PLoS One, 5*(1), e8888.

Iliadou, A. N., Janson, P. C., & Cnattingius, S. (2011). Epigenetics and assisted reproductive technology. *J Intern Med.*

Illingworth, R., Kerr, A., Desousa, D., et al. (2008). A novel CpG island set identifies tissue-specific methylation at developmental gene loci. [Research Support, Non-U.S. Gov't]. *PLoS Biol, 6*(1), e22.

Iqbal, K., Jin, S. G., Pfeifer, G. P., & Szabo, P. E. (2011). Reprogramming of the paternal genome upon fertilization involves genome-wide oxidation of 5-methylcytosine. *Proc Natl Acad Sci U S A, 108*(9), 3642-3647.

Ito, S., D'Alessio, A. C., Taranova, O. V., et al. (2010). Role of Tet proteins in 5mC to 5hmC conversion, ES-cell self-renewal and inner cell mass specification. *Nature, 466*(7310), 1129-1133.

Ito, S., Shen, L., Dai, Q., et al. (2011). Tet Proteins Can Convert 5-Methylcytosine to 5-Formylcytosine and 5-Carboxylcytosine. *Science.*

Jair, K. W., Bachman, K. E., Suzuki, H., et al. (2006). De novo CpG island methylation in human cancer cells. *Cancer Res, 66*(2), 682-692.

Jin, S. G., Guo, C., & Pfeifer, G. P. (2008). GADD45A does not promote DNA demethylation. *PLoS Genet, 4*(3), e1000013.

Jin, S. G., Kadam, S., & Pfeifer, G. P. (2010). Examination of the specificity of DNA methylation profiling techniques towards 5-methylcytosine and 5-hydroxymethylcytosine. *Nucleic Acids Res, 38*(11), e125.

Jones, P. A., & Laird, P. W. (1999). Cancer epigenetics comes of age. *Nat Genet, 21*(2), 163-167.

Jones, P. L., Veenstra, G. J., Wade, P. A., et al. (1998). Methylated DNA and MeCP2 recruit histone deacetylase to repress transcription. *Nat Genet, 19*(2), 187-191.

Jurkowski, T. P., Meusburger, M., Phalke, S., et al. (2008). Human DNMT2 methylates tRNA(Asp) molecules using a DNA methyltransferase-like catalytic mechanism. *RNA, 14*(8), 1663-1670.

Kaminskas, E., Farrell, A. T., Wang, Y. C., Sridhara, R., & Pazdur, R. (2005). FDA drug approval summary: azacitidine (5-azacytidine, Vidaza) for injectable suspension. *Oncologist, 10*(3), 176-182.

Kaneda, M., Okano, M., Hata, K., et al. (2004). Essential role for de novo DNA methyltransferase Dnmt3a in paternal and maternal imprinting. *Nature, 429*(6994), 900-903.

Kato, Y., Kaneda, M., Hata, K., et al. (2007). Role of the Dnmt3 family in de novo methylation of imprinted and repetitive sequences during male germ cell development in the mouse. *Hum Mol Genet, 16*(19), 2272-2280.

Khosla, S., Dean, W., Brown, D., Reik, W., & Feil, R. (2001). Culture of preimplantation mouse embryos affects fetal development and the expression of imprinted genes. *Biol Reprod, 64*(3), 918-926.

Kim, J. M., Hong, K., Lee, J. H., Lee, S., & Chang, N. (2009). Effect of folate deficiency on placental DNA methylation in hyperhomocysteinemic rats. *J Nutr Biochem, 20*(3), 172-176.

Kimura, H., & Shiota, K. (2003). Methyl-CpG-binding protein, MeCP2, is a target molecule for maintenance DNA methyltransferase, Dnmt1. *J Biol Chem, 278*(7), 4806-4812.

Klimasauskas, S., Kumar, S., Roberts, R. J., & Cheng, X. (1994). HhaI methyltransferase flips its target base out of the DNA helix. *Cell, 76*(2), 357-369.

Kothari, R. M., & Shankar, V. (1976). 5-Methylcytosine content in the vertebrate deoxyribonucleic acids: species specificity. *J Mol Evol, 7*(4), 325-329.

Kouidou, S., Malousi, A., & Maglaveras, N. (2006). Methylation and repeats in silent and nonsense mutations of p53. *Mutat Res, 599*(1-2), 167-177.

Kriaucionis, S., & Heintz, N. (2009). The nuclear DNA base 5-hydroxymethylcytosine is present in Purkinje neurons and the brain. *Science, 324*(5929), 929-930.

Kumar, S., Cheng, X., Klimasauskas, S., et al. (1994). The DNA (cytosine-5) methyltransferases. *Nucleic Acids Res, 22*(1), 1-10.

Kunert, N., Marhold, J., Stanke, J., Stach, D., & Lyko, F. (2003). A Dnmt2-like protein mediates DNA methylation in Drosophila. *Development, 130*(21), 5083-5090.

Latos, P. A., & Barlow, D. P. (2009). Regulation of imprinted expression by macro non-coding RNAs. [Research Support, Non-U.S. Gov't Review]. *RNA biology, 6*(2), 100-106.

Ley, T. J., Ding, L., Walter, M. J., et al. (2010). DNMT3A mutations in acute myeloid leukemia. *N Engl J Med, 363*(25), 2424-2433.

Li, E., Bestor, T. H., & Jaenisch, R. (1992). Targeted mutation of the DNA methyltransferase gene results in embryonic lethality. *Cell, 69*(6), 915-926.

Li, J. Y., Lees-Murdock, D. J., Xu, G. L., & Walsh, C. P. (2004). Timing of establishment of paternal methylation imprints in the mouse. *Genomics, 84*(6), 952-960.

Li, X., Ito, M., Zhou, F., et al. (2008). A maternal-zygotic effect gene, Zfp57, maintains both maternal and paternal imprints. *Dev Cell, 15*(4), 547-557.

Lin, H. (2011). S-Adenosylmethionine-dependent alkylation reactions: When are radical reactions used? *Bioorg Chem.*

Liu, K., Wang, Y. F., Cantemir, C., & Muller, M. T. (2003). Endogenous assays of DNA methyltransferases: Evidence for differential activities of DNMT1, DNMT2, and DNMT3 in mammalian cells in vivo. *Mol Cell Biol, 23*(8), 2709-2719.

Lu, Q., Kaplan, M., Ray, D., et al. (2002). Demethylation of ITGAL (CD11a) regulatory sequences in systemic lupus erythematosus. *Arthritis Rheum, 46*(5), 1282-1291.

Lucifero, D., Mann, M. R., Bartolomei, M. S., & Trasler, J. M. (2004). Gene-specific timing and epigenetic memory in oocyte imprinting. *Hum Mol Genet, 13*(8), 839-849.

Maiti, A., & Drohat, A. C. (2011). Thymine DNA glycosylase can rapidly excise 5-formylcytosine and 5-carboxylcytosine: Potential implications for active demethylation of CpG sites. *J Biol Chem.*

Mann, J. R. (2001). Imprinting in the germ line. *Stem Cells, 19*(4), 287-294.

Mann, M. R., Lee, S. S., Doherty, A. S., et al. (2004). Selective loss of imprinting in the placenta following preimplantation development in culture. *Development, 131*(15), 3727-3735.

Margot, J. B., Aguirre-Arteta, A. M., Di Giacco, B. V., et al. (2000). Structure and function of the mouse DNA methyltransferase gene: Dnmt1 shows a tripartite structure. *J Mol Biol, 297*(2), 293-300.

Margot, J. B., Ehrenhofer-Murray, A. E., & Leonhardt, H. (2003). Interactions within the mammalian DNA methyltransferase family. *BMC Mol Biol, 4*, 7.

Matje, D. M., Coughlin, D. F., Connolly, B. A., Dahlquist, F. W., & Reich, N. O. (2011). Determinants of precatalytic conformational transitions in the DNA cytosine methyltransferase M.HhaI. *Biochemistry, 50*(9), 1465-1473.

Mayer, W., Niveleau, A., Walter, J., Fundele, R., & Haaf, T. (2000). Demethylation of the zygotic paternal genome. *Nature, 403*(6769), 501-502.

McGrath, J., & Solter, D. (1984). Completion of mouse embryogenesis requires both the maternal and paternal genomes. *Cell, 37*(1), 179-183.

McKay, J. A., Wong, Y. K., Relton, C. L., Ford, D., & Mathers, J. C. (2011). Maternal folate supply and sex influence gene-specific DNA methylation in the fetal gut. *Mol Nutr Food Res.*

Meissner, A. (2010). Epigenetic modifications in pluripotent and differentiated cells. *Nat Biotech, 28*(10), 1079-1088.

Meissner, A., Mikkelsen, T. S., Gu, H., et al. (2008). Genome-scale DNA methylation maps of pluripotent and differentiated cells. *Nature.*

Metivier, R., Gallais, R., Tiffoche, C., et al. (2008). Cyclical DNA methylation of a transcriptionally active promoter. *Nature, 452*(7183), 45-50.

Morgan, H. D., Dean, W., Coker, H. A., Reik, W., & Petersen-Mahrt, S. K. (2004). Activation-induced cytidine deaminase deaminates 5-methylcytosine in DNA and is expressed in pluripotent tissues: implications for epigenetic reprogramming. *J Biol Chem, 279*(50), 52353-52360.

Morris, K. V. (2009a). Long antisense non-coding RNAs function to direct epigenetic complexes that regulate transcription in human cells. *Epigenetics,* 4(5), 296-301.

Morris, K. V. (2009b). RNA-Directed Transcriptional Gene Silencing and Activation in Human Cells. *Oligonucleotides, 19*(4), 299-306.

Munro, S. K., Farquhar, C. M., Mitchell, M. D., & Ponnampalam, A. P. Epigenetic regulation of endometrium during the menstrual cycle. *Mol Hum Reprod, 16*(5), 297-310.

Munzel, M., Globisch, D., Bruckl, T., et al. (2010). Quantification of the sixth DNA base hydroxymethylcytosine in the brain. *Angew Chem Int Ed Engl, 49*(31), 5375-5377.

Nagae, G., Isagawa, T., Shiraki, N., et al. (2011). Tissue-specific demethylation in CpG-poor promoters during cellular differentiation. *Hum Mol Genet, 20*(14), 2710-2721.

Nakamura, T., Arai, Y., Umehara, H., et al. (2007). PGC7/Stella protects against DNA demethylation in early embryogenesis. *Nat Cell Biol, 9*(1), 64-71.

Nan, X., Ng, H. H., Johnson, C. A., et al. (1998). Transcriptional repression by the methyl-CpG-binding protein MeCP2 involves a histone deacetylase complex. *Nature, 393*(6683), 386-389.

Ng, H. H., Zhang, Y., Hendrich, B., et al. (1999). MBD2 is a transcriptional repressor belonging to the MeCP1 histone deacetylase complex. *Nat Genet, 23*(1), 58-61.

Ng, R. K., Dean, W., Dawson, C., et al. (2008). Epigenetic restriction of embryonic cell lineage fate by methylation of Elf5. *Nat Cell Biol, 10*(11), 1280-1290.

Oberlander, T. F., Weinberg, J., Papsdorf, M., et al. (2008). Prenatal exposure to maternal depression, neonatal methylation of human glucocorticoid receptor gene (NR3C1) and infant cortisol stress responses. *Epigenetics, 3*(2), 97-106.

Oelke, K., Lu, Q., Richardson, D., et al. (2004). Overexpression of CD70 and overstimulation of IgG synthesis by lupus T cells and T cells treated with DNA methylation inhibitors. *Arthritis Rheum, 50*(6), 1850-1860.

Ohgane, J., Yagi, S., & Shiota, K. (2008). Epigenetics: the DNA methylation profile of tissue-dependent and differentially methylated regions in cells. *Placenta, 29 Suppl A*, S29-35.

Ohinata, Y., Payer, B., O'Carroll, D., et al. (2005). Blimp1 is a critical determinant of the germ cell lineage in mice. *Nature, 436*(7048), 207-213.

Oka, M., Rodic, N., Graddy, J., Chang, L. J., & Terada, N. (2006). CpG sites preferentially methylated by Dnmt3a in vivo. *J Biol Chem, 281*(15), 9901-9908.

Okano, M., Bell, D. W., Haber, D. A., & Li, E. (1999). DNA methyltransferases Dnmt3a and Dnmt3b are essential for de novo methylation and mammalian development. *Cell, 99*(3), 247-257.

Okano, M., Xie, S., & Li, E. (1998a). Cloning and characterization of a family of novel mammalian DNA (cytosine-5) methyltransferases. *Nat Genet, 19*(3), 219-220.

Okano, M., Xie, S., & Li, E. (1998b). Dnmt2 is not required for de novo and maintenance methylation of viral DNA in embryonic stem cells. *Nucleic Acids Res, 26*(11), 2536-2540.

Osmond, C., Barker, D. J., Winter, P. D., Fall, C. H., & Simmonds, S. J. (1993). Early growth and death from cardiovascular disease in women. *BMJ, 307*(6918), 1519-1524.

Oswald, J., Engemann, S., Lane, N., et al. (2000). Active demethylation of the paternal genome in the mouse zygote. *Curr Biol, 10*(8), 475-478.

Paulsen, M., Davies, K. R., Bowden, L. M., et al. (1998). Syntenic organization of the mouse distal chromosome 7 imprinting cluster and the Beckwith-Wiedemann syndrome region in chromosome 11p15.5. *Hum Mol Genet, 7*(7), 1149-1159.

Paulsen, M., & Ferguson-Smith, A. C. (2001). DNA methylation in genomic imprinting, development, and disease. *J Pathol, 195*(1), 97-110.

Penn, N. W., Suwalski, R., O'Riley, C., Bojanowski, K., & Yura, R. (1972). The presence of 5-hydroxymethylcytosine in animal deoxyribonucleic acid. *Biochem J, 126*(4), 781-790.

Pierard, V., Guiguen, A., Colin, L., et al. DNA cytosine methylation in the bovine leukemia virus promoter is associated with latency in a lymphoma-derived B-cell line: potential involvement of direct inhibition of cAMP-responsive element (CRE)-binding protein/CRE modulator/activation transcription factor binding. *J Biol Chem, 285*(25), 19434-19449.

Polyzos, N. P., Polyzos, I. P., Mauri, D., et al. (2009). Effect of periodontal disease treatment during pregnancy on preterm birth incidence: a metaanalysis of randomized trials. *Am J Obstet Gynecol, 200*(3), 225-232.

Pradhan, S., Bacolla, A., Wells, R. D., & Roberts, R. J. (1999). Recombinant human DNA (cytosine-5) methyltransferase. I. Expression, purification, and comparison of de novo and maintenance methylation. *J Biol Chem, 274*(46), 33002-33010.

Prendergast, G. C., & Ziff, E. B. (1991). Methylation-sensitive sequence-specific DNA binding by the c-Myc basic region. *Science, 251*(4990), 186-189.

Probst, A. V., Dunleavy, E., & Almouzni, G. (2009). Epigenetic inheritance during the cell cycle. *Nat Rev Mol Cell Biol, 10*(3), 192-206.

Quddus, J., Johnson, K. J., Gavalchin, J., et al. (1993). Treating activated CD4+ T cells with either of two distinct DNA methyltransferase inhibitors, 5-azacytidine or procainamide, is sufficient to cause a lupus-like disease in syngeneic mice. *J Clin Invest, 92*(1), 38-53.

Ramsahoye, B. H., Biniszkiewicz, D., Lyko, F., et al. (2000). Non-CpG methylation is prevalent in embryonic stem cells and may be mediated by DNA methyltransferase 3a. *Proc Natl Acad Sci U S A, 97*(10), 5237-5242.

Razin, A., Webb, C., Szyf, M., et al. (1984). Variations in DNA methylation during mouse cell differentiation in vivo and in vitro. *Proc Natl Acad Sci U S A, 81*(8), 2275-2279.

Reik, W. (2007). Stability and flexibility of epigenetic gene regulation in mammalian development. *Nature, 447*(7143), 425-432.

Rhee, I., Bachman, K. E., Park, B. H., et al. (2002). DNMT1 and DNMT3b cooperate to silence genes in human cancer cells. *Nature, 416*(6880), 552-556.

Rich-Edwards, J. W., Kleinman, K., Michels, K. B., et al. (2005). Longitudinal study of birth weight and adult body mass index in predicting risk of coronary heart disease and stroke in women. *BMJ, 330*(7500), 1115.

Richardson, B., Scheinbart, L., Strahler, J., et al. (1990). Evidence for impaired T cell DNA methylation in systemic lupus erythematosus and rheumatoid arthritis. *Arthritis Rheum, 33*(11), 1665-1673.

Robertson, K. D. (2005). DNA methylation and human disease. *Nat Rev Genet, 6*(8), 597-610.

Rozenberg, J. M., Shlyakhtenko, A., Glass, K., et al. (2008). All and only CpG containing sequences are enriched in promoters abundantly bound by RNA polymerase II in multiple tissues. *BMC Genomics, 9*, 67.

Saito, M., & Ishikawa, F. (2002). The mCpG-binding domain of human MBD3 does not bind to mCpG but interacts with NuRD/Mi2 components HDAC1 and MTA2. *J Biol Chem, 277*(38), 35434-35439.

Saitou, M. (2009). Germ cell specification in mice. *Curr Opin Genet Dev, 19*(4), 386-395.

Sancar, A., Lindsey-Boltz, L. A., Unsal-Kacmaz, K., & Linn, S. (2004). Molecular mechanisms of mammalian DNA repair and the DNA damage checkpoints. *Annu Rev Biochem, 73*, 39-85.

Santos, F., & Dean, W. (2004). Epigenetic reprogramming during early development in mammals. *Reproduction, 127*(6), 643-651.

Santos, F., Hendrich, B., Reik, W., & Dean, W. (2002). Dynamic reprogramming of DNA methylation in the early mouse embryo. *Dev Biol, 241*(1), 172-182.

Saxonov, S., Berg, P., & Brutlag, D. L. (2006). A genome-wide analysis of CpG dinucleotides in the human genome distinguishes two distinct classes of promoters. *Proc Natl Acad Sci U S A, 103*(5), 1412-1417.

Schaefer, M., & Lyko, F. (2009). Solving the Dnmt2 enigma. *Chromosoma, 119*(1), 35-40.

Seki, Y., Hayashi, K., Itoh, K., et al. (2005). Extensive and orderly reprogramming of genome-wide chromatin modifications associated with specification and early development of germ cells in mice. *Dev Biol, 278*(2), 440-458.

Shah, M. Y., & Licht, J. D. (2011). DNMT3A mutations in acute myeloid leukemia. *Nat Genet, 43*(4), 289-290.

Shen, L., Kondo, Y., Guo, Y., et al. (2007). Genome-wide profiling of DNA methylation reveals a class of normally methylated CpG island promoters. *PLoS Genet, 3*(10), 2023-2036.

Song, C. X., Szulwach, K. E., Fu, Y., et al. (2010). Selective chemical labeling reveals the genome-wide distribution of 5-hydroxymethylcytosine. *Nat Biotechnol, 29*(1), 68-72.

Straussman, R., Nejman, D., Roberts, D., et al. (2009). Developmental programming of CpG island methylation profiles in the human genome. *Nat Struct Mol Biol, 16*(5), 564-571.

Sunahori, K., Juang, Y. T., & Tsokos, G. C. (2009). Methylation status of CpG islands flanking a cAMP response element motif on the protein phosphatase 2Ac alpha promoter determines CREB binding and activity. *J Immunol, 182*(3), 1500-1508.

Surani, M. A., Barton, S. C., & Norris, M. L. (1984). Development of reconstituted mouse eggs suggests imprinting of the genome during gametogenesis. *Nature, 308*(5959), 548-550.

Szwagierczak, A., Bultmann, S., Schmidt, C. S., Spada, F., & Leonhardt, H. (2010). Sensitive enzymatic quantification of 5-hydroxymethylcytosine in genomic DNA. *Nucleic Acids Res, 38*(19), e181.

Tahiliani, M., Koh, K. P., Shen, Y., et al. (2009). Conversion of 5-methylcytosine to 5-hydroxymethylcytosine in mammalian DNA by MLL partner TET1. *Science, 324*(5929), 930-935.

Takai, D., & Jones, P. A. (2002). Comprehensive analysis of CpG islands in human chromosomes 21 and 22. *Proc Natl Acad Sci U S A, 99*(6), 3740-3745.

Tang, L. Y., Reddy, M. N., Rasheva, V., et al. (2003). The eukaryotic DNMT2 genes encode a new class of cytosine-5 DNA methyltransferases. *J Biol Chem, 278*(36), 33613-33616.

Valinluck, V., & Sowers, L. C. (2007). Endogenous cytosine damage products alter the site selectivity of human DNA maintenance methyltransferase DNMT1. *Cancer Res, 67*(3), 946-950.

Valinluck, V., Tsai, H. H., Rogstad, D. K., et al. (2004). Oxidative damage to methyl-CpG sequences inhibits the binding of the methyl-CpG binding domain (MBD) of methyl-CpG binding protein 2 (MeCP2). *Nucleic Acids Res, 32*(14), 4100-4108.

Van den Wyngaert, I., Sprengel, J., Kass, S. U., & Luyten, W. H. (1998). Cloning and analysis of a novel human putative DNA methyltransferase. *FEBS Lett, 426*(2), 283-289.

Watanabe, D., Suetake, I., Tada, T., & Tajima, S. (2002). Stage- and cell-specific expression of Dnmt3a and Dnmt3b during embryogenesis. *Mech Dev, 118*(1-2), 187-190.

Waterland, R. A., Dolinoy, D. C., Lin, J. R., et al. (2006). Maternal methyl supplements increase offspring DNA methylation at Axin Fused. *Genesis, 44*(9), 401-406.

Waterland, R. A., & Jirtle, R. L. (2003). Transposable elements: targets for early nutritional effects on epigenetic gene regulation. *Mol Cell Biol, 23*(15), 5293-5300.

Waterland, R. A., & Michels, K. B. (2007). Epigenetic epidemiology of the developmental origins hypothesis. *Annu Rev Nutr, 27*, 363-388.

Weaver, J. R., Susiarjo, M., & Bartolomei, M. S. (2009). Imprinting and epigenetic changes in the early embryo. *Mamm Genome, 20*(9-10), 532-543.

Weber, M., Hellmann, I., Stadler, M. B., et al. (2007). Distribution, silencing potential and evolutionary impact of promoter DNA methylation in the human genome. *Nat Genet, 39*(4), 457-466.

White, G. P., Watt, P. M., Holt, B. J., & Holt, P. G. (2002). Differential patterns of methylation of the IFN-gamma promoter at CpG and non-CpG sites underlie differences in IFN-gamma gene expression between human neonatal and adult CD45RO- T cells. *J Immunol, 168*(6), 2820-2827.

Williams, K., Christensen, J., Pedersen, M. T., et al. (2011). TET1 and hydroxymethylcytosine in transcription and DNA methylation fidelity. *Nature, 473*(7347), 343-348.

Wossidlo, M., Nakamura, T., Lepikhov, K., et al. (2011). 5-Hydroxymethylcytosine in the mammalian zygote is linked with epigenetic reprogramming. *Nat Commun, 2*, 241.

Wu, J. C., & Santi, D. V. (1985). On the mechanism and inhibition of DNA cytosine methyltransferases. *Prog Clin Biol Res, 198*, 119-129.

Wu, J. C., & Santi, D. V. (1987). Kinetic and catalytic mechanism of HhaI methyltransferase. *J Biol Chem, 262*(10), 4778-4786.

Wu, S. C., & Zhang, Y. (2010). Active DNA demethylation: many roads lead to Rome. *Nat Rev Mol Cell Biol, 11*(9), 607-620.

Wu, Y., Strawn, E., Basir, Z., Halverson, G., & Guo, S. W. (2006). Promoter hypermethylation of progesterone receptor isoform B (PR-B) in endometriosis. *Epigenetics, 1*(2), 106-111.

Xie, S., Wang, Z., Okano, M., et al. (1999). Cloning, expression and chromosome locations of the human DNMT3 gene family. *Gene, 236*(1), 87-95.

Yoder, J. A., & Bestor, T. H. (1998). A candidate mammalian DNA methyltransferase related to pmt1p of fission yeast. *Hum Mol Genet, 7*(2), 279-284.

Yoder, J. A., Soman, N. S., Verdine, G. L., & Bestor, T. H. (1997). DNA (cytosine-5)-methyltransferases in mouse cells and tissues. Studies with a mechanism-based probe. *J Mol Biol, 270*(3), 385-395.

Yoo, J., & Medina-Franco, J. L. (2011). Trimethylaurintricarboxylic acid inhibits human DNA methyltransferase 1: insights from enzymatic and molecular modeling studies. *J Mol Model*.

Yu, Z., Genest, P. A., ter Riet, B., et al. (2007). The protein that binds to DNA base J in trypanosomatids has features of a thymidine hydroxylase. *Nucleic Acids Res, 35*(7), 2107-2115.

Yuen, R. K., Jiang, R., Penaherrera, M. S., McFadden, D. E., & Robinson, W. P. (2011). Genome-wide mapping of imprinted differentially methylated regions by DNA methylation profiling of human placentas from triploidies. *Epigenetics Chromatin, 4*(1), 10.

Yuen, R. K. C., Neumann, S. M. A., Fok, A. K., et al. (2011). Extensive epigenetic reprogramming in human somatic tissues between fetus and adult. *Epigenetics and Chromatin, 4*(7).

Zhu, B., Zheng, Y., Angliker, H., et al. (2000). 5-Methylcytosine DNA glycosylase activity is also present in the human MBD4 (G/T mismatch glycosylase) and in a related avian sequence. *Nucleic Acids Res, 28*(21), 4157-4165.

Zhu, B., Zheng, Y., Hess, D., et al. (2000). 5-methylcytosine-DNA glycosylase activity is present in a cloned G/T mismatch DNA glycosylase associated with the chicken embryo DNA demethylation complex. *Proc Natl Acad Sci U S A, 97*(10), 5135-5139.

Zhu, J. K. (2009). Active DNA demethylation mediated by DNA glycosylases. *Annu Rev Genet, 43*, 143-166.

Zimmermann, C., Guhl, E., & Graessmann, A. (1997). Mouse DNA methyltransferase (MTase) deletion mutants that retain the catalytic domain display neither de novo nor maintenance methylation activity in vivo. *Biol Chem, 378*(5), 393-405.

Zucker, K. E., Riggs, A. D., & Smith, S. S. (1985). Purification of human DNA (cytosine-5-)-methyltransferase. *J Cell Biochem, 29*(4), 337-349.

Part 3

Perspectives in Embryology

8

Stem Cell Therapies

D. Amat[1], J. Becerra[2,3], M.A. Medina[4],
A.R. Quesada[4] and M. Marí-Beffa[2,3]
[1]*Department of Human Anatomy,*
Faculty of Medicine, University of Malaga, Malaga,
[2]*Department of Cell Biology, Genetics and Physiology,*
Faculty of Science, University of Malaga, Malaga,
[3]*Networking Research Center on Bioengineering,*
Biomaterials and Nanomedicine, (CIBER-BBN), Malaga,
[4]*Department of Molecular Biology and Biochemistry,*
Faculty of Science, University of Málaga, Malaga,
Spain

1. Introduction

Both development and adult homeostatic maintenance of different tissues and organs are dependent on the activity of a specific cell type named stem cell (SC). In recent years, SCs has been tested for clinical therapies against complex diseases. For many of these diseases, SC treatments offer a glimmer of hope. Fortunately, in some other instances, these therapies are an interesting reality.

In this chapter, we aim to introduce the readers to the nature and diversity of SC. We will describe their origin and location in the human body and the main SC therapies used in clinical practice. Finally, we will propose a standard goal of current applications to convince readers about future avenues in these treatments.

2. Stem cells

2.1 Concept and types

A stem cell is an undifferentiated cell that is able to proliferate, giving rise to various types of differentiating cell lineages. Through unequal divisions, SCs may predominantly give rise to two different cells: one SC and one progenitor cell which continues cell division to finally initiate cell differentiation. These are the two inherent features of the lineages of these cells, self-renewal and differentiation. Depending on their location and un-differentiation state, these cells are able to generate/restitute specific tissues, organs or complete embryos (Figure 1). This capacity is named potency. A stem cell shows a high potency when many different cell lines can be obtained in vitro or in vivo. SCs show reduced potency when very few cell lines can be obtained.

According to this potency concept, SC can be classified into the following groups:

Totipotent stem cells: Cells able to differentiate into any embryonic or extra-embryonic cell line. Blastomeres from zygote to morula are totipotent cells.

Pluripotent stem cells: Cells able to differentiate into any cell line derived from any of the three embryonic sheets. Cells of the Inner Cell Mass, the embryoblast, and cells of the germinal ridges are pluripotent SCs.

Multipotent stem cells: Cells able to differentiate into any cell type derived from only one embryonic sheet. Ectoderm, endoderm or mesoderm stem cells are multipotent.

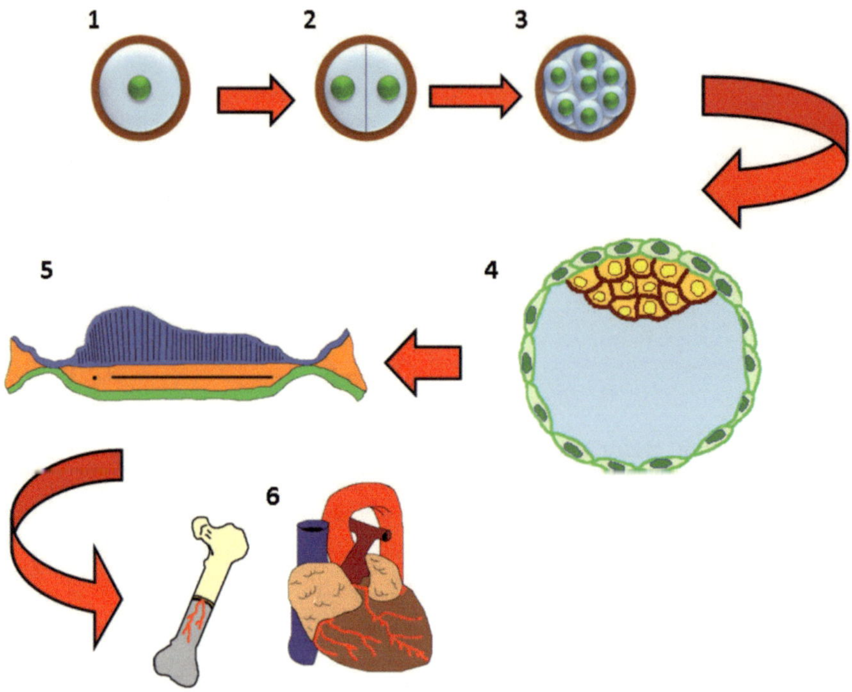

Fig. 1. **Cell potency varies during human embryogenesis and organogenesis**.
The zygote (1) and blastomeres at two cell stage (2) and morula (3) are totipotent cells. Cells in the inner mass of the blastocyst (4) are pluripotent. Trilaminar disc (5) cells are multipotent. Stem cells at heart or large bones (6) are either multipotent or unipotent. Green and yellow cells in the blastocyst respectively are trophoblast and inner cell mass.

Under this classification, the embryo appears as a very important source of many different types of stem cells. Cells of the Inner Cell Mass are named **Embryonic stem cells (ESCs,** Thomson et al., 1998) and show a high potency. Besides, **Adult (or Somatic) Stem Cells (ASCs)** are resident cells in adult tissues all along the complete life span. These cells show a smaller potency than ESCs but lesser risks and bioethical problems (see below) during their clinical use.

ASCs are responsible of adult tissue **homeostasis**. This homeostasis is established as a balance between cell death and cell proliferation. A clear example can be found in the skin. Superficial cell layers in the epidermis are continuously dying and being lost. In order to maintain epidermis stability, basal cell layer continuously proliferate. This proliferation is carried out by a population of epidermal stem cells.

In order to maintain tissue homeostasis, ASC show specific cell cycle properties. These cells are in a quiescent state. Under a wound stimulus, they proliferate and give rise to differentiated cell lineages. Cell proliferation and differentiation are compensated in such a way that new differentiated cells are proportionate to the original induction by wound signals. Following a tissue injury and under new inducing signals released from the wound, ASC leave the quiescent state, proliferate and differentiate to replace damaged or dead cells. This natural process can restore absent or wounded tissues in few days.

Modern Cell Biology technologies have also been able to induce potency in adult differentiated cells. By specific in vitro culturing conditions, differentiated cells can dedifferentiate and proliferate to initiate several newly-differentiating cell lines. These cells are named **induced pluripotent stem cells (iPS**, Takahashi et al., 2007). All above-mentioned stem cells can be used in a potential cell therapy.

2.2 Locations and functions

Since the earliest stages of embryogenesis, stem cells can be inferred to occur. Blastomeres in all pre-morula stages are able to generate a complete organism when isolated from the rest. Thus, these blastomeres are the source of totipotent embryonic stem cells. A well-known example of this is both groups of blastomeres isolated to form two monozygotic twins during early development. Both resulting individuals are genetically identical and generate independent chorion, placenta and amniotic cavities. Thus, these monozygotic twins are also diamniotic, dichorionic and diplacental.

Just before implantation, blastocysts form pluripotent stem cells in the Inner Cell mass or embryoblast. At this stage, a first cell commitment can be observed during human embryogenesis, the embryoblast is able to generate a complete organism but not the extraembryonic tissues. During embryoblast formation, the isolation of two inner cell masses also generates two monozygotic and diamniotic twins, but they are monochorionic and monoplacental. These unique structures are generated from a different embryonic tissue named trophoblast.

Following this stage, the inner cell mass further specifies and moves their cells to form the embryonic sheets in two consecutive structures, the bilaminar and the trilaminar discs (see Figure 1). These three embryonic sheets, the ectoderm, the endoderm and the mesoderm, are formed by multipotent stem cells, committed to generate specific cell fates. Stem cells from any of these sheets are unable to differentiate into the typical fates of cells from other sheets (Thomson et al., 1998).

The ectoderm forms the nervous system and the integument. In the adult, a group of ectoderm-derived stem cells can be found in the basal layer of the epidermis, the epidermal stem cells, and in sub-ventricular positions within the central nervous system, the neural stem cells. The former regulate a continuous epidermal renewal, whereas the latter maintain the cellular structure of the nervous tissue against a classical paradigm of nervous proliferative quiescence. Also associated to ectoderm, a derived embryonic source of stem cells is found in the neural crest. During neural crest fusion, a group of ectodermal cells trans-differentiate into migratory derived cells and generate different cell types all along the body. These cell fates are skin melanocytes, facial bones and sensory neurons among others. This large migratory pathway can provide ectoderm-derived stem cells in peculiar locations such as carotid body, from which stem cells can be obtained to use them in therapies for Parkinson's disease.

On the other hand, the endoderm gives rise to the respiratory and digestive systems. Endoderm-derived stem cells are multipotent and can be found in mucosae basal layer in both systems. These cells show a high cell proliferation rate and are continuously in risk to be affected by carcinogenic reagents. These reasons may have led these systems to show the highest neoplasia frequency of all human organs. In the human liver, another endoderm-derived stem cell type can be found, the oval cells. These cells have been recently shown to be involved in hepatic-regeneration. A number of cell types can be differentiated from these oval cells, such as the hepatocytes or biliary duct cells.

Finally, mesoderm is involved in the formation of many different apparatus or systems: the skeletal apparatus (bones, cartilage, tendons), the vascular system (blood, blood vessels and heart), excretory (nephric systems and kidney) and reproductive (gonads) systems and the whole connective tissue in the organism. Thus, many mesoderm-derived stem cells are also located in many locations in the body to form all these tissues. From a clinical point of view, optimal stem cells are those forming the blood, the haematopietic stem cells in the bone marrow, and mesenchymal stem cells which differentiate into muscle (myoblast), bone (osteoblast), cartilage (chondroblast), fat tissue (adipoblast) or fibroblast progenitor cells. These cell types can be found in any mesoderm-derived tissue. A special type of mesenchymal stem cells is umbilical cord stem cells that can be differentiated into blood or mesenchymal cells. These stem cells from umbilical cord are a handy and bloodless source of highly proliferative stem cells useful for clinical purposes.

2.2.1 Stem cell genetic program

The knowledge of a genetic program for stem cells was initiated by two German embryologists, Theodor Boveri and Hans Driesch. In the late XIXs, both scientists respectively showed that the program was both located inside the nucleus and dependent on cell-to-cell interactions. Hans Driesch also proposed that the early embryo is a harmonic system in which all, or many, cells show the same potency which is dependent on cell-to-cell interactions. Modern Molecular and Developmental Biology are providing important information on the nuclear location of the genetic program and the cell-to-cell interaction-dependence of animal embryogenesis.

Any animal or human cell is also under the control of a genetic program. This genetic program is regulated by a hierarchy of transcription factors that either enhance or silence the activity of a single enzyme named the RNA polymerase II (Figure 2). This enzyme generates a heterogeneous RNA using as a substrate a DNA template in the open reading frame of the gene. This heterogeneous RNA is processed by RNA splicing and editing to generate the messenger RNA. In the ribosome, this messenger RNA is translated into protein. Protein synthesis requires a complete set of other specific proteins and RNAs to either form the ribosome or regulate its activity.

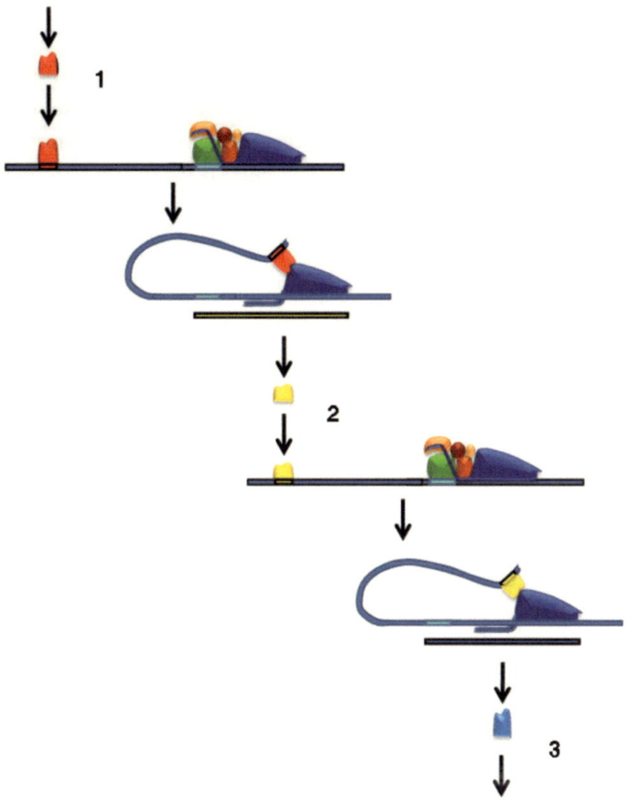

Fig. 2. **A gene transcriptional hierarchy controls cell specification.**
At each stage, blue proteins to the right are RNA polymerase II enzymes. Protein complex in various colors is the general transcription complex. At each stage, isolated color boxes to the left (1 to 3) are specific transcription factors. Longer blue narrow rectangles are DNA strands and shorter yellow and dark blue narrow rectangles are messenger RNAs codifying for transcription factors in the same color. Explanations can be found in the text.

The activity of the RNA polymerase II depends upon two different regulatory protein-complexes, the general and the specific transcription complexes. The latter complexes are formed by the transcription factors (Figure 2) and co-factors. These proteins are able to bind DNA and step-by-step activate new downstream genes. Some of these new genes may codify for other transcription factors that activate further new downstream genes. Some of these transcription factors which act during animal embryogenesis are codified by genes named *Hox, achaete, scute, engrailed, myoD, Dorsal, cubitus interruptus, apterous, Pannier* or *Iroquois*. These animal proteins are ordered in a transcriptional hierarchy, the so-called genetic program (Figure 2). This hierarchy is so complex that almost all embryonic cells can be supposed to be different at the transcription level (See Alberts et al., 2007). Finally, this genetic program also transcribes mRNAs codifying for other protein types. These new proteins may be structural or collaborate in the regulation of many cell properties, such as cell division, apoptosis or differentiation. This completes the genetic control of embryogenesis or tissue maintenance.

Two additional features can be observed in animal development, cell-to-cell interaction dependence and *pleiotropism*. In animals, transcription regulation is balanced among all cells. This balanced regulation is able to generate cell diversity, organ size or pattern formation during embryogenesis. By this process, the transcription regulation of a given cell depends upon the transcription in surrounding cells. In order to balance these transcription activities, a number of different signal transduction pathways have been found (see Figure 3). Wnt, Sonic hedgehog, FGFs and BMPs are ligands that activate some of these transduction pathways. This balanced transcription can be traced back to the genome from the mother to regulate follicle cells signaling. Moreover, these proteins can interact with many other proteins in other cells to regulate many cell processes. Thus, the same ligands may *pleiotropically* regulate cell differentiation in ectoderm, endoderm and mesoderm-derived stem cells. The embryonic-sheet specificity is provided by the genetic program of each cell.

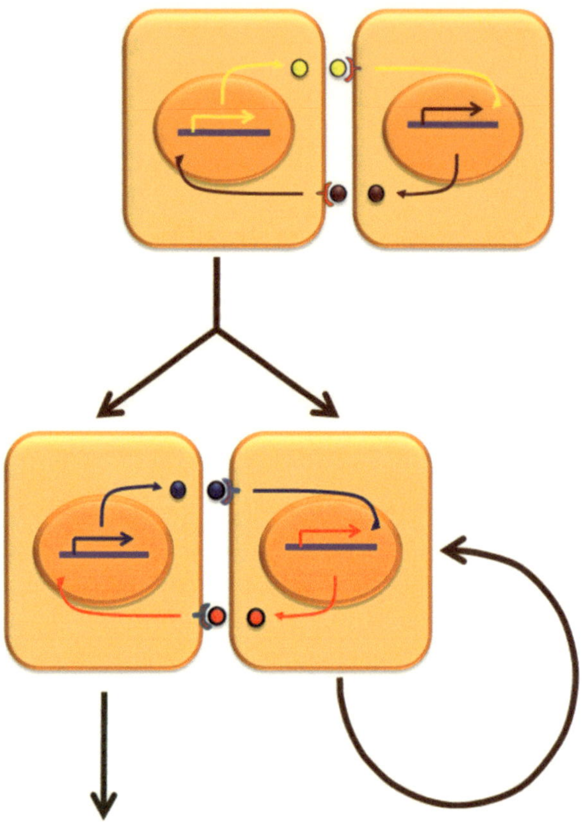

Fig. 3. **Cell-to-cell interactions regulate gene transcription within each human cell**. Signals can travel along varying distances. The upper left cell is a stem cell. The lower left and right cells respectively are a differentiating cell and a new stem cell. Details are described in text.

These studies in animal species provide a good tool to now find human genes candidate to be regulating any cell process. By sequence similarity between any animal and human gene, an orthologous human gene can be found. When compared with the animal gene, this human gene may also show similar functions in the human organism.

When studying human stem cells, all these regulatory features can also be found. Notch or Wnt signaling pathways have been shown to control the cell-to-cell-interaction-dependent decision between cell proliferation and cell differentiation.In humans, additional functions have been shown for these signaling pathways. The same can be found when other signaling pathways are studied. Transcription factors acting in stem cells are Bmi1, Oct4, SOX2, Klf-4, cMyc, FoxD3, or EGR1. An interesting effort might completely relate SC gene transcription to embryonic genetic program.

In summary, the differentiation of any stem cell type depends on a hierarchy of transcription, traced back to maternal information. This hierarchical regulation also influences, in a balanced way, the differentiation of other surrounding cell types. These neighboring influences, such as the so-called stem cell niche, are under profound analysis.

2.2.2 Stem cell niches

Stem cells require a very specific chemical and physical environment for a correct function. This environment is provided by **the Niche, a group of cells providing signals to stem cells for a balanced activity within the tissue** (see Becerra et al., 2011).

SC niche is defined as the microenvironment formed by SC, non-SC niche resident cells and specific signaling and ECM molecules surrounding them. The niche controls SC divisions. This SC proliferation can be regulated to fulfill the homeostatic needs (Becerra et al., 2011).

Although the information on niche structure is still scarce, a recent increase in scientific publication on this topic is being issued. Drosophila testis or germarium, hair follicle, intestine crypts, bone marrow and brain sub-ventricular zone are niches under intense study (Fig. 4).

Thus, the molecular/cellular paradigm is providing a new solution to one of the oldest unsolved questions in current developmental biology. Potency is now being understood under the concepts of genetic program and cell-to-cell interactions, and this may also help clinical practices.

3. Cell therapies

3.1 Introduction and concept

An ancestral question, traced back much before ancient Greece, is to understand why human body does not regenerate. A Greek example is Prometheus legend. Prometheus was immortal but stole fire from Zeus and was condemned to be devoured by an eagle. The eagle ate his liver every day, but Prometheus continuously regenerated his organ back and back again. The ancient observation of living beings led to the idea of animal regeneration. Indeed, sea urchin, triton or fishes are able to regenerate their limbs to restitute previous amputations. This probably inspired the legend.

The discovery and study of human stem cells brought about new hopes of recovering absent or injured organs. This was the birth of Cell therapy. **Cell therapy is the treatment of**

human diseases by application of cells. Although this discipline is at its beginnings, many current therapies aim to use stem cells to treat an important variety of human diseases.

Fig. 4. **Simplified schematic representation of the niche of stem cells in the bone marrow**. MSC and HSC interact in the endosteal niche, around the bone, and the vascular niche, close to the blood vessels. A lot of factors have been identified related to the regulation of hematopoiesis and the fate of the MSC. (1) spicules of bone, (2) osteoblasts, (3) hematopoietic stem cells (HSC), (4)Hematopoietic progenitor cells, (5) mesenchymal stem cells (MSC), (6) fenestrated capillaries, (7) progenitors inside capillaries, and (8) is an osteoclast.

The above-mentioned cells, neural, respiratory or digestive mucosae stem cells, haematopoietic, epidermal, mesenchymal or umbilical cord stem cells are currently used to treat diseases. However, the methods for obtaining these cells and treatment effectiveness vary depending upon cell type or potency.

In 1998, embryonic stem cells were first discovered (Thomson et al., 1998). Inner cell mass cells from a blastocyst were cultured in vitro. An empirical device was used to succeed in this task. This empirical device was being used since 1960s for in vitro culture of animal cells. Fibroblasts were killed by irradiation inducing extensive mutations in their nuclei DNA. A lawn of irradiated fibroblasts was deposited in a Petri dish, the feeder layer, in order to feed the non-irradiated embryonic cells. During culturing, these cells spontaneously dissociated from one another. These living cells were sub-cultured in another feeder layer to obtain embryonic stem cell colonies. These cells showed an important pluripotency when induced to differentiate in vitro under a variety of different stimuli and eventually produced teratomas when grafted into a mouse.

More modern techniques have been used to culture stem cells by in vitro amplification. Some of these cells can be obtained by egg enucleation and subsequent nuclear implantation from a differentiated somatic cell. If a pre-treatment of nutrient deprivation is carried out, nuclear re-programming may occur. This re-programming is able to de-differentiate the somatic cell nucleus and re-induce early embryogenesis in vitro. Under these conditions, a population of cultured cells from the patient can be obtained. Following these techniques, **Asexually Produced Totipotent (APT) Cells** were obtained. This technique shows a great advantage as these cells are not immunologically rejected when clinically used.

Two experimental trials are used to verify the quality of embryonic stem cells:

i. Teratome formation

If ESCs are experimentally implanted in animals, they generate tumors. These tumors may differentiate structures such as gastric glands, teeth or hairs. The nature of these structures suggests that these tumors may be originated from any embryonic sheet.

ii. Chimera induction

If embryonic stem cells are implanted into a genetically different inner cell mass, a chimeric organism is formed. Although classical terminology states that chimeric organism are mixture of different species,ESC-derived chimerism only implies genetic differences within a given species. In these mammalian organisms, the cells show two different genomes. This may be useful to trace the location of cells by simple histological observations, but it can also be used as a cellular treatment of diseases.

All these manipulations of human embryos or genes have precluded a direct application of these techniques due to bioethics reasons. Due to continuous argumentations in the society, an alternative to ESC use in cell therapies have arisen since the beginning. This technique use multipotent stem cells obtained from the adult, the above-mentioned ASC. Cell therapies using ASC has been widely expanded in clinical use around the world. However, although ASC are safe and very useful in many therapies, they show several practical problems that have to be considered.

A first problem is related to methods to obtain ASCs. Almost any stem cell type has to be obtained by a different experimental protocol. Thus, there are simple ways and very difficult protocols of obtaining stem cells. The easiest protocol is bone marrow aspiration to obtain mesenchymal and haematopoietic stem cells. This is clinically used in everyday treatment of leukemia or blood cell diseases. The most difficult protocol is to obtain neural stem cells from sub-ependymal layers. The clinical protocols using this latter technique are far away from being regularly used at hospitals due to its low benefit/risk level.

Beside stem cells, these techniques normally require further auxiliary material. Among them, bioactive molecules and artificial scaffold are currently necessary for growth and cell differentiation to obtain tissue integration. All these auxiliary techniques are included in the so-called Tissue Engineering.

3.2 Current applications

Many different cell therapy techniques are currently under legalization process to be applied to treat modern diseases. Some of these diseases affect bone, cartilage, central nervous and immune systems, skin, heart or blood. Bone Marrow therapies to treat blood or skeletal diseases are widely spread treatments used all around the world. However, other cell therapies are still under intense research. These stem cell techniques are treating self-immune or neurological diseases, such as Parkinson, ELA or medullar traumas.

The most important factor boosting improvement of these treatments is the close-knit collaboration between basic and applied research. From basic research, model systems must be proposed to evaluate the effectiveness of cell therapy. These model systems require the definition of precise model species, organ, tissues and/or cell type to verify responses to treatments. Experimental conditions, such as transgenic mice, implanted animals or in vitro culturing, may also be necessary for this model system definition. In some instances, some of these techniques may also be applied to hospitals, where material useful for the definite cell treatment is necessary. During this process, biological research grants are substituted with pre-clinical trials, always under medical rigorousness and refereeing. Since most applications require ex vivo manipulation of cells collected from the donor, the European Union has regulated its use as a drug.

In order to provide examples of this tortuous journey, we will discuss three examples. These three cell treatment examples use the same stem cell type, the mesoderm-derived mesenchymal stem cells (MSC). We will show that these three cases are elements in a single routing process that can be used as stepping stones to be followed when similar strategies are expected to be applied (Figure 5).

The first therapy is widely used around the world. Every hospital has enough experience and tradition in its application. Moreover, citizens in modern societies have a clear knowledge of its existence and utility. This technique is Haematopietic stem cells implantation, or the well-known bone marrow transplantation. The second therapy is also well implanted in hospitals but it is rarer than previous ones. Although it has roots in ancient Egypt, there is not such a consensual application as HSC implantation. This is the use of mesenchymal stem cells in treatments of bone diseases. The third case is at pre-clinical trials stage in some hospitals but it is widely unknown in clinical environments. This is the use of MSC to treat psoriasis or other angiogenic diseases. The three cases can be understood as a gradual process of accepting a unique clinical success. At this stage, the reconstruction of the original environment of stem cells in the body is reproduced and then these cells are able to restitute homeostasis in the organism.

3.2.1 Cell therapy and haematopoiesis

The tissue that generates blood cells is the Bone Marrow. In the bone marrow, millions of HSC and MSC can be found. If after an accident, an important quantity of blood is lost,

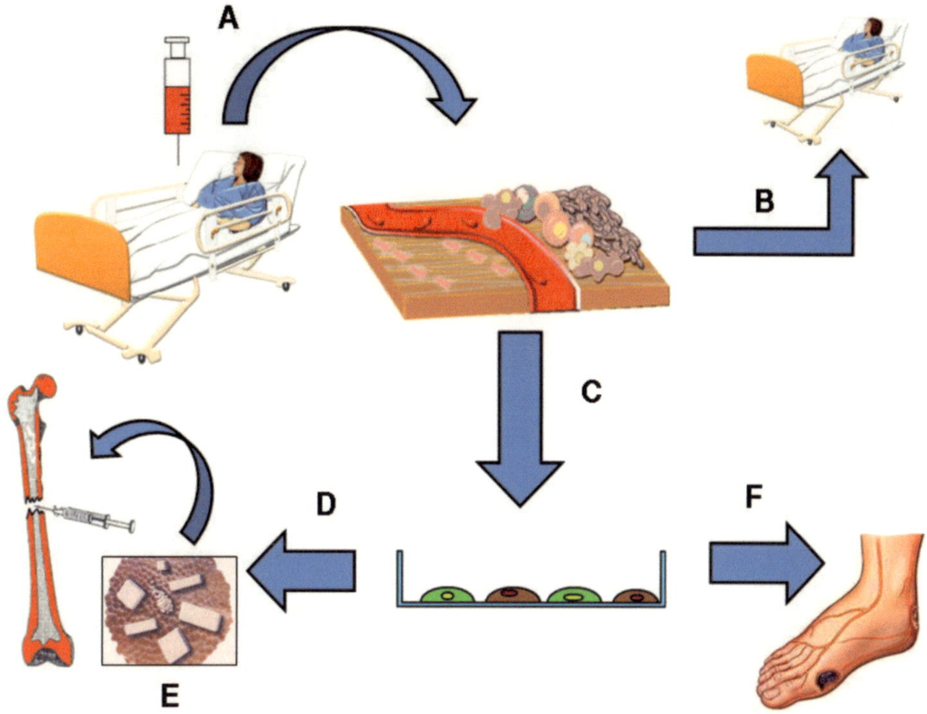

Fig. 5. **Bone marrow aspiration and liposuction provide mesenchymal stem cells.** (A) bone marrow aspiration or liposuction, (B) direct grafting of mesenchymal stem cells (mononuclear cell fraction), (C) in vitro amplification of progenitor cells, (D) grafting of amplified stem cells, adsorbed on pieces of biomaterial (E) and (F) potential treatment of a diabetic foot.

signals are released at bone marrow to activate HSC. This activation induces cell proliferation and differentiation of the haematopoietic lineage to restore all blood cells. The restoring of blood cells is mandatory for reconstitution of blood in the injured patient.

Following experiments with irradiated dogs, bone marrow transplantation was shown to be useful to restitute blood homeostasis. The physiological knowledge of the process has directed a clear improvement of treatments against hematological diseases. This treatment obtains bone marrow from healthy donors and implants them in compatible patients. Even though blood transfusions are commonly used due to mobilization of marrow progenitors to peripheral blood, bone marrow sources are iliac crests, sternum, femur and humerus. Bone marrow is the residence of millions of HSCs able to restitute these cells.

Effectiveness of bone marrow transplantation can be easily studied in hematological diseases, such as congenital aplastic anemia, thalassemia, platelet defects or coagulopathy. In these instances, HSC at transplanted bone marrow restitute absent or defective blood cells in the patient (Figure 5). However, leukemia treatments provide additional information on this clinical strategy. This disease mostly affects white blood cells. In these diseases, white blood cells are present in the patient but they are in variable quantities suggesting loss of regulation by uncontrolled proliferation. Bone marrow transplantation has also been important to treat leukemia. This cell therapy restores red blood cells and platelets previously reduced by chemical treatments. But this therapy also regulates abnormal HSC by substituting them with healthy HSC from donors. These healthy HSCs regulate patient haematopoiesis to homeostatic conditions.

Millions of MSC, adipocytes or pericytes are still present in the irradiated host bone marrow. Some of these auxiliary cells may also help to generate a correct signaling ambient for the proliferation of cells to be up-regulated (Figure 4). Indeed, some of these cell types have been proposed to act as HSC Niche to provide a fine regulation of the restituting process. In the end, this fine regulation is crucial for homeostatic renewal and health restitution. This fine regulation of stem cells is the real standard goal of this technique. A tentative hypothesis would extend this standard goal to any other cell therapy method. Since this type of cell therapy does not perform ex vivo manipulation, they do not impose regulatory standards of drug quality.

3.2.2 Cell therapy and chondrogenesis/osteogenesis

In modern societies, a number of diseases have appeared associated to the new life style. The increase in life span brings about additional degenerative diseases, such as osteoporosis or arthritis, which affect bones and joints of an increasingly number of citizens. Moreover, traffic accidents or the massive play of sports also generates an important number of bone and joint diseases. These ostearticular patients are thus widely these ostearticular patients are widely distributed in the population and very different age groups can be affected by similar diseases. Although some evidence of limited digit regeneration has been reported in children, treatments for these diseases must be adapted to the special conditions of the patient.

Cell therapy of partial bone or cartilage-defects, requires the in situ administration of osteoblast and chondroblast progenitor cells. These cells can also be obtained from bone marrow or other origins, amplified and induced in vitro into a osteo- or chondrogenic lineage. Then, these cells can easily be applied to the injured tissue (Figure 5). But, when surgeons transplanted these cells, they did not obtained proper results. Scaffolds to support cells and appropriate signals were not present and the skeletal tissue was not completely restored. A classical clinical practice was to obtain bone powder or small bone fragments and apply them to the operation field. Following this technique, a very important increase in bone regeneration was obtained. Then, surgeons transplanted a mixture of this powder and osteoblast progenitor cells to significantly improve bone regeneration. Even in these treatments, small or even larger pieces of bones were also transplanted in the mixture to gradually obtain better results. At this stage, a clear conclusion can also be obtained. Stem cells cannot fulfill a correct treatment by themselves, and a potential stem cell niche, either natural or artificial, to provide a necessary cell induction, must be present.

Indeed, artificial biocompatible materials have also been extensively developed in this task. These biomaterials show superficial properties of special adherence to progenitor cells. Moreover, a number of different proteins, many in the TGF-beta super-family, can also be applied. These molecular treatments aim to reproduce the natural regulation of progenitor cells in the body and are combined with the cell therapy. This type of Tissue Engineering is one of the most promising techniques in this area. Pre-clinical and clinical trials are well advanced and are showing significant results. As expected, these "artificial" treatments will be compared with "natural" ones after bone fragments transplantation, where cells, signals and extracellular matrix components are present. As the latter cannot always be applied due to the type of bone lesion, artificial scaffold treatments could be a candidate cell therapy in most hospitals in a short future.

In general, these therapies take advantage of easy methods to obtain MSC. Bone Marrow aspiration is a well-founded medical technique. This technique can be applied to the patient and an important number of MSC obtained in vitro. These cells and the appropriate natural or artificial scaffold can hopefully collaborate in the regeneration of osteocartilage injuries. Recent alternatives to bone marrow aspiration can be found in liposuction techniques. Vasculo-stromal fraction from fat tissue obtained by this technique can be use as an in vitro source of MSCs. Potency and quality of these cells are enough to obtain osteoblasts or chondroblasts under the appropriate stimuli. But liposuction is less invasive than bone marrow aspiration and can further help the cosmetic surgery of the patient. In any case, cell therapy for skeletal repair is one of the most promising applications of regenerative medicine.

3.2.3 Cell therapy and angiogenesis

Angiogenesis is the generation of new capillaries by a process of sprouting of pre-existing microvessels. In health, vessel proliferation is under stringent control and occurs only during embryonic development, endometrial regulation, reproductive cycle and wound repair. On the contrary, a persistent and deregulated angiogenesis is related to diseases such as proliferative retinopathies, psoriasis and rheumatoid arthritis, and seems to be essential for tumor growth and metastasis (Carmeliet, 2005). On the other hand, the concept of *therapeutic angiogenesis* has emerged as an approach to ischemic diseases in which stimulation of new vessels growth is intended to restore blood supply to ischemic tissue (Carmeliet, 2005; Tirziu and Simons, 2005).

Promising results in pre-clinical studies prompted the initiation of a number of clinical trials in patients with advanced coronary and peripheral arterial disease who had no other treatment option (Lachmann and Nikol, 2007; Tse et al., 2007). In spite of the recent advances in medical therapy, coronary artery disease remains the major cause of morbidity and mortality in the developing countries. In patients with severe coronary artery disease, persistent myocardial ischemia in hibernated myocardium can result in progressive loss of cardiomyocytes with development of heart failure. On the other hand, peripheral arterial disease is a common manifestation of systemic atherosclerosis that is associated with a significant limitation in limb function due to ischemia and high risk of cardiovascular mortality. The lower limb manifestations of peripheral arterial disease are classified into the categories of chronic stable claudication, critical leg ischemia, and acute limb ischemia. Lower limb ischemia is a major health problem since, in the absence of effective pharmacological, interventional or surgical treatment, amputation becomes the only solution to unbearable symptoms at the end-stage.

A part of therapeutic angiogenesis approaches to treat coronary and peripheral artery diseases is based on cell therapy. Bone marrow consists of multiple cell populations, including endothelial progenitor cells, which have been shown to differentiate into endothelial cells and release several angiogenic factors and thereby enhance neovascularisation in animal models of hind limb ischemia. The promising results from various preclinical studies provide the basis for clinical trials using bone marrow-derived cells or non-bone marrow cells, like cells from the peripheral blood or other tissues (Figure 5). However, the mechanisms by which these cells exert their positive effects are poorly understood (Lachmann and Nikol, 2007). Furthermore, although the initial pilot clinical trials showed potential clinical benefit of bone marrow derived cell therapy for therapeutic angiogenesis, the long-term safety, the optimal timing and treatment strategy remains unclear (Tse et al., 2007). This might explain why, up to the moment, some controversial results have been obtained. While a meta-analysis of randomized, controlled clinical trials of therapeutic angiogenesis published in 2009 concluded that patients with peripheral arterial disease -and in particular those with critical ischemia- improved their symptoms when treated with cell therapy with acceptable tolerability, a more recently published article concluded that stem cell or progenitor cell therapy did not reveal clinical benefit in patients with peripheral artery disease (De Haro et al., 2009; Nikol, 2011). Additional controlled clinical trials are in progress. When data from large randomized placebo-controlled trials will be available, it will be possible to evaluate properly the actual impact of this therapeutic approach.

4. Risks and bioethics problems

As any other medical treatment, cell therapy techniques may show intrinsic risks or lead to unexpected damages. First, to obtain adult MSC is bloody. Although it varies with the used technique, all methods show morbidity. Improvement can be found from Bone marrow aspiration to liposuction. The first technique is painful and shows potential surgical risks, whereas the second operation shows lesser morbidity although certain potential surgical complications can still arise. In any case, other alternatives show even worse side effects. Neural stem cells or liver oval cells can be obtained from an adult, but a complete neurosurgical or hepatic operation is necessary with many additional drawbacks.

Obtaining the ASCs is not the whole problem. Stem cell implantation is another risky part of the protocol. When manipulating biological material during the surgical operation, the infection risk is very high. In order to reduce this risk, the surgeon and auxiliary manipulators has to follow strict conditions and actuations. All these risk does not preclude the application of these techniques. However, **all implicated manipulators have to strictly follow many security protocols at each stage.** Good Manufactured Practices (GMP) must be observed at all times.

Moreover, postsurgical risks must also be considered. When heterologous grafts (donor and host are different persons) are carried out, immunological rejection is always important. Indeed, very severe diseases can appear such as Graft-against-Host disease or Systemic Inflammatory Response Syndrome which can drive to death to the patient. A continuous risk of cell treatments is to produce an uncontrolled growth in grafted cells to generate neoplasia. Statistically, the risk to suffer neoplasia is much lower when treated with ASC than using ESC. In ASC treatments, this probability is similar to that observed in untreated patients due to other carcinogenic factors. Recent findings indicate the MSCs produce recovery by trophic and immunomodulatory effects (Caplan, 2009). This has boosted numerous clinical trials that may hopefully lead to new healing therapies.

Another limiting question is posed by bioethics commissions. A big controversy arose with embryonic stem cells. To obtain these cells, embryos must be destroyed or their genetic material manipulated. This leads to moral problems which involves both the scientific community and the non-scientific society. Some countries approved these protocols whereas others did not. So, besides the moral questions, other economical-political reasons also influenced the natural development of these protocols. All this has led scientists and physicians to avoid the use of these cells in favor of ASCs.

5. Future perspectives

Besides all these improvements, Cell therapy has a long journey ahead. Among all obvious perspectives, to obtain Stem Cells through a lesser bloody operation, to find the appropriate stem cell source and to investigate the best signaling cocktail for any tissue differentiation, are clear options. In order to reach this, a continuous interaction between scientists and physicians must occur. If these therapies are approved for use in every hospital, GMP labs would be installed near the operating room. On the contrary, if marketing criteria are imposed, "therapeutic cells are drugs", the situation will be very different.

6. Acknowledgements

This work was partially supported by the Spanish Network on Cell Therapy (Red TerCel), PS09/02216 and BIO2009-13903-C01-01 (MICINN and FEDER), P07-CVI-2781 and P08-CTS-3759 (Autonomous Government and FEDER). CIBER-BBN is an initiative funded by the VI National R&D&I Plan 2008-2011, *Iniciativa Ingenio 2010, Consolider Program, CIBER Actions* and financed by the Instituto de Salud Carlos III with assistance from the *European Regional Development Fund.*

7. References

Alberts, B., Johnson, A., Lewis, J., Raff, M., Roberts, K., & Walter, P. (2007). *Molecular Biology of the Cell* (fifth edition), Garland Science, ISBN 978-0-8153-4100-5, New York.

Becerra, J., Santos-Ruiz, L., Andrades, J.A., & Marí-Beffa, M. (2011).The stem cell niche should be a key issue for cell therapy in regenerative medicine. *Stem Cell Rev. and Rep.* Vol.7 No.2, pp. 248-255.

Caplan, A.I. (2009). Why are MSCs therapeutic? New data: new insight. *J. Pathol.* Vol. 217, pp. 318-324.

Carmeliet, P. (2005). Angiogenesis in life, disease and medicine. *Nature* Vol. 348, pp. 932-936.

De Haro, J., Acin, F., López-Quintana, A., Flórez, A., Martínez-Aguilar, E., & Varela, C. (2009). Meta-analysis of randomized, controlled clinical trials in angiogenesis: gene and cell therapy in peripheral arterial disease. *Heart Vessels* Vol.24, pp. 321-328.

Duong Van Huyen, J.P., Smadia, D.M., Bruneval, P., Gausssem, P., Dal-Cortivo, L., Julia, P., Fiessinger, J.N., Cavazzana-Calvo, M., Aiach, M., & Emmerich, J. (2008). Bone marrow-derived mononuclear cell therapy induces distal angiogenesis after injection in critical leg ischemia. *Mod. Pathol.* Vol.21, pp. 837-846.

Lachmann, N., & Nikol, S. (2007). Therapeutic angiogenesis for peripheral artery disease: stem cell therapy. *Vasa* Vol.36, pp. 241-251.

Nikol, S. (2011). Therapeutic angiogenesis using gene transfer and stem cell therapy in peripheral artery disease. *Dtsch. Med. Wochenschr.* Vol.136, pp. 672-674.

Takahashi, K., Tanabe, K., Ohnuki, M., Narita, M., Ichisaka, T., Tomoda, K., & Yamanaka, S. (2007). Induction of Pluripotent Stem Cells from Adult Human Fibroblasts by Defined Factors. *Cell* Vol.131, No.5, pp. 861-872.

Thomson, J. A., Itskovitz-Eldor, J., Shapiro, S. S., et al. (1998). Embryonic stem cell lines derived from human blastocysts. *Science*, Vol.282, pp. 1145–1147.

Tirziu, D., & Simons, M. (2005). Angiogenesis in the human heart: gene and cell therapy. *Angiogenesis* Vol.8, pp. 241-251.

Tse, H.F., Yiu, K.H., & Lau, C.P. (2007). Bone marrow stem cell therapy for myocardial angiogenesis. *Curr. Vasc. Pharmacol.* Vol.5, pp. 103-112.

9

Self-Organization, Symmetry and Morphomechanics in Development of Organisms

Lev V. Beloussov
Department of Embryology, Faculty of Biology Moscow State University
Russia

1. Introduction

This chapter may look strange for a text-book. While the usual text-books expose firmly established facts and theories, the main aim of this essay is to tell about what we *do not* know and *do not* understand and to show that this "dark area" is probably greater than the elucidated one. No less strange may look that our arguments are based to a great extent upon the data obtained long ago and for many times described but, as we try to argue, up to now adequately non-interpreted. On the other hand, the main pathway which we suggest to move along, that is the application of a self-organization theory to developmental events, is missed in conventional text-books. Taking into consideration a strange genre of this essay, the author have to apologize the potential readers for its inevitable shortages: some points may be discussed too briefly, while others too much emphasized. Nevertheless, my goal will be achieved if just single readers will realize that in the science about organic development much more than some small details are unknown and unexplained; and that the young generation of researchers has ahead a fascinating field for further studies.

2. Do we understand development?

Being an aged Professor of Moscow State University, within several decades I am reading Embryology lectures for a large class of Biology students. I was a witness of an exciting transformation of this science (which, in the hope to look more modern changed its traditional name to "Developmental Biology") from a minor and poorly known affiliation of zoology or histology to a powerful and highly respectable branch of life sciences, closely linked with genetics and molecular biology and becoming an indispensable part of stem cell research, regenerative medicine, and so on. At the first glance, everybody even to a small extent related to this science should be proud of its achievements. But nevertheless, several times during my lecture course I feel myself uneasy with my students, as if I do not tell them the whole truth. And the truth is that, in spite of all the technological achievements, we the specialists do not understand the development of organisms not so much in details, as in its main outlines. Yes, we can produce by our willing in artificial conditions some types of cells and multicellular structures, but we have to take these and other results as given, without really explaining them. Actually, we cannot answer a question which looks naïve, but is actually very deep: *why in the course of normal development a given stage* (that is, a given set of embryonic structures) *is exchanged by another one, no less definite; or why, what looks even more miraculous, a variable set of structures comes towards quite a definite end-result* (Fig. 1).

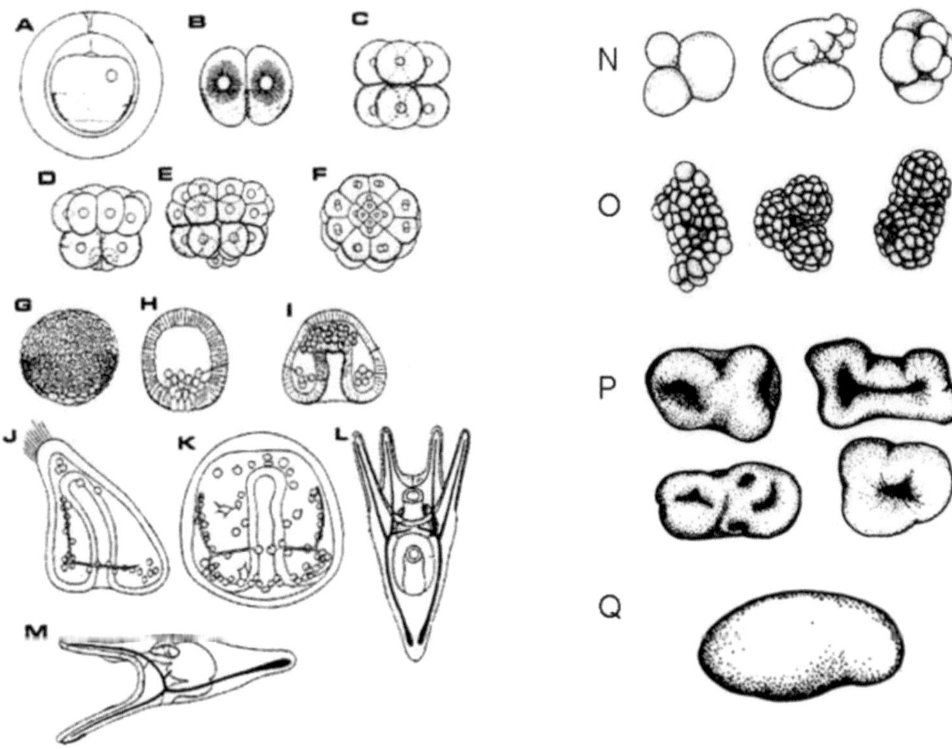

Fig. 1. Two examples of developmental successions. A-M: succesive stages of sea-urchin development.In this case the structure of each next stage is strictly determined. N-Q: development of a hydroid polyp from a cleaving egg to larva stage.The early and intermediate stages (N-P) have quite a variable structure, but the end-stage Q is the same in all the cases (From Beloussov, 2008).

True, a response to a question "why" which can satisfy us is itself in no way definite and unambiguous, especially in biology. If you ask, why a given embryonic structure is appeared at this time and location, at least three different kinds of "explanation" can be given.

First, some people will be satisfied by claiming that a given structure is arisen here and at that time moment because this is required for fulfilling its subsequent physiological functions, and/or obtaining some selective advantages, and so on. All such statements, which exchange the question "why" by "for what purpose" belong to so called teleology - a view which is looking into future for finding the reasons for what has happened just now or in the past. Teleology cannot be completely withdrawn from life sciences –for a biologist to look for goals is a respectable business. However, if we want to follow the main way taken by other natural sciences, we have to search the answers to the "why" questions in the immediate past of a given event, rather than in its future.

Just this idea became a basis for a classical causality, which is often called Laplacian, because it was formalized by the great French mathematician Pierre Simon Laplace at the beginning of XIX century and was considered for a long time to be the only one compatible with a real science. By this approach, the main aim of a science is to analyze the observed world to such an extent that it could be presented as (being split to) a chain of one-to-one cause-effects links, a single cause being able to produce no more than one event. The main task of investigator is to compile a complete list of the causes. If fulfilling this task, the surrounding world will become completely predictable: nothing new (unexpected) can happen in it.

Paradoxically, the Laplacian approach became, in the course of time, much more deeply rooted in biology, than in physical sciences, which Laplace had into mind. In particular, it has been introduced in embryology by the German embryologist Wilhelm Roux already at the end of XIX century. Up to now it remains to be the leading ideology of this science (although most of experimenters do not even suspect this).

Meanwhile, in physics since Galileo and Newton times another approach, which may be called law-centered one, took the leading positions. In a certain sense, it is opposite to the causal one, although the both developed hand by hand. While classical causality is directed towards detalization (by splitting a world into a set of as detailed as possible cause-effects relations), the law-centered approach tends to generalize, by establishing *invariable* relations between as much as possible events. For example, if two physical bodies are moving along different trajectories, this approach invites us to formulate a common law describing the both movements, while if following the classical causality we should search the specific causes for each of the movements, and even for the small parts of the trajectories. It was the law-centered approach who gave to the physical sciences a predictive power, that is, any power at all. Our main question will be - should we use this approach in developmental biology, or we are completely satisfied by a classical causality? The answer will depend mainly upon whether the successions of developmental events are underlain by perfect causal chains, determined in all their links. Let us look, whether this is the case. In doing this, we shall explore the possibilities of two mostly used versions of a causal approach. The first one claims, that the main causes of the developmental events are genes, while another ascribes a leading role to the influences of the earlier arisen embryonic structures upon the subsequent ones.

A genocentric approach seems, at the first glance, firmly substantiated, because the genes, or, if speaking more precisely, so called signaling pathways, that is the relays of protein-protein interactions, triggered by so called ligands (in most cases, products of genes activity) and switching on other genes are indispensable participants of virtually all the biological processes, including developmental ones. If blocking (knockouting) certain genes and/or signaling pathways, many developmental events will be abolished and distorted. Does it mean however that there exists one-to-one relation between a gene/signaling pathway on one hand and a given embryonic structure on the other?

Many years ago the biologists believed, that this was just the case. One of the milestones of a first half of XX century biology was a claim: "One gene - one character" (a character was taken at that time as something static, related to an adult state). More recently, however, when the amazing technical progress permitted to trace the expression of single genes in the course of development, quite unexpected results have been obtained: it turned out that the products of activity of the same or closely homologous genes and/or the same signaling pathways were involved in quite different developmental events and vice versa. Text-books in developmental and cell biology are full of such examples. Here are just few of them:

- "The interactions betweeen msx-1 and msx-2 homeodomain proteins characterize the formation of teeth in the jaw field, the progress zone in the limb field, and the neural retina in the eye" (Gilbert, 2010).
- The transcription factor Pax-6 is expressed at different times and at different levels in the telencephalon, hindbrain and spinal cord of the central nervous system; in the lens, cornea, neural and pigmented retina, lacrimal gland and conjunctiva of the eye; and in the pancreas (Alberts et al., 2003).
- In Drosophila embryos a gene Engrailed is involved in segmentation of a germ band, development of intestine, nervous system and wings. In mouse same gene participates in brain and somites development. In Echinodermata it takes part in skeleton and nervous system development (Alberts et al., 2003).
- Delta-Notch signaling pathway regulates: neuro-epithelial differentiation in insects, feather formation in birds,fates of blastomeres in Nematodes, differentiation of T-lymphocytes etc (Alberts et al., 2003).
- Hunchback gene is involved at the early stage of Drosophila development as one of so-called gap genes and at the later stages participates in development of neural system.

By summarizing: if we know everything about the genes/signaling pathways being in work in the given space/time location, we can tell nothing about what embryonic process is going on, and vice versa. This is enough for concluding that the genes/signaling pathways in spite of all their importance cannot be considered as "causes" of development, much better to say that they are tools, which can be utilized by a developing organism for quite different purposes. Certainly, the tools deserve to be studied, and such studies can be very important and useful, but they do not help us to answer our main question. Accordingly, the results of our studies will have no predictable power - we are doomed to investigate each next experimental point separately.

Let us pass to the second version of the causal approach, ascribing the main role to the interactions between embryonic rudiments. Just this version was used by Roux and his followers.

The first task which Roux decided to solve by his approach was a long standing controversy between two general views upon development. The first of them, called preformism, claimed that each structure of an adult has its own material representative from the very beginning of development, the latter being localized somewhere inside an egg or spermatozoon. By this view, from the very beginning of development an embryo is no less spatially complicated than the adult organism. The alternative view, called epigenesis, negated this idea, suggesting that an early embryo is less complicated than the advanced one, and may be even homogeneous. Roux attempted to resolve this alternative by dividing an embryo into parts: if the preformism were true, an isolated part of embryo will produce, under subsequent development, nothing else than that set of organs, which will be normally produced from this very part; if, meanwhile, a part, after its isolation, will produce another or, moreover, the larger set of organs, preformism should be rejected. Roux himself performed this procedure by killing one of two first cells (blastomeres) of a frog egg with a heated needle. As a result, the remaining blastomere produced roughly a half of embryo. It looked, as if preformism was true. However, such a situation lasted less than for a decade. Several years later another German embryologist, Hans Driesch, separated the blastomeres of another animal, sea urchin, by more delicate technique: by using sea water lacking calcium ions he separated the blasstomeres, but kept all of them alive. The result was quite another: each of two first blastomeres, and even each of the first four ones gave rise to *entire, almost normal embryos* (although of a correspondingly diminished sizes), rather than to the parts which had to be normally developed from the isolated blastomeres (Fig. 2). This effect was called *embryonic regulations*.

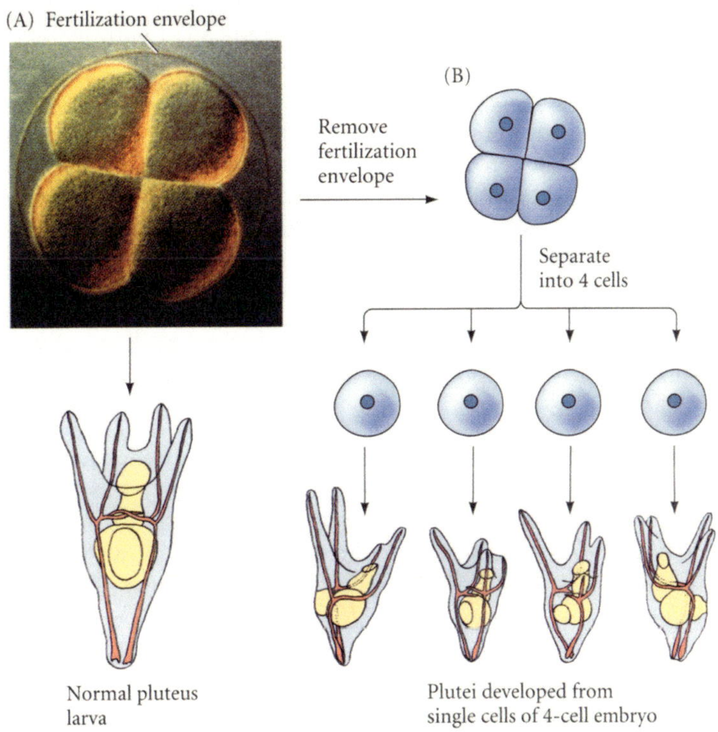

Fig. 2. A scheme of Driesch's experiment demonstrating embryonic regulations: so-called plutei larvae developed from the single blastomeres separated at 4-cell stage have roughly the same structure (at the diminished size) as the normal larva (from Gilbert, 2010, modified).

A similar result was obtained in another set of Driesch experiments, in which the blastomeres were rearranged (changed their neighbors). In spite of rearrangement, the subsequent development was going in a normal way. Since Driesch times, hundreds of such experiments have been performed at the different animal species and stages of development, obtaining in general (if omitting details, unnecessary in out context) quite similar results.

What should we derive from these experiments as related to the concept of one-to-one causal chains?

If we continue to assume that each structure of an adult or of an advanced embryo possesses its own causal chain traceable from the very beginning of development, embryonic regulations enforce us to conclude that each one of say four blastomeres contains at the same time ¼, ½ or a full set of the causal chains and, moreover, the portions of these sets contained in the same blastomere may be different in the different experiments. Taking into mind, that a blastomere "do not know" in advance, whether he will be isolated or not, and what neighbors will he have, we have to accept that embryonic regulations make the idea of one-to-one causal chains contradictory and absurd: at least during the developmental period when the regulations are taking place, any causal chains should be smoothed and lost.

Besides embryonic regulations, there are also other arguments against the existence of one-to one causal chains. One of the main ones is a so-called equifinality, illustrated by Fig. 1N-Q. It is the attainment of the same end-result of development by quite various, sometimes purely stochastic developmental pathways. A stochasticity of embryonic processes firstly emphasized already a century ago by the Russian biologist Alexander Gurwitsch, looks to be a background of many morphogenetic events, first of all those associated with branched rudiments (blood vessels, lungs, leaf veins). These structures are of a fractal nature and are hence generated in chaotic regimes, to which the notion of specific causes is completely inapplicable.

Now let us look, what conclusions from his experiments made Driesch himself. He expressed them in a laconic statement, known as Driesch law: "A fate of a part of embryo depends upon its position within a whole" [let us add: rather than upon its internal properties].

By this formulation Driesch wanted to interpret embryonic regulations in the following way. At the first step, the shape of a normal early embryo is in rough outlines and in diminished size restored. Next, each cell of a regulated embryo "recalculates" its position according to its coordinates within a new "whole" and develops according to this recalculated position, rather than follows its normal destiny. Formally such interpretation may be true, but several important questions remain unanswered. First of them is: by what means a roughly normal shape of an early embryo is restored? This process is not explained by Driesch law. Moreover, well after Driesch it was shown that a normal shape can be restored from the cells arranged in a completely chaotic manner (Fig. 3). The second question is: what are the

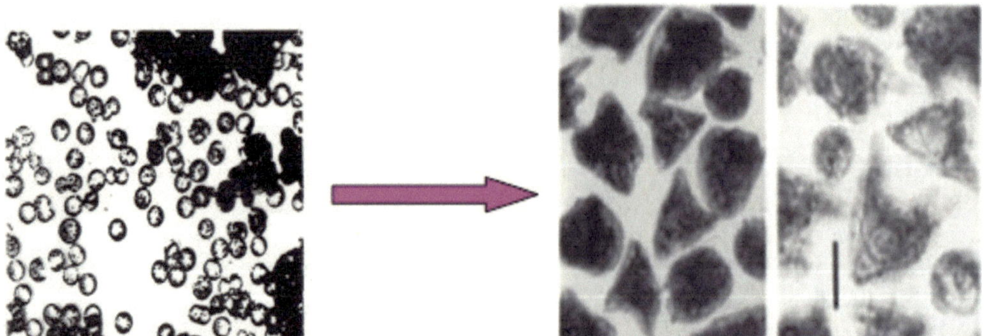

Fig. 3. Normal sea-urchin larvae can arise from completely random aggregations of embryonic cells.

reference points for the recalculation procedure? Driesch formulation – "according to a whole"- is too vague, although, as we'll see later, such a vagueness has its own justifications. Meanwhile, in a new and a most popular version of Driesch law – a concept of "positional information" (PI) (Wolpert, 1996) - the answer was another: cell positions are referred to certain special predetermined points, often defined as "sources" and "sinks" of some diffusible substances, the morphogenes. But it is easy to demonstrate that the existence of such predetermined points is incompatible with embryonic regulations. The matter is that under either partial removal or rearrangement of embryonic material all of its elements (including those which are suggested to be the reference points) take the positions, geometrically non-homologous to those occupied by the same points in normally developing embryos (Fig. 4). Moreover, so far as the early embryos are capable to

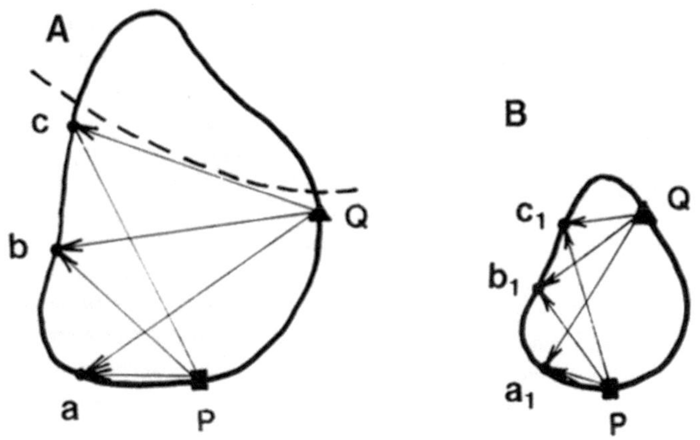

Fig. 4. Embryonic regulations are incompatible with the assumption of any prelocalized specific material elements (say, P andQ), regarded as the sources of a positional information (PI). After the dissection of an embryo part upper from a dotted line shown in A and closure of the wound (B) the positions of the elements P and Q (as well as all the others) will become geometrically non-homologous to the same elements' positions in A. As a result, any points of an embryo which occupy in A and B homologous positions (say, a and a1, b and b1, c and c1) will perceive quite different PI signals which is incompatible with embryonic regulations (from Beloussov, 1998).

regulations after the removals and rearrangements of quite different embryonic areas, any predetermined elements will take in different experiments quite different (but each time geometrically non-homologous) positions. Consequently, we have only two formal possibilities to "save" a concept of predetermined reference points: either to suggest that such a role is passed each time to the elements, occupying geometrically homologous positions, or to assume that *all* the elements of a partial or normal embryo play a role of the reference points. However, the both versions (out of which the second one looks more consistent) imply that either only the reference points or even all the embryonic elements should somehow "feel" the shape of a whole. Here we see that a vagueness of the Driesch formulation had its reasons: intuitively he felt that the regulations which he discovered cannot be explained without implying the action of something related to irreducible whole to the minor elements. In his time, when the Laplacian determinism still was in full power, such a claim looked as something inappropriate for a real science. Today it is impossible to negate such things; moreover, the verbal expressions like "top-down causation", "emergent behavior" and "context-dependency" have been coined for describing them. However, mere words are not enough; what we need is a coherent law-centered theory explaining the arising of complexity, holistic regulations and so on. Remarkably, such a theory has been emerged already several decades ago quite outside of biology as a result of convergence of several branches of mathematics and physics. This is a self-organization theory (SOT). Before addressing to SOT directly, we must however get some knowledge of another topics - a symmetry theory (even irrespective to SOT it is very useful for any biologist).

3. Elements of a symmetry theory as related to development

In short, a symmetry theory is dealing with invariable transformations of geometric bodies, that is, with those kinds of movements which superpose a body with itself. The emphasis upon invariability makes this theory closely related to the very essence of law-centered approach.

We'll restrict ourselves to the elementary part of a symmetry theory, which is dealing with three kinds of movements: rotations, reflections and translations and explore some simple examples. It is easy to see, for example, that a rectangle superposes with itself under rotations around its center to 90^0, 180^0, 270^0 and 360^0 (that is, in four positions), a regular triangle do the same under rotations to 120^0, 240^0 and 360^0 (3 positions), while a disc superposes with itself under rotation to an infinite (∞) numbers of angles. A number of positions superposing a body with itself is defined as an *order of symmetry* (in the above cases it is a *rotational* symmetry). If, as in the case of a disc, the number of such rotations is infinite, one speaks about a *power* of a symmetry order. Passing from 2-dimensional disc to 3-dimensional sphere, we obtain a rotational symmetry power ∞/∞, what means that the rotations may go around an infinite number of central axes intersected at any angles.

The bodies possessing any order of a rotational symmetry may have or not have *reflection* (mirror) *symmetry*, its plane denoted as m. Thus, a combined rotation/reflection symmetry of a sphere is $\infty/\infty \cdot m$. The bodies having no mirror symmetry at all can exist in two mutually reflected modifications, which can be arbitrarily defined as left and right.

The *translational* symmetry is that of linear repeated patterns. Its order is characterized by a smaller linear shift n, which superposes a shifted pattern with non-shifted one. If a pattern is homogeneous along the shift direction (n is infinitesimal), the translational symmetry order is ∞.

After learning these definitions, we can easily see that the development of an egg towards the adult state is associated with a stepwise reduction of symmetry order (or a series of symmetry breaks, as is often told). Thus, an egg before the establishment of its polar axis has a symmetry order of a sphere ($\infty / \infty \cdot m$), after the axis establishment it is reduced up to $\infty \cdot m$, while after determination of a saggital plane (into which the antero-posterior axis of a future organism is located) it becomes $1 \cdot m$ (we ignore a right-left asymmetry, which is of a molecular origin and seems to persist throughout the entire life cycles without any fundamental perturbations). The development of advanced embryos is mostly associated with reduction of a translational symmetry order, that is, with establishment of the finite (rather than infinitesimal) n values. Most obvious examples are the formation of mesodermal somites out of a roughly homogeneous cell mass, or a subdivision of an initially smooth neural tube into brain vesicles.

We pay so much attention to symmetry breaks, because they are closely associated with the entire problem of causation. This linkage has been formulated by a classical principle claimed by the French physicist Pierre Curie exactly at the time when Driesch made his regulations experiments (although the both scientists did not know anything about each other). In his principle, Curie gave for the first time a strict definition of an effect and its cause. By his idea, any observable event is associated with the reduction of a symmetry order (by his words, "This is a dissymmetry, which creates an event"). Next, by Curie principle, no symmetry break can take place spontaneously, that is, without a somewhere located "dissymmetrizer", an object with the already reduced symmetry order. It is a dissymmetrizer, which fits a notion of a "cause".

By applying this concept to developmental events, we have to conclude that any step of the above mentioned symmetry breaks, according to Curie principle, demands a dissymmetrizer, located either outside or inside of an entire egg/embryo. Let us start from the earliest developmental events. At the first glance, they require external dissymmetrizers. For example, an egg polarity in the eggs of brown algae can be established by a directed illumination of an egg and the polarity of many animal eggs by the surrounding structures of an ovary. The position of a saggital plane in amphibian eggs is determined by the point of a sperm entrance, and so on. However, very accurate observations have shown, that the external agents are not necessary: the algae eggs acquire polarity under absolutely isotropic illumination and amphibian eggs can select a plane of saggital symmetry out of an infinite bunch of planes even in the absence of a spermatozoon (parthenogenesis), or if it was inserted accurately into the egg pole (where it cannot act as a dissymmetrizer).

Even less are the chances to find any dissymmetrizers for the events taking place in more advanced embryos. Here, as known from embryology text-books, in very many cases one rudiment plays a role of a so called inductor which triggers the development of another one, and in most cases this process is directly or indirectly mediated by chemical agents, emitted by inductor. Usually the inductors are regarded as the "causes" of the induced organs formation, but is it so in the terms of Curie principle? It is easy to show, that virtually in all these cases the symmetry order (as a rule, translational) of an induced morphological structure is considerably reduced in relation to that of an inductor; for the cases of purely chemical induction this is obvious without any comments. In the terms more customary for biologists this means that the morphological structure of an induced rudiment cannot be derived in one-to-one manner from that of an inductor: certain factors, increasing the complexity of the induced organ and non related to inductor itself should be involved.

In general, both embryonic regulations and symmetry breaks without dissymmetrizers leads us to conclude, that in the course of development more complicated (less symmetric), although if perfectly ordered entities are emerged from less complicated (more symmetric) ones. This is incompatible with a classical causal approach, but perfectly fits to what is called self-organization. Is such a process unique for the living beings?

4. Self-organization in inanimate matter

Already more than century ago the first examples of the similar events proceeding in non-biological systems has been described by the French physicist Benard. This was the formation of cell-like structures (Benard cells) from a homogeneous viscous liquid, intensively heated from below (Fig. 5A, D). These structures immediately disappeared after heating was stopped, or became less pronounced. As proved later by one of SOT founders, Ilya Prigogine, these structures appeared because under enough intense flow of energy the convection streams (upward shifts of heated liquid particles and downward shifts of the cooled ones) pass from a random to so-called *coherent regime,* characterized by collective movements along some common trajectories which became now energetically more advantageous than the random movements along individual tracks (Fig. 5B, C). What we see here is a real emergence of an ordered complexity from a homogeneous state or, in other words, a spontaneous (non-embedded from outside) reduction of a symmetry order: the initial infinite order of a translational symmetry is reduced up to that of *n* order, where *n* is a Benard cell diameter.

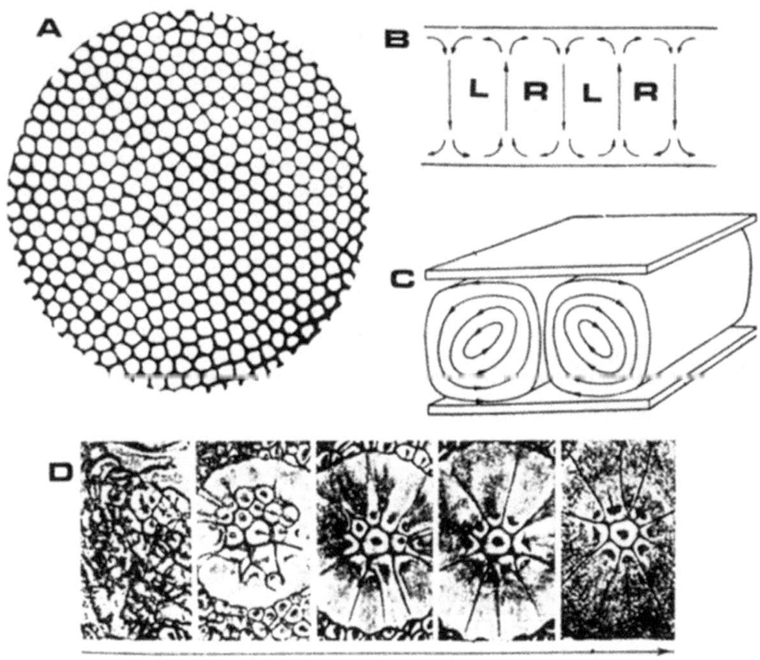

Fig. 5. Benard cells. A: general view from the top. B, C: schemes Of coherent convection streams. D: evolution of Benard cells patterns under constant heating (from left to right)

While the phenomenon of Benard cells formation did not pay much interest and was not considered as a breakthrough event, quite another was a public reaction to the occasional discovery of a fluctuating chemical reaction by the Russian chemist Boris Belousov in 1950ieth. Although firstly it was rejected by the editorial board of a scientific journal (the referee wrote that it violates the second law of thermodynamics and hence should not exist) very soon an entire research team from the Institute of Biophysics, Russian Acad. Sci. extensively elaborated this reaction, transformed it into space-unfolded "autowaves" and gave its complete theory. Because of its vividness, the reaction became very popular throughout the world: everybody could see that within a couple of minutes a series of ever complicated spiral waves appear from "nowhere", that is from a completely homogeneous state (Fig. 6).

Fig. 6. Successive structures (1-8) arisen during Belousov-Zhabotinsky chemical reaction.

5. A theory of something emerged from nothing

Now we'll give a very brief and elementary review of SOT principles (for much more complete, but still popular SOT account see Capra, 1996; for a developmentally related account see Beloussov, 1998). Let the readers only slightly familiar with math be not afraid: the math will be minimal. As other great specialist in this field, Rene Thom said - "this is not the math, this is a mere drawing". Our drawings will be also minimal - most will be expressed by words.

The first point to be noticed is that contrary to classical mathematics, SOT is about a real world, which is full of so called unexpected perturbations, or a noise. Without noise none of the effects, predicted and described by SOT, will take place. For us biologists this is quite obvious: all the organisms are living in a very noisy world, which they have to resist and/or assimilate, preserving their individual, or a species-specific way of living. Such a property of a dynamic, or functional (not static!) resistance is also one of the main components of a self-organization. It is called robustness. All the natural systems are to a certain extent robust - otherwise they would not exist at all. However, robustness always has its limits, and when they are exceeded, a system abruptly passes into another state, which is as a rule also robust.

Let us express the above said by mathematical symbols. We shall see that such a transformation will very much clarify what was told before. Our main tool will be differential equations, firstly one variable linear, and then two variables nonlinear ones. Why is it necessary? The matter is, that even the simplest differential equation like

$$dx/dt = kx \tag{1}$$

has the following properties, lacking, say, in algebraic equations:

1. It describes a process, rather than a static state;
2. It contains a feedback loop: not just the right part variables affect the left part ones, but vice versa as well. The feedback may be either positive, or negative, or, in the case of two variables equations, positive-negative (±). The latter is mostly useful for self-organization.
3. Most important in our context: differential equations combine the values of quite a different order - dx/dt is an infinitely small part of x. Thus, x represents a whole, while dx/dt its small part. Correspondingly, the action of x upon dx/dt is the action of a whole upon its parts. Just formally, differential equations imply a holistic causation, which we beforehand derived from experiments. .

Let us now add a free member "-*A*" to eq. (1), obtaining

$$dx/dt = kx - A \tag{1a}$$

and make a graph (Fig. 7A), depicting by arrows the directions of dx/dt . We'll get what is called the vector field (in this case 1-dimensional). Owing the presence of a free member, we have a stationary point $dx/dt = 0$ at $x = A/k$, from which the vectors dx/dt are *diverged*. Correspondingly, if we reverse the signs of the right part members, getting

$$dx/dt = -\ kx + A \tag{1b}$$

the vectors will be *converged* towards the point with the same coordinates (Fig. 7B). This is enough for coming to the main notions of SOT: those of a dynamic (or Lyapunov) *stability/instability*. The solution (stationary point) in eq (1a) is unstable, because any infinitesimal shift from this point will bring us away without any chance to return back. On the contrary, in the framework of (1b) equation after any shift we'll come back to the stationary point, which is unlimitedly stable. We call this kind of stability/instability dynamic because it relates to the variable x, which dynamics is just traced in the equations. Besides these *dynamic variable(s),* in all the equations another kind of values is always present and plays a leading role: those are so called *parameters,* which either do not change at all their values, or change them in an order more slowly than the dynamic variables. A distinction between dynamic variables and parameters is very important, because it relates to the fundamental concept of the *structural-dynamic levels.* This concept, belonging to so called systems theory, claims that a surrounding world (both animate and non-animate) is stratified into a number of more or less discrete levels distinguishing from each other by characteristic times (Tch) and characteristic dimensions (Lch) of the related events; the both hierarchies are, as a rule, roughly parallel to each other. In this language, the parameters, at least by Tch criteria, should be attributed to a much higher level than the dynamic variables. As concerning Lch it is crucial that in the developing organisms the dynamic variables are always the collective entities: all the developmental events are based upon the action of many cells, or many molecules, occupying different positions. As a rule, all the members of this collective share the same parameters values (otherwise this would not be a common system). Correspondingly, the area of the parameters action (the parameters Lch) is also greater than Lch for each one dynamic variable.

In eq (1a, 1b) the parameters are represented by k and A values. Even in these simplest equations they are playing the main role and, remarkably, do it in quite a robust manner. Namely, there are the signs (+ or -) rather than the absolute values of the both parameters which decide whether the solution will be stable or unstable: it is easy to see, for example, that only eq (1b) rather than (1a) has stable solutions. At the same time, in the immediate vicinity of $k = 0$ just very small shifts of k values are enough for switching a solution from stable to unstable, and vice versa. In other words, in relation to the shift of, say, parameter k eq (1) is unstable at $k \approx 0$ and stable in all the other areas. This is another kind of stability/instability, which is called *parametric*, or *structural*. A notion of a structural stability very adequately represents such biological realities as, for example, a morphology of a taxon, because it reflects at the same time a preservation of a general "Bauplan" and some considerable, but nevertheless limited fluctuations.

The notions of stability/instability (both dynamic and parametric) and their regulation are of a primary importance for understanding the developmental transformations and their relations to causality. When we notice, as mentioned above, that at least some of the symmetry breaks look to be proceeded "spontaneously", this actually means that the preceded symmetry order has lost its dynamical stability and hence can be broken by negligibly small perturbations (of a noise intensity), to which a developing organism is insensible during stable periods. By the way, this means that Curie principle formally keeps its validity during instability periods as well, but at that time the "causes" are so small that cannot be distinguished from the ever presented noise.

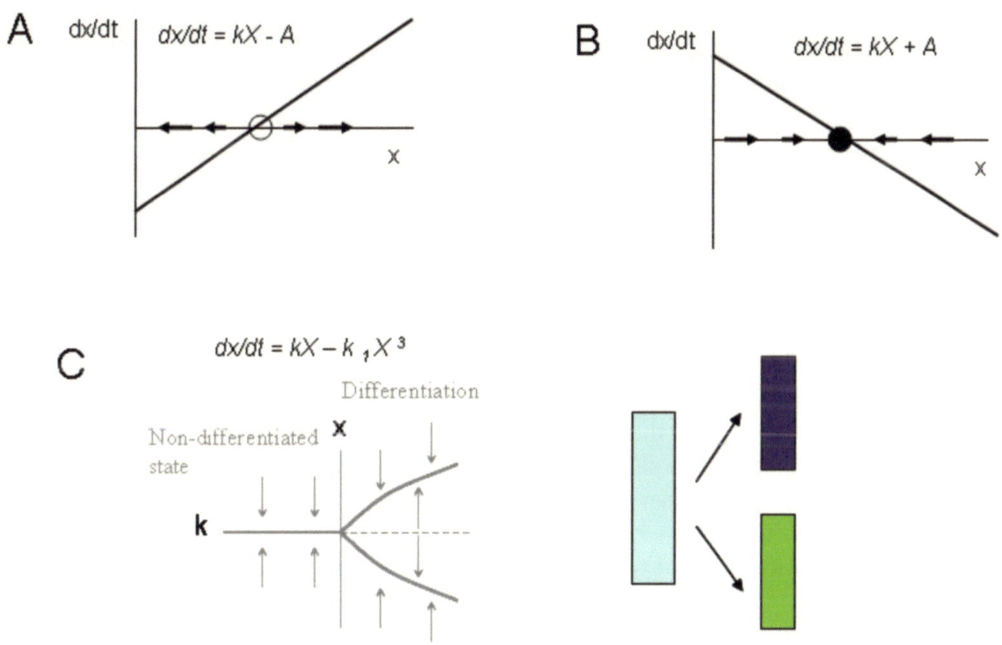

Fig. 7. Vector fields and solutions of some simple differential equations. A, B: linear equations. In A the solution is unstable (empty circle, vectors diverging), while in B it is stable (filled circle, vectors converging). C: under $k < 0$ there is only one stable solution $x = 0$, while under $k > 0$ this solution becomes unstable while two stable solutions appear in exchange. A transition from negative to positive k values corresponds to that from a single non-differentiated to a differentiated state (colored scheme to the right). For more details see text.

Under these conditions, it is meaningless to look for the "causes" in their classical sense. What we need to know instead is why at the given moment an embryonic structure has lost its previous stability. Meanwhile, we know already, that the loss (or acquiring) of stability is provided by the changes of the parameters values. This conclusion is of a direct methodological usage: instead of splitting the studied systems into ever smaller parts in pursuit of ever escaping causes, we should concentrate our interest onto macroscopic levels, to which the parameters belong.

For illustrating the role of parametric regulation in more details, we have to go from linear to non-linear equations. This shift is not just formal: it means that we pass from independently acting elements to interacting ones (those which either enhance, or inhibit each other, or make the both things together). In other words, non-linearity means *cooperative interactions*, most of all important for developmental events. A simple example: suggest that N identical elements affect the event A. If these elements are independent of each other, their action upon A is proportional to N, while if N elements enhance each other, it is proportional to $N(N-1) \approx kN^2$.

We miss a quadratic non-linearity and come directly to the 3-rd order one, as providing one of the best illustrations of developmental processes. Consider the equation

$$dx/dt = kx - k_1x^3 (k_1 > 0) \quad (2)$$

which describes a first order positive feedback between kx and dx/dt and 3-rd order negative feedback between k_1x^3 and dx/dt . It can be easily tested, that under $k < 0$ eq (2) has only one rational solution $x_1 = 0$ which is stable, while under $k > 0$ this solution becomes unstable, while two new stable symmetric solutions appear:

$$x_2, x_3 = \pm \sqrt{k/k_1}$$

(Fig. 7C). The main property of this model is that when k parameter in his rightwards movement reaches positive values, a number of stable solutions increases from one to two. Accordingly, it is a simplest model of the complexity increase, or of the reduction of a symmetry order (under $k < 0$ the sole stable solution $x_1 = 0$ is the axis of a rotational and reflection symmetry, while under positive k none of the stable solutions x_2, x_3 can play this role). In biological language this is just what we define as differentiation. The most important lesson from this model is that such a crucial step is under a full parametric control. That does not mean that the dynamic variables play no role at all, but this role is parametrically dependent. Namely, only under $k > 0$ the dynamic variables can select one of two vacant stable solutions. But they do it in quite a robust manner, without being obliged to take precise values: under any $x > 0$ the positive solution is selected, while under any $x < 0$ the negative one. Therefore:

(1) there are the parameters which determine the number and the values of stable solutions, that is, stable states, *potentially achievable* by a system; (2) among these, the actual states are selected by the dynamic variables in quite a robust manner: each stable state is a "basin of attraction" of infinitesimal number of the dynamic variables values.

By the way, these conclusions undermine a myth of an extremely precise organization of the living beings: the main condition for surviving and keeping their individuality is a small number of potentially achievable and highly robust stable states, rather than a precise arrangement of the dynamic variables, which never exists.

Remarkably, most of the above said can be easily translated into embryological language.

One of the most important notions of embryology is indeed that of a *competence*: briefly speaking, this is a capability of a given embryo region at a given stage to develop into more than one direction. Now we may see that in SOT language it corresponds to that region of the parameters values, which has more than one stable solution. Therefore, the existence or the absence of a competence should be regulated parametrically. The next step after reaching the region of competence will be to come into the "attraction basin" of a definite solution. This event corresponds to what is defined in embryology as *determination*, and we can conclude that it is a matter of dynamic regulation. Same will be true for the final reaching of a stable state; in embryological language this is *differentiation*.

At the end of this section, let us briefly describe, even missing formulae, some more complicated self-organizing systems. Biologically very important class of systems is described by 2-variables (*X* and *Y*) non-linear differential equations, where Tch for *Y* are in an order smaller than for *X*: therefore, if including the parameters, these systems are at least three-leveled. In addition, the variables are interconnected by (+, -) feedbacks: a slower variable *X* inhibits a fast variable *Y*, while the latter enhances the first one. As a result, in a wide range of the values of a single controlling parameter we get so called autooscillations, that is, non-damped regular fluctuations of the both variables values. Complementing this system by a linear dependence between *Y* and *dx/dt*, we transform ever persisting autooscillations into a so-called trigger regime with two stable states, exchanging each other after finite perturbations of one of the variables. The arisen structures may be either only time-dependent, or in addition space-unfolded. In the latter case one has to assume that at least one of the variables is diffusing through space (it may exemplify not only a chemical substance, but also a certain physical state). In any case, all of these either purely temporal, or spatial-temporal structures are able to create, under a proper range of controlling parameters, quite stable patterns out a completely homogeneous state; note however that the patterns are stable until the supply of dynamic variables will continue.

6. Application of SOT to embryonic development

The first person to be mentioned here is Conrad Waddington, a British scientist who, even before SOT emerged in its present form, suggested a very stimulating allegory of development, that of a mountain landscape, consisting of valleys (which symbolize stable developmental trajectories) and crusts (imaging unstable states between valleys) (for recent account see Goldberg et al., 2007). There is also a tale that it was Waddington who asked a famous mathematician Alan Turing whether it is possible to construct a model generating a macroscopic order out of a completely homogeneous state. Turing did so postulating feedback interactions and diffusion of two reagents (Turing, 1952). His model became quite famous, even if it had no relations to any real biological process. An entire series of models, aiming to imitate biological realities have been constructed later on by Gierer and Meinhardt (Meinhardt, 1980). In general, the models postulated feedback interactions between two chemical substances, one of them (the activator) stimulating the development of a certain structure, while another (the inhibitor) suppressing the activator. Necessary was also the inequality of the both components diffusion rates: the inhibitor should diffuse much more rapidly than the activator.

These models permitted to reproduce a number of biomorphic patterns, mostly periodic ones and in particular those related to surface designs. Also, they introduced an important principle: "short-range activation – long range inhibition" - which seems to be a wide-spread tool for pattern formation, although not necessarily connected with diffusible chemical substances. Meanwhile, the authors were fully satisfied by reproducing some single steps of development, without expressing any interest to model more or less prolonged chains of events. As a result, the initial conditions and the relations between postulated chemical substances and morphological structures which they assume to "activate" or "inhibit" had to be taken each time in quite an arbitrary way.

7. Mechanically-based self-organization (morphomechanics)

Much closer to biological realities and less connected with arbitrary assumptions became another class of models, emerged since 1980ieth. The main acting agents in these models were *mechanical stresses* (MS), generated by embryonic cells. Even a priori MS looked to be good candidates for being involved into regulatory circuits by the following reasons at least:

- they belong to universal (largely non-specific) natural agents;
- they are acting at the same time on quite different structural levels, from molecular to that of whole organisms;
- MS create very effective feedbacks with geometry of stressed bodies: any changes in MS pattern affect geometry in a well-predicted way, and vice versa. As D'Arcy Thompson told in his classical book "On Growth and Form" (last edition: Thompson, 1961) "Form is a diagram of forces".

As discovered during several last decades, embryonic tissues of all the studied animals, from lower invertebrates to human beings are mechanically stressed (same, even to a greater extent is true for plants). Embryonic MS are of different origin. In early development the main stressing force is turgor pressure in embryonic cavities (blastocoel, subgerminal cavity), which is born due to ion pumping and which stretches the surrounding cell layers. At the advanced stages most of stresses are caused by collective movements of many dozens of cells. Cell proliferation also contributes to MS. It is of a particular importance, that MS are arranged along ordered patterns, remaining topologically invariable during successive developmental periods and drastically changing in between. They never are uniformly spread throughout the developing embryos, but are generated in a certain part and transmitted by rigid structures to others.

Already several decades ago the German anatomist Bleschmidt described a large set of MS patterns emerging in human development, and claimed that "the general rules… that are applicable to man … have much in common with the rules of the developmental movements that take place in animals and even in plants" (Bleschmidt and Gasser, 1978). In advanced embryos he distinguished 8 different kinds of MS fields which participate in development of practically all the organs. Some of them are depicted in Fig.8A-C. Modulations of MS patterns (relaxation, reorientation, changes in MS values) in amphibian and chicken embryos lead to grave developmental anomalies. A number of fetus pathologies are also mechano-dependent.

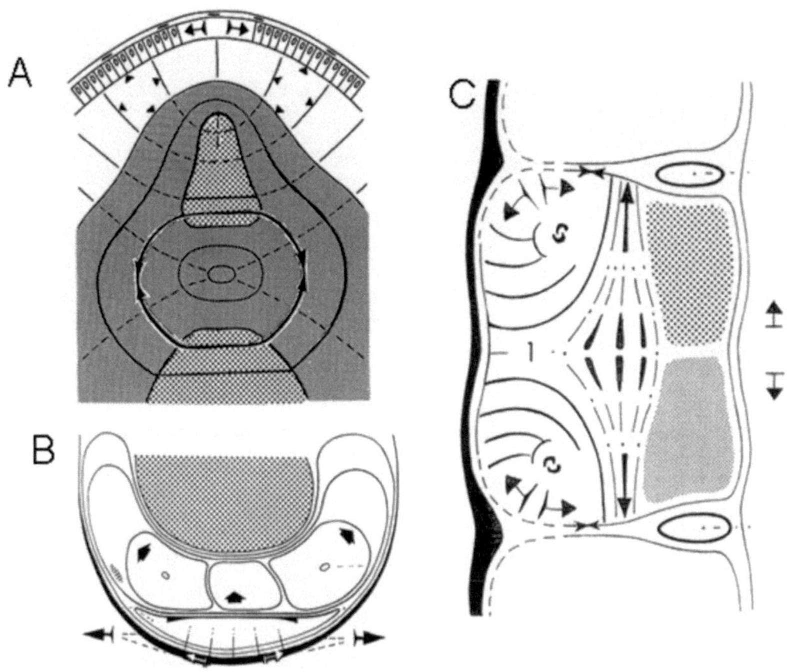

Fig. 8. Some examples of "biokinetic" schemes of human embryos anlagen by Blechschmidt and Hasser (1978). A: a rudiment of a finger; B: heel pad of 5 months fetus; C: a somite with surrounding tissues. Diverging and converging arrows depict stretching areas and those resisting to stretch, correspondingly. The main idea is that all the anlagen have their own patterns of mechanical stresses.

Most important, self-generated MS affect each other, creating feedbacks. Harris and coworkers (1984) evidenced the presence of such feedbacks by observing cell cultures seeded onto highly elastic substrates which the cells were able to stretch by their own contractile forces (Fig. 9A, B). As a result, homogeneously seeded cells became rearranged into regular clusters (Fig. 9C). This is a real self-organization (reduction of symmetry order) created by a feedback between short range adhesive interactions, tending to clump cells together into a tight cluster and long range stretching forces which extend the substrate and hence decrease cell density. Within the model framework, the adhesive forces correspond to short range activation, while the stretching forces to the long range inhibition of Gierer-Meinhardt models. Therefore, mathematics is roughly the same, but physics quite another – mechanics instead of chemistry! Quite similar, although if independently developed approach has been used in Belintzev et al. (1987) model, aiming to reproduce a segregation of initially homogeneous epithelial layers into the domains of columnar and flattened cells. In this model a role of short range activation was played by so called contact cell polarization (CCP) – cell-cell transmission of a tendency to become columnar. At the same time, long range inhibition, similarly to Harris et al. model, was provided by mechanical tension, arisen in the epithelial layer with fixed ends just because of CCP. Hence, again we have here a mechanically based (+, -) feedback. This model is of a special interest, because (unexpectedly to the authors) it became able to reproduce some main properties of embryonic regulations, namely preservation of proportions under different absolute dimensions of a layer. This became possible, because the model equation contains a member, referring to holistic (independent from individual elements) property of a layer: this is the average cell polarization throughout the entire layer. Thus, the model can be considered as a mathematical expression of the Driesch law.

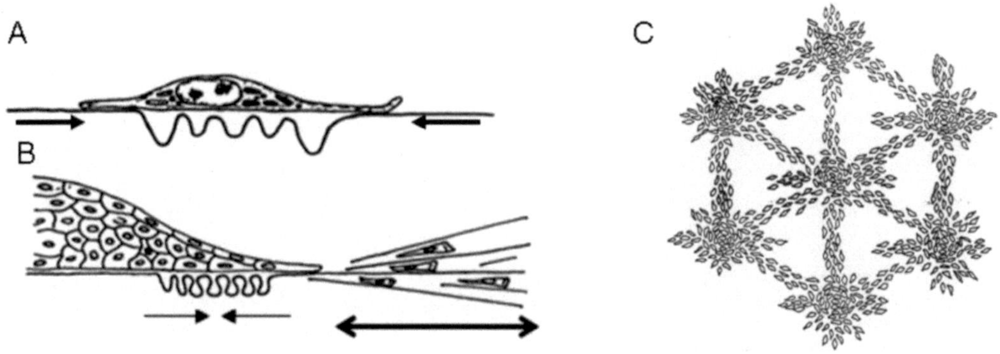

Fig. 9. Formation of regular cell clusters onto an elastic substrate. A: a single crawling cell shrinks the underlain substrate and hence stretches that located outside. B: a large cluster of adhered cells (to the left) stretches the cells located outside (to the right). C: a regular cell pattern arisen in the dermal layer of chicken embryo under the similar mechanical conditions (from Harris et al., 1984, with the authors' permission).

The described models made the first steps away from a purely static view upon development: we began to understand, why a given stage is exchanged by the next one. However, the modeled chains were too short and very soon abrupt. Is it possible to use a mechanically-based approach for reproducing much more prolonged developmental successions, including the above models as particular cases? Such an attempt has been performed by our research group about two decades ago (at the initial stage of this enterprise very important contribution was made by Dr. Jay Mittenthal from Illinois University, USA).

Our main idea was very simple. It is well known that any organism, deviated by any external perturbation (including certainly mechanical forces) from its normal functioning, tries to diminish the results of perturbation up to their complete annihilation. We modify this almost trivial statement by adding that any part of a developing organism affected by a mechanical force (coming normally from another part of the same embryo) not only tends to restore its initial stress value, but do it with a certain overshoot. This assumption, called the hypothesis of MS hyper-restoration (HR), permitted to make several predictions, opened for experimental and model testing (Beloussov and Grabovsky, 2006; Beloussov, 2008).

For example, according to this model, a stretching of a tissue piece by an external force should produce the active reaction which firstly diminishes stretching and then, as a part of HR response, generates the internal pressure force, directed along a previous stretching (Fig. 10A). As a rule, this is done by so called cell intercalation, that is, cells insertion between each other in the direction, perpendicular to stretching (Fig. 10C). Accordingly, if a tissue piece is relaxed or, the more, compressed, its cells should actively contract in the direction of relaxation/compression, tending to produce tension is this very direction (Fig. 10B, D). If applying these predictions to a cell sheet bent by external force, we should expect that its concave (compressed) side will be actively contracted, while the convex (stretched) one extended. As a result, their cooperative action will actively increase the folding, just triggered by external force.

Importantly, these reactions are connected by feedbacks with each other. Among them, one of the main can be called "contraction-extension" (CE) feedback. As any other self-organizing event, it starts from a fluctuation - in this case, of a stretching/compression stress along cell layer. For example, if a part A of a layer is stretched slightly more than a neighboring part B and the layers edges are firmly fixed, at the next time period part A will be actively extended and hence compresses the part B. The latter will respond to this by active contraction, even more stretching A, and vice versa. The modeling showed (Beloussov & Grabovsky, 2006), that the results of these interactions crucially depend upon one of the parameters, a so called threshold stretching stress (TSS), that is the minimal stretching stress required for generating the internal pressure. If TSS is taken large enough, a layer will be segregated into single alternated domains of columnar and flattened cells. Just this situation corresponds to Belintzev et al. model. Under TSS decrease the number of alternative cell domains is increased while under very small TSS no stationary structures are produced: instead, a series of running waves is generated. This exemplifies a parametric dependence of morphogenesis.

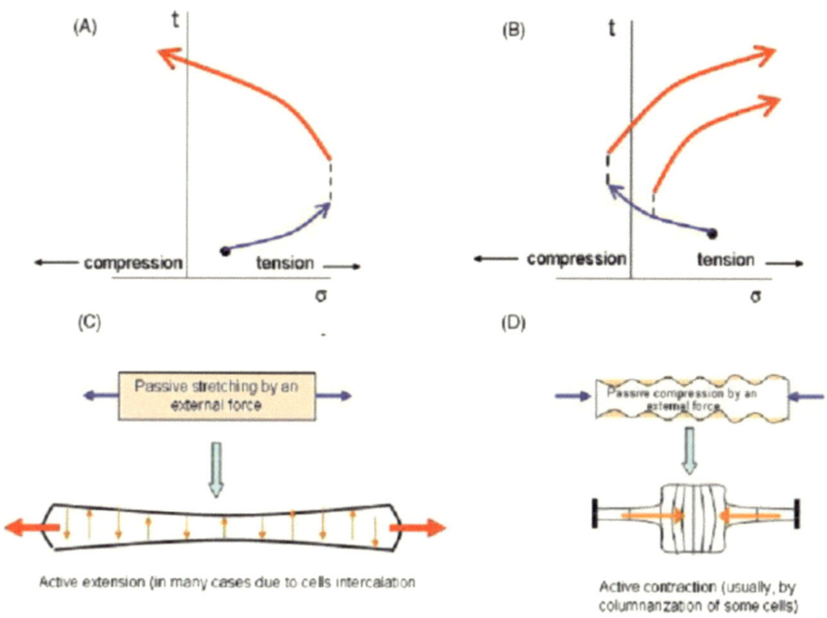

Fig. 10. Model of hyperrestoration of mechanical stresses. A, B: schemes of the responses to stretching and to relaxation/compression, correspondingly. Horizontal axis: mechanical stress (compression to the left, tension to the right). Vertical axis: time. C: a typical way for a response to stretching (cell intercalation). D: response to relaxation by tangential contraction (columnarization) of some neighboring cells. Vertical bars: firmly fixed edges.

Now let us reproduce in very broad outlines a more or less prolonged (but still uncomplete) chain of morphogenetic events. We start from so called "idealized blastula" stage, a spherically symmetric body with the walls of equal thickness which surround a concentric cavity and are stretched by the turgor pressure within the latter. Why during normal development a blastula stage embryo will not stop at this stage but passes instead towards a more complicated (less symmetric) form? By our suggestion, this is because, according to CE-feedback, a spherical symmetry of blastula is unstable: even small local variations in its wall thickness will produce the corresponding differences of tensile stresses: thinner parts will be stretched to a greater extent and hence produce the greatest internal pressure, actively extending themselves and compressing the resting part(s) of the cavity wall. At the next step the mostly compressed part will generate, according to above said, the active contraction force. This delimits the start of the next stage, called gastrulation. In general, such a contraction can be achieved by different ways. The first of them is emigration of some cells from the compressed part inside the blastula cavity. This is typical for some lower Invertebrates (Cnidaria). Another one is the folding of a compressed part of a cell sheet; it is more elaborated type of gastrulation, called invagination. However, the folding itself may go in different geometric ways: the extreme ones are exemplified by a creation of a straight slit (Fig. 11A), and a circular fold (Fig. 11E). Take a sheet of paper and try to reproduce each of them. You will see how easy is to make a slit-like fold, while to make a circular one is virtually impossible - so much radial folds are arisen around! In order to smooth out the folds, the excessive cells should be removed (emigrated) from the folded area. In principle, CE-feedback can provide such a mechanism, but it is to be well tuned. On the contrary, a slit-like folding does not demand so refined regulation.

Nature employed the both ways: the first one is typical, for example, to Annelides and Arthropoda (belonging to a large group called Protostomia), while the second one for Echinodermata and Chordata (belonging to so called Deuterostomia). Interestingly, some lower Invertebrates, belonging to the type Cnidaria, took a variable intermediate way (Fig. 11B). In any case, the geometry of gastrulation very much affects subsequent development. The laterally compressed slit-like Protostomia blastopores should actively elongate themselves along the slit axis, compresses their polar regions (which later on transform to the oral and anal openings) but to a very small extent affects mechanically the rest of embryo (Fig. 11C, D). On the contrary, a gradually contracted hoop-like blastopore of Deuterostomia embryos creates around it a diverged radial tensile field, being extended over the entire embryonic surface and thus involving it into a coordinated morphogenesis (Fig. 11F).

Meanwhile, a uniform mode of a circular blastopore contraction is also unstable: similarly to what took place at the blastula stage, even small local irregularities in contraction rate along the blastopore periphery should subdivide it into the compression and extension zones. As a result, an ideal circular symmetry will be sooner or later broken and the blastopore together with its surroundings will acquire either a radial symmetry of *n* order, depicted by a symbol $n \cdot m$, or a mirror symmetry $1 \cdot m$. The latter mode of symmetry means formation of dorso-ventrality, still rudimentary in Echinodermata but fully expressed in Chordata. The dorsal line is that of a maximal active extension of embryonic body. Along this line another CE-feedback is created: its posterior part becomes extended, while the anterior one relaxed/compressed. This latter region transforms into a transversely extended head (Fig. 11H, h).

Later on the body of Vertebrate embryos becomes segregated into more or less independently developing territories ("fields of organs") into which the similar events are taking place in diminished scales. It will be a fascinating and as yet almost untouched field of studies to construct self-organizing models for all of them.

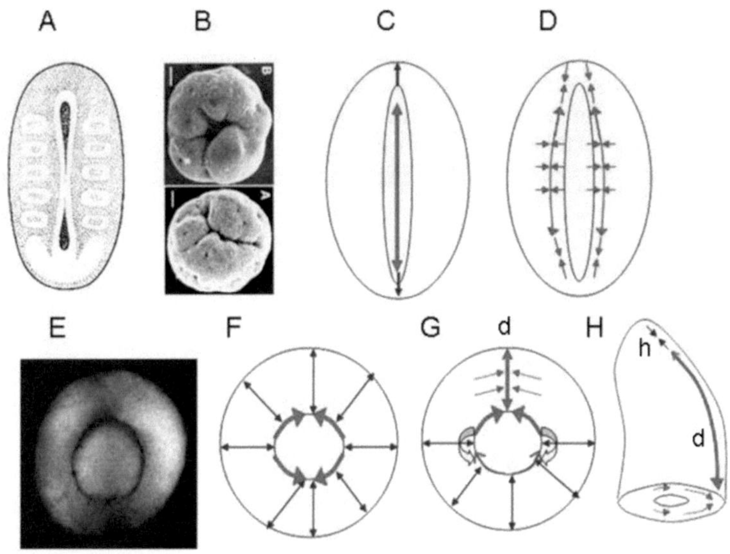

Fig. 11. Formation and mechanical role of slit-like and circular blastopores. A: typical blastopore of Protostomia. B: irregular blastopores of Cnidarian embryos. C, D: tensile fields in the vicinity of slit-like blastopores. E: circular blastopore of amphibian embryo. F-H: transformation of a radially symmetric tensile field around a circular blastopore (F) into *1 m* symmetry field with a dominating dorsal axis (G, H). H is a view from the left. d: dorsal side, h: head region.

8. Few words in conclusion

The main message of this essay is that the classical causal approach, continued to be used (even if unconsciously) by overwhelming majority of investigators cannot explain the main properties of development – the arising of more complex entities from less complex and even homogeneous ones and the associated "top-down causation" – influence of an irreducible whole upon its parts. As a result, a present-day developmental biology looks, even in the best cases, as a list of separate "instructions" of how to make this or that structure rather than a science with its own laws and predictive power. On the other hand, SOT, one of the leading branches of the modern knowledge, turned out to be quite suitable for promoting developmental biology to perform a desired transformation. To the present time, however, only the first steps on this way have been made. What we need now, is not so much a construction of specific models, but a general understanding of the nature of feedbacks, parameters and dynamic variables mostly involved in development. To a great extent this is related to a widely used notion of "genetic information". We have already seen, that it is far from being specific (an old concept "one gene - one morphological structure" is fully incompatible with modern data). Meanwhile, SOT opens a much more rational way for qualify it more properly: so far as genomes belong to the most constant components of the life cycles and the genetic factors are dealing, first of all, with the rates of molecular processes, it looks reasonable to attribute to genetic factors the role of the highest order parameters (those having the largest Tch and Lch). Within such a framework, the role of "genetic information" is quite powerful, but non-addressed: knowing the parameters' values but being non-informed about the structure of the feedback loops into which they are involved, we cannot tell anything about the role which is played by the first ones (as a rule, these roles should be quite multiple). The parameters themselves, if taken out of the context of an associated equation or, at least, of a feedback contour are, so to say, blind – they have no definite meaning at all. If continuing to apply the notion of information to developmental events (although this never can be done with a desired degree of precision), we have to conclude, that it is smoothed between several structural levels: not only DNA and proteins, but also morphological structures and embryo geometry bear an essential amount of "information", irreducible to other levels events. Or, in other words, the biological information is embedded in wide contexts, rather than in single elements. By believing that the solution of all the developmental mysteries can be reached by splitting embryos in ever diminished parts, we may miss the very essence, which is resided onto meso- and macroscopic, rather than microscopic level.

Our last question will be of a utilitarian nature: why do we need to pay any efforts by transforming biology into a law-centered science, if already in its present-day state it gives so many results, useful for medicine and biotechnology? True, as a great physicist Boltzman told, nothing is more practical than a good theory. But is this citation applicable to biology? Nobody can be sure of it, but the attempt is worth to be performed. In any case, one can hardly be content to see the science about the most complicated, ordered and aesthetically perfect natural processes using a methodology not very different from that of a medieval alchemy.

9. References

Alberts R. et al. (2003) Molecular Biology of the Cell. Garland Science. Taylor & Francis Group.

Belintzev B.N., Beloussov L.V., Zaraiskii A.G. (1987) Model of pattern formation in epithelial morphogenesis. J. theor. Biol. 129: 369-394.

Beloussov L.V. and Grabovsky V.I. (2006). Morphomechanics: goals, basic experiments and models. Int. J. Dev Biol. V. 50. P. 81-92.184.

Beloussov L.V., Luchinskaia N.N., Ermakov A.S. and Glagoleva N.S. (2006). Gastrulation in amphibian embryos, regarded as a succession of biomechanical feedback events. Int. J. Dev Biol. V. 50. P. 113-122.

Beloussov L.V. (2008). Mechanically based generative laws of morphogenesis. Physical Biology. V. 5. № 1. 015009

Beloussov L.V. (1998). The Dynamic Architecture of a Developing Organism. Kluwer Acad. Publishers. Dordrecht/Boston/London.

Blechschmidt E., Gasser R.F. (1978) Biokinetics and biodynamics of human differentiation. Ch.C.Thomas Publ. Springfield Illinois.

Capra, F. (1996). The Web of Life. Anchor Books. N.Y.

Gilbert S.F. (2010) Developmental Biology. Sinauer Ass. Sunderland, MA.

Goldberg A.D., Allis C.D., Bernstein E (2007). Epigenetics: a landscape takes shape. Cell 128: 635-638.

Harris A.K., Stopak D. and Warner P. (1984) Generation of spatially periodic patterns by a mechanical instability: a mechanical alternative to the Turing model. J. Embryol. Exp. Morphol. 80: 1-20.

Meinhardt, H. (1982) Models of Biological Pattern Formation. Acad Press. N.Y., London

Thompson, D'Arcy (1961). On Growth and Form. Cambridge University Press, Cambridge.

Turing, A.M. (1952). The chemical basis of morphogenesis. Phil.Trans.Roy.Soc.L. ser B, 237: 37-72.

Wolpert L. (1996). One hundred years of positional information. Trends in Genetics 12: 359-364.